Renewable Hydrogen Production

Renewable Hydrogen Production

Ibrahim Dincer
Ontario Tech. University, Canada

Haris Ishaq
Ontario Tech. University, Canada

Elsevier
Radarweg 29, PO Box 211, 1000 AE Amsterdam, Netherlands
The Boulevard, Langford Lane, Kidlington, Oxford OX5 1GB, United Kingdom
50 Hampshire Street, 5th Floor, Cambridge, MA 02139, United States

Notices

Knowledge and best practice in this field are constantly changing. As new research and experience broaden our understanding, changes in research methods, professional practices, or medical treatment may become necessary.

Practitioners and researchers must always rely on their own experience and knowledge in evaluating and using any information, methods, compounds, or experiments described herein. In using such information or methods they should be mindful of their own safety and the safety of others, including parties for whom they have a professional responsibility.

To the fullest extent of the law, neither the Publisher nor the authors, contributors, or editors, assume any liability for any injury and/or damage to persons or property as a matter of products liability, negligence or otherwise, or from any use or operation of any methods, products, instructions, or ideas contained in the material herein.

Library of Congress Cataloging-in-Publication Data
A catalog record for this book is available from the Library of Congress

British Library Cataloguing-in-Publication Data
A catalogue record for this book is available from the British Library

ISBN: 978-0-323-85176-3

For information on all Elsevier publications visit our website at
https://www.elsevier.com/books-and-journals

Publisher: Candice Janco
Acquisitions Editor: Peter Adamson
Editorial Project Manager: Leticia M Lima
Production Project Manager: Niranjan Bhaskaran
Cover Designer: Victoria Pearson

Typeset by TNQ Technologies

Contents

Preface

We are in an era where the societal energy dimensions are not sustainable enough because of the key issues, such as that global energy demand has risen significantly with a primary dependence on fossil fuels and that local and global environmental challenges have become prominent through stratospheric ozone depletion, acid rain, and global warming. Collective efforts have been placed on fossil-fuels-free solutions to meet the energy needs and protect the environment. Further efforts have, in this regard, been exerted on carbon-free fuels and energy solutions that primarily dwell on hydrogen and hydrogen energy. Hydrogen appears to eradicate the described energy and environment-related challenges. The usage of hydrogen minimizes environmental pollution enormously and reduces fossil fuel dependency. Hydrogen and oxygen can be employed in the fuel cells to generate electricity, and heat and water are the only by-products undergoing zero greenhouse gases. Hydrogen can be produced using numerous renewable energy sources either centrally and formerly distributed or onsite where it is required. Hydrogen is a zero-emission fuel when burned with oxygen. Hydrogen can be used for multiple applications namely, oil refineries, synthesis of ammonia, and synthetic fuels, as an energy carrier, as fuel, as energy storage media, and can also be employed in the fuel cells for power generation. This book provides clean, sustainable, and environmentally benign routes of renewable energy-driven hydrogen production systems that are expected to make a global transition to a hydrogen economy possible.

The first chapter comprises a detailed introduction of global energy demands and supplies, carbon emissions, and sectoral energy utilization and their consequences. After providing the historical background of hydrogen with its advantages and disadvantages, a comparative evaluation of hydrogen and other fuels is presented. Renewable energy-based sustainable hydrogen production and storage methods along with its infrastructure, transportation, and distribution as well as utilization aspects are discussed. Furthermore, it concerns hydrogen fuel cell applications, and the detailed analysis and modeling of different fuel cell types, proton exchange membrane fuel cells, phosphoric acid fuel cells, ammonia fuel cells, solid oxide fuel cells, and alkaline fuel cells are provided and discussed in detail.

Chapter 2 presents various types of hydrogen production methods, including conventional (particularly, natural gas reforming and coal gasification) and cleaner methods with renewable energy sources of solar, wind, hydro, ocean thermal energy conversion, tidal, geothermal, and biomass. This chapter also covers the thermochemical cycles, chlor-alkali electrochemical process, and water electrolysis process including proton exchange membrane electrolyzer, solid oxide electrolyzer, and alkaline electrolyzer.

Chapter 3 concerns the solar energy-driven hydrogen production methods, including thermochemical, photochemical, photoelectrochemical, and electrolysis processes. Two case studies are presented to investigate solar energy using solar PV panels, solar heliostat field, and solar concentrating collectors for clean hydrogen production, and both theoretical and experimental results are in this regard presented and discussed. The designed case studies also analyze both steady-state and dynamic (time-dependent) cases as explained and investigated to explore the intermittent nature of solar energy sources.

Chapter 4 encompasses the wind energy-driven hydrogen production methods. Onshore and offshore wind turbines are explained comprehensively, including the horizontal-axis and vertical-axis wind turbines. Each part of the wind turbine system is explained in the wind turbine configuration, and a case study is designed to investigate the steady-state and dynamic situations of wind hydrogen production systems.

Chapter 5 dwells on geothermal energy-driven hydrogen production, geothermal reserves/capacities, and geothermal utilization along with the advantages and disadvantages of geothermal energy. The approaches of geothermal power plants are considered, including with and without reinjection techniques and geothermal heat pumps. Dry steam, flash steam, and binary cycle geothermal power plants as well geothermal heat pumps including closed-loop (vertical, horizontal, lake or pond), open-loop, and hybrid systems are presented and discussed in detail. The different flashing types of single, double, and triple-stage geothermal-assisted hydrogen production plants are presented, and a case study is designed to illustrate how a geothermal hydrogen production system is implemented.

Chapter 6 comprehensively discusses the hydro energy-based hydrogen production systems, including the pros and cons of hydro energy. Some significant topics, such as the classification of the hydro power plant, hydroelectric turbine and generator, hydroelectric power plant, and pumped storage are discussed in detail. This chapter also covers the types of hydropower turbines along with the modeling of the single penstock, surge tank, and wave travel time.

Chapter 7 is aimed to present some methods that are applicable to the ocean thermal energy conversion (OTEC)-assisted hydrogen production system and also discusses different types of OTEC systems such as closed-cycle, open-cycle, and hybrid. It further includes the ocean energy devices and designs considering the pros and cons of different types of ocean energy such as ocean thermal energy, osmotic power, and tides and currents. A case study is designed to explore the OTEC-based hydrogen production application.

Chapter 8 deals with the biomass-based hydrogen production methods and discusses the advantages and disadvantages of biomass as a renewable energy resource. It also discusses the different types of biological methods including dark fermentation, photo fermentation, microbial electro-hydro-genesis cell, direct/indirect biophotolysis, and thermochemical methods including gasification, high-pressure aqueous, and pyrolysis. Various types of pyrolysis reactions and gasifier types including updraft, downdraft, fluidized bed, and cross-draught gasifier are discussed in detail. A case study is presented to illustrate a biomass gasification-based hydrogen production application.

Chapter 9 is about the integrated systems for hydrogen production for buildings, hydrogen, and combined heating and power to elaborate the significance of integrated energy systems. This chapter also covers the sustainable energy supply aspects including power-to-gas, power-to-heat, and battery storage. Three case studies are accommodated to investigate the different integrated configurations of solar PV, solar heliostat field, solar concentrating collector, wind, hydro, geothermal, OTEC, and biomass for clean and sustainable hydrogen production.

Chapter 10 states some key closing remarks and future directions for the potential development and deployment of renewable hydrogen production technologies and highlights the advancements in hydrogen-based energy technologies for global transition.

Ibrahim Dincer
Haris Ishaq

Nomenclature

A area (m^2)
cnv conductivity (S/m)
C_p power coefficient (kJ/kg.K)
C carbon
DA day angle (°)
E actual voltage of cell, V
$\mathbf{E_{act,c}}$ activation overpotential of cathode, V
$\mathbf{E_{act,a}}$ activation overpotential of anode, V
$\mathbf{E_{ohmic}}$ ohmic overpotential, V
$\mathbf{E_{conc}}$ concentration overpotential, V
en specific energy, kW
$\dot{E}x_{dest}$ exergy destruction (kW)
ex specific exergy (kJ/kg)
ex^0_{cg} standard chemical exergy (kJ/mol)
E_{ecc} eccentricity factor
f frequency (μm)
F Faraday constant
FF fill factor
G Gibbs free energy (kJ)
g gravitational acceleration (m/s^2)
h specific enthalpy (kJ/kg)
HHV higher heating value (kJ/kg)
H hydrogen
I direct normal irradiance (kW/m^2)
I current (A)
$\dot{I}_{beam}$ incoming beam radiation (kW/m^2)
$\dot{I}_{normal}$ direct normal radiation (kW/m^2)
J current density (A/m^2)
$\mathbf{J_{oc}}$ exchange current density at cathode, A/m^2
$\mathbf{J_{oa}}$ exchange current density at anode, A/m^2
k thermal conductivity (W/m^2K)
LHV lower heating value (kJ/kg)
$\dot{m}$ mass flowrate, kg/s
MW molecular weight
Ms moisture content
N nitrogen
n number of cells
$\dot{N}$ molar flow rate (mol/s)
O oxygen
P power (kW)
$\dot{Q}$ heat rate (kW)
$\dot{Q}_{solar}$ solar input (kW)
s specific entropy (kJ/kg K)
$\dot{S}_{gen}$ specific entropy (kW/K)

t time (s)
T temperature (°C)
V voltage (V), velocity (m/s)
V_{OC} open circuit voltage (V)
$\dot{W}$ work rate (kW)
x_j mole fraction
Z elevation (m)

Greek letters

η energy efficiency
ε_{act} activation energy
I_L limiting current density
ζ ionic conductivity
δ declination angle
β biomass parameter
ω moisture content, hour angle
θ temperature ratio
τ scattering transmittance
ψ exergy efficiency
Ω resistance
ϕ latitude
ω hour angle
θ_{zenith} zenith angle
ρ density
ε_i voidage
μ viscosity (kg/m.s)
ψ shape factor
κ number of electrons transferred

Subscripts

act activation polarization
a anode
c cathode
conc concentration overpotential
C# component name
ch chemical
Comp compressor
d destruction
dest destruction
di distillate
el electrolyzer
en energy
ex exergy
f_w feed water
FC fuel cell, flash chamber
gen generator, generation
MED multieffect desalination
oc open circuit

ohm Ohmic polarization
ov overall
ph physical
PHA phosphoric acid
PV photovoltaic
rev reversible
s steam
SC short current
SEP separator
SP state point
SOFC solid oxide fuel cell
ST solar time
sw sea water
turb turbine
W work
WT wind turbine
z zenith

Acronyms

AC air compressor
ALK alkaline
CHP combined heating and power
CLG chemical looping gasification
Cu-Cl copper chlorine
DAFC direct ammonia fuel cell
DNI direct normal irradiance
EES engineering equation solver
FC fuel cell, flash chamber
GHG greenhouse gas emissions
GT gas turbine
HCU hydrogen compression unit
HEX heat exchanger
HST hydrogen storage technology
ICE internal combustion engine
IEA International Energy Agency
LHV lower heating value
LPG liquefied petroleum gas
MED multi-effect desalination
NGLs natural gas liquids
OECD organization for economic co-operation and development
ORC organic rankine cycle
OTEC ocean thermal energy conversion
PEM proton exchange membrane
PHA phosphoric acid
PSA pressure swing adsorption
PV photovoltaic
RO reverse osmosis
RRC reheat rankine cycle
SEP separator

SHF solar heliostat field
SMR steam methane reforming
SO solid oxide
SOFC solid oxide fuel cell
SOFC solid oxide fuel cell
ST solar time, steam turbine
TEG thermoelectric generator
WF wind form
WGSR water gas shift reactor

CHAPTER

Introduction

1

A significant growth in economic development and population occurring around the globe has been a key reason behind the increased energy demand. It is a well-known fact that power generation plays a vital role in the industrial revolution of any country. The major portion of this energy demand is set to be covered by the traditional energy production methods employing fossils fuels, and a part of global energy demand is covered by renewable energy sources that are growing significantly. The global energy production can be divided into the categories of the crude sector, natural gas liquids (NGLs) and feedstocks, coal sector, natural gas sector, biofuels and waste sector, nuclear sector, hydro sector, solar/wind/other sectors, geothermal sector, and other sectors. Fig. 1.1 exhibits the global energy demand by different

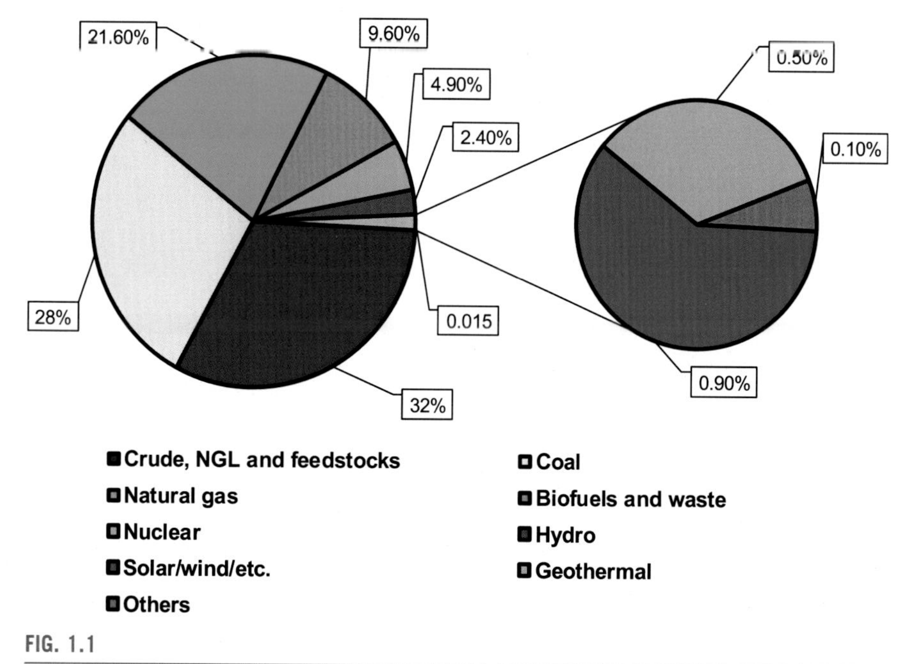

FIG. 1.1

Global energy production by sectors.

Data from [1].

Renewable Hydrogen Production. https://doi.org/10.1016/B978-0-323-85176-3.00001-9

sectors. It, therefore, displays that the crude, NGLs, and feedstocks lead the global energy production by 32% followed by the coal sector as 28%. The natural gas sector comes in the third position with 21.6% followed by biofuels and waste at 9.6%. A good percentage of 3.8% of global energy production is covered by renewable energy sources divided into 2.4% from hydro, 0.9% from solar/wind/other, and 0.5% from geothermal energy sources.[1]

Fossil fuels cover a significant portion of this increasing energy demand, and these traditional sources are fronting extreme challenges after the quick depletion. The key drawback of consuming these traditional sources is CO_2 emissions and increased global warming.[2] The global CO_2 emissions can be divided into different major sectors of electricity and heat producers, other energy industries, industry sector, transportation sector, residential sector, commercial and public services, agriculture sector, and fishing sector. Fig. 1.2 displays the CO_2 emissions distribution is different sectors in the last few years from 1990 to 2015. The primary *y*-axis represents the electricity and heat producers, other energy industries, industry sector, and transportation sectors, while the residential sector, commercial and public services, agricultural sector, and fishing sectors are displayed on the secondary *y*-axis. The electricity and heat production sector increased from 7625 to 13,405 Mt from 1990 to 2015, other energy industries sector increased from 977 to 1613 Mt, the industrial sector increased from 3959 to 6158 Mt, transportation sector increased from 4595 to 8258 Mt, the residential sector increased from 1832 to 2033 Mt, commercial and public services sector increased from 774 to 850 Mt, and agriculture sector increased from 398 to 428 Mt[3]].

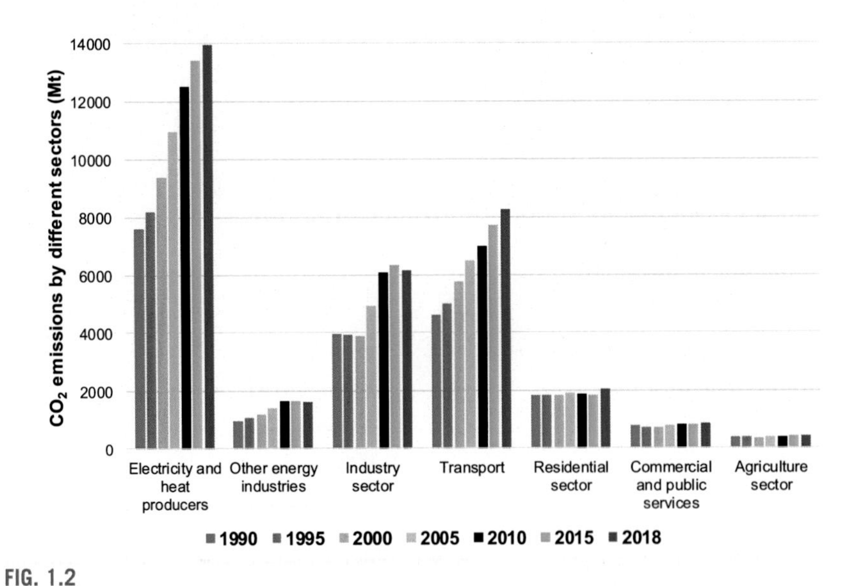

FIG. 1.2

Global CO_2 emissions by sector.

Data from [3].

The renewable energy sources stand the finest applicant to replace these traditional sources due to the increased environmental problems. The increasing greenhouse gas emission, environmental problems, and carbon emissions taxes justify the need for the transition of traditional energy sources to renewable energy sources. Renewable energy sources, namely solar, wind, hydro, geothermal, ocean thermal energy conversion (OTEC), and biomass sources appears to be the most suitable candidacies to replace the traditional sources. Due to the intermittent nature of some renewable energy sources, sources and hydrogen can take the full benefit from renewable energy sources and to be used as storage media. Hydrogen has recently been getting more serious recognition globally as a potential fuel and a unique energy solution as it proposes advantages accompanied by its usage and carbon-free solutions availability. A recent study[4] offered a comparative assessment of different renewable energy-driven hydrogen production methods. Moreover, it is a well-known fact that the existing fuel storage and transportation infrastructures that are employed for other chemical fuels can be used for hydrogen storage and transportation. Table A-1 in the appendix displays the distribution of CO_2 emissions per capita globally.

The intermittent nature of solar and wind renewable energy sources increases the requirement of an energy storage media, and hydrogen is the chief candidate that can be used for multiple purposes such as fuel, energy storage media, synthesis of methanol and ammonia, and energy carrier. Hydrogen is getting more recognition globally as a potential fuel and a unique energy solution, as it proposes advantages accompanied by its usage and carbon-free solutions availability. Moreover, it is a well-known fact that the existing fuel storage and transportation infrastructures that are employed for other chemical fuels can be used for hydrogen storage and transportation.

A recent study[5] has included a solar power and natural gas-driven hydrogen production system to minimize the carbon fuel taxes. A significant focus was to integrate solar energy sources with natural gas (steam methane) reforming for hydrogen production. They introduced a new concept of an improved steam methane reforming (SMR) system that varies the level of endothermic nature of steam methane reforming through the integration of steam and CO in SMR feed. The system was presented in which the previously mentioned resources are generated internally and recycled as well to create an SMR-based system for the hydrogen production system employing methane and driven by solar and methane energy supply. They also reviewed the environmental carbon tax legislative, and its probable influence on SMR and improved SMR was quantified for economic viability.

Fig. 1.3 exhibits the global hydrogen demand from 1975 to 2018. The significant global hydrogen demand is divided into three different categories of refining, ammonia, and other applications. The global transition from commercial energy sources to renewable energy sources is increasing the global hydrogen for hydrogen fuel-cell vehicles. In the refining sector, pure hydrogen demand increased from 6.2 Mt to 38.24 Mt from 1975 to 2018, followed by the global hydrogen demand in the ammonia synthesis sector ranged from 10.88 Mt to 31.46 Mt and hydrogen demand in other sectors increased from 1.08 Mt to 4.19 Mt.

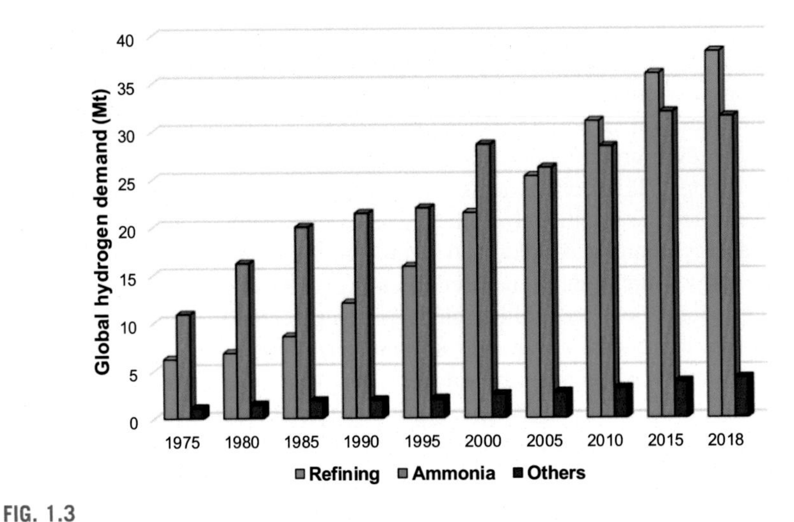

FIG. 1.3

Global demand for pure hydrogen, 1975–2018.

Data from [8].

A review paper[6] was published on the steam reforming process-based on hydrogen production and conducted the economic analysis where their objectives aimed to offer a comprehensive economic and environmental study for producing hydrogen employing steam reforming of different raw materials, namely biomass, ethanol, biogas, and natural gas. It was found through the literature review that natural gas steam reforming offers lesser installed capital in comparison with other hydrogen production methods owing to the existence of high unconverted hydrocarbon quantities in produced gas (also known as tar) during other hydrogen production methods, namely biogas steam reforming. A recent study[7] published a review article on methane solar thermal reforming for syngas and hydrogen production. The solar-driven steam methane reforming is accepted as a transition that offers a feasible and viable approach to generate a transition route toward solar-hydrogen economy and for decarbonization of fossil fuels. The commercially recognized traditional natural gas reforming concept that is extremely endothermic can be reduced using a solar-driven system and also offers reduced cost to introduce renewable energy-based hydrogen production techniques. These two technologies also share parallel technical problems considering thermodynamics and thermochemistry linking to exploit solar energy efficiently. In this viewpoint, they offered a comprehensive review study on the advancement and current standing of solar-driven reforming systems by keeping the significant focus on reactor technologies and the techniques that are employed until now to integrate SMR heat requirement principles with concentrated solar power. A detailed review was presented that addressed the solar reactors from each scale and also suggested future work directions. Fig. 1.4 displays the share of renewables in the total energy consumption for 2018.

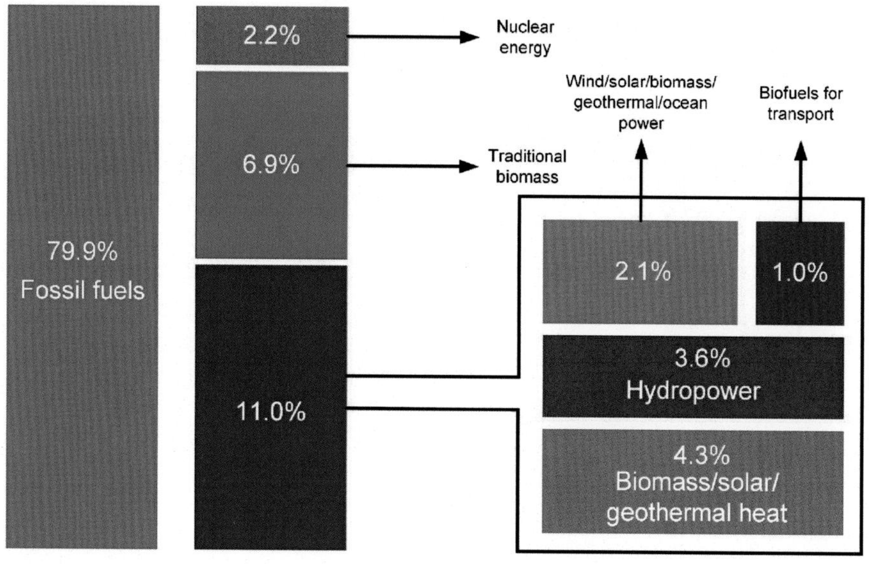

FIG. 1.4

The share of renewables in the total energy consumption for 2018.

Data from [9].

The hydrogen required to produce ammonia can be cracked into hydrogen again as ammonia production is a reversible process and ammonia can be cracked into hydrogen and nitrogen. Unlike other fuels, hydrogen offers a high heating value that increases the benefit of using hydrogen fuel accompanied by the advantage of the environmentally benign solutions offered with hydrogen as it releases zero harmful emissions when combusted. Fig. 1.5 shows the comparison of lower heating values (LHV) of different fuels. Hydrogen leads in the charts with LHV of 119.9 MJ/kg followed by the propane LHV of 45.6 MJ/kg and methanol offers the lowest LHV of 18 MJ/kg among hydrogen, propane, methane, gasoline, methanol, ammonia, and diesel fuels.

In the coming decades, hydrogen is the best candidate to become a part of emissions mitigation efforts. Today, one-third of the total hydrogen production (120 million tonnes) can be found in the mixture while two-thirds is pure hydrogen. Water is the most significant source of hydrogen that can split into oxygen and hydrogen by supplying electrical power. Currently, commercial natural gas reforming is considered the most significant method of hydrogen production, and a part of hydrogen is produced through renewable sources and this trend is expected to change soon; the transition of traditional sources to 100% renewable energy sources is taking place due to the increasing environmental problems, greenhouse gas (GHG) emissions, and carbon emissions taxes. The massive amount of produced hydrogen is used on-site in industry. Oil refining and ammonia production are the leading sectors that account for almost two-third of the produced hydrogen. Ammonia is used for the production of different

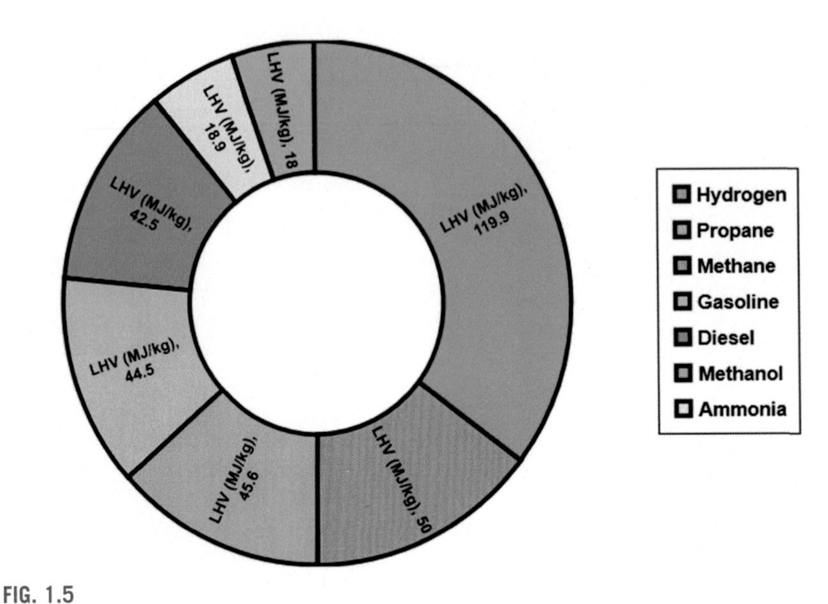

FIG. 1.5

Comparison of lower heating values of different fuels.

chemicals such as nitric acid, urea, etc. as well as used as nitrogen fertilizer, while hydrogen is added to heavy oil to ease the fuel transportation in petroleum refineries. Due to the limited relevance of hydrogen usage for energy evolution, hydrogen handling has resulted in ample experience. The pipeline system for hydrogen transportation has already spanned hundreds of kilometers in numerous regions and countries and operates without incidents followed by a long background of hydrogen transportation in dedicated trucks. The hydrogen fuel-cell vehicles are expected to employ a significant portion of global hydrogen, as it provides a clean energy solution eliminating combustion. Hydrogen-fueled aviation, heavy vehicles, drones, boats, and sustainable communities are also being focused on clean energy solutions. For the energy transition, hydrogen has to be produced through environmentally benign systems using renewable energy sources and by decarbonizing the hydrogen supply and transportation.

Hydrogen can be used for multiple purposes as a promising feedstock in addition to being a fuel and an energy carrier. The demand for hydrogen is distributed among various sectors. For example, hydrogen is employed for numerous chemical processes, namely hydrocracking, hydro-desulfurization, hydro-dealkylation, fats and oils dehydrogenation, welding, and metallic ores reduction, and also hydrogen is added to heavy oil to ease fuel transportation. A major portion of hydrogen is used in oil refineries, synthesis for synthetic fuels, and the formation of hydrochloric acid and ammonia synthesis. Oil refineries and ammonia synthesis sectors are two significant sectors of hydrogen demand. For ammonia synthesis, hydrogen is produced using a traditional natural gas reforming method followed by the Haber Bosch

Table 1.1 Comparison of Hydrogen Properties With Other Conventional Fuels.

Fuel	Hydrogen	Propane	Methane	Gasoline	Diesel	Methanol
LHV (MJ/kg)	119.9	45.6	50	44.5	42.5	18
HHV (MJ/kg)	141.6	50.3	55.5	47.3	44.8	22.7
Stoichiometric Air/Fuel ratio (kg)	34.3	15.6	17.2	14.6	14.5	6.5
Combustible Range (%)	4–75	2.1–9.5	05–15	1.3–7.1	0.6–5.5	6.7–36
Flame Temperature (°C)	2207	1925	1914	2307	2327	1870
Minimum Ignition Energy (MJ)	0.017	0.3	0.3	0.29	–	0.14
Auto Ignition Temperature (°C)	585	450	540–630	260–460	180–320	460

process for ammonia synthesis. Table 1.1 tabulates the comparison of hydrogen properties with other conventional fuels of propane, methane, gasoline, diesel, and methanol. To present a comparison, the significant properties of lower heating values (MJ/kg), hydrogen heating values (MJ/kg), stoichiometric air/fuel ratio (kg), combustible range (%), flame temperature (°C), minimum ignition energy (MJ), and autoignition temperature (°C).

The current global annual energy consumption is $\sim 445 \times 10^{18}$ J and rising continuously with the growth in population and living standards. Residential energy consumption predictions per capita in foremost economic regions of OECD (Organization for Economic Cooperation and Development)-Asia, OECD-America, OECD-Europe, Non-OECD Europe, Non-OECD Asia, Africa, Middle East, and Central and South America. North America and European countries are the chief candidates for the highest per capita energy consumption in the residential sector.

Hydrogen primarily offers the highest burning velocity as compared to all other fuels whether liquids or gases. A high-performance indicator is offered by the hydrogen fuel cells in terms of efficiencies as fuel cells do not follow the efficiency limits of the Carnot cycle. All over the world, a massive amount of H_2 fuel stations is projected to install in this year to facilitate hydrogen fuel cell and fuel-cell hybrid vehicles. Hydrogen carries several advantages, such as efficient energy conversion production through water splitting with zero carbon emissions, synthesis of different fuels such as synthetic fuels and ammonia followed by different chemical reactions, the abundance of sources, availability of storage options, existing infrastructure can be utilized for long-distance transportation, high LHV as compared with traditional fossil fuels and provides clean energy solutions for energy sector through renewable energy sources to eradicate the environmental problems.

1.1 Fuels Utilization

Hydrogen is the most abundant, common and simplest chemical element and a building block of all matter. In comparison with other atoms that include electrons, neutrons, and protons, hydrogen only consists of one proton and one electron. Hydrogen is the most abundant element in the universe and covers almost three quarters of all universal matter. Hydrogen is an odorless and colorless nonmetal and extremely combustible in the common form and tends to burst into a flame that makes it a dangerous and valuable resource. Robert Boyle[10], a British scientist, first discovered the hydrogen in 1671. He dipped different metals in acid for experimented investigations. A single-displacement reaction occurs on dipping a pure metal in acid. For instance, the addition of potassium to the hydrochloric acid solution causes a single-displacement reaction. The hydrochloric acid reacts with potassium metal to form potassium chloride salt and hydrogen atoms form hydrogen gas by the following reaction:

$$2K + 2HCl \rightarrow 2KCl + H_2 \quad (1.1)$$

Henry Cavendish[11], a British scientist, confirmed hydrogen as a distinct element in 1776. Cavendish and Boyle observed hydrogen as a flammable gas. Explicitly, hydrogen rapidly undergoes the combustion process in the presence of oxygen.

$$2H_2 + O_2 \rightarrow 2H_2O + \text{Heat} \quad (1.2)$$

Van Troostwijk and Deiman[12] discovered water electrolysis in 1789. James Dewar[13] produced the first liquid hydrogen in 1898. According to the stoichiometric exothermic reaction, oxygen and hydrogen molecules combine to form water and generate heat. Cavendish and Boyle along with flammability also detected that hydrogen gas is lighter than air. Hydrogen, similar to helium, is suitable for filling balloons and even better as compared to helium at lifting things. In the early 1900s, hydrogen was used as lifting gas for large airships. Nevertheless, this craze of hydrogen airship lifting did not last long. A tragedy occurred in 1937 when German airship (Hindenburg) was caught in a fire and blasted at Lakehurst, and killed 36 people. The designers of the airship were aware of the flammability of hydrogen and recognized helium was a better option but it was expensive and rare. After this Hindenburg disaster, hydrogen was abandoned as lifting gas for large airships.

The pH scale was introduced in 1909 by P.L. Sorensen.[14] J.N. Bronsted[15] defined an acid as a proton donor in 1923. Deuterium was discovered by Harold Urey[16] in 1931. The chemical $LiAlH_4$ was prepared by H.I. Schlesinger-Chicago University[17] in 1947. Further achievements established in the coming years were detonation of H-Bomb in 1954,[18] super acid (BF3−HF) in 1960 by G.A. Olah[19] Nobel 1994, H.C. Brown Nobel Prize,[17] Purdue University in 1978, first stable T.M. dihydrogen compound discovered by G. Kubas[20] in 1984, and metallic hydrogen was prepared in 1996.[21]

The Space Shuttle launched by the National Aeronautics and Space Administration (NASA), which was canceled in 2011, was hydrogen-fueled, and the engine was powered by liquid oxygen and hydrogen burning. NASA engineers understood the dangers of using hydrogen and decided to take maximum advantage of the whole raw power carefully.

Lately, governments and people became progressively interested in reducing greenhouse gases due to the increasing environmental problems. In the transportation sector, the concept of hydrogen fuel-cell-powered cars was introduced, which gained high interest, and unlike other fuels, using hydrogen causes zero greenhouse gases and releases water only. Hydrogen storage was one of the major problems accompanied by employing hydrogen as a car fuel source. Although hydrogen offers more energy by weight but offers less energy by volume as compared to gasoline. That reveals that a bigger hydrogen tank is needed to drive a reasonable distance once filled with fuel but gas tanks will not be sufficient. Scientists and researchers are trying to convert hydrogen into a solid because of the low energy density. Human hydrogen understanding has traveled a long journey since it was first discovered in 1671. Hydrogen has been employed for airship lifting and to get people into space. Hydrogen is expected to be the power source for the transportation sector that will fuel the cars of tomorrow.

Hydrogen offers the highest burning velocity as compared to all other fuels whether liquids or gases. The probable reduction of high-level greenhouse gas emissions including other hazardous emissions is a significant considerable factor to compare the fuel alternatives. Fig. 1.6 displays a detailed historical background of hydrogen since 1671 including significant landmarks such as hydrogen discovery, hydrogen detection as a flammable gas, discovered of water electrolysis, liquid hydrogen discovery, detected that hydrogen gas is lighter than air, use as lifting gas for large airships, pH scale establishment, deuterium discovery, $LiAlH_4$ preparation, detonation of H-Bomb, super acid (BF3—HF) preparation, dihydrogen compound discovery, metallic hydrogen, heat and power production systems, and hydrogen fuel-cell vehicles. Fig. 1.7 displays the hydrogen utilization as fuel, feedstock, and energy carrier. Hydrogen can be employed as a fuel for numerous applications such as fuel cells, combustion, power, heating and cooling, ammonia synthesis, and synthetic fuels. Fig. 1.8 shows the carbon to hydrogen ratios of numerous transportation fuels. The fuels considered in this figure are coal, diesel, gasoline, liquefied petroleum gas, ethanol, natural gas, ammonia, and hydrogen. In comparison with other fuels employed for combustion applications, NH_3 offers the highest energy density, contains zero carbon that offers zero global warming potential, and produces water and nitrogen over combustion.

The advanced hydrogen deployment end-use applications require a combined ramped-up hydrogen supply chain, along with the supplementary purification, production, and pressurizing for transportation and distribution capacity. Fig. 1.9 exhibits the global energy transition required for the hydrogen industry. According to the theoretical standpoint, extensive spectrum options are possible, ranging from on-site production to centralized production and tanker trucks based on long-distance delivery through dedicated hydrogen pipelines.

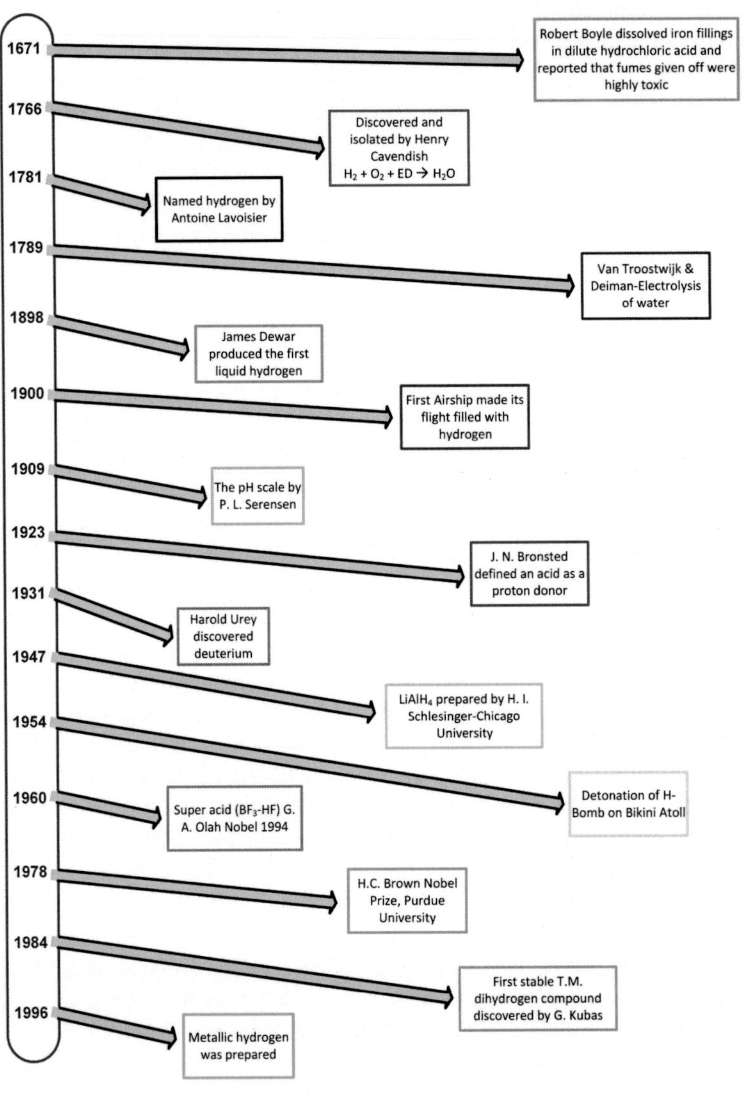

FIG. 1.6

A historical milestone list of hydrogen.

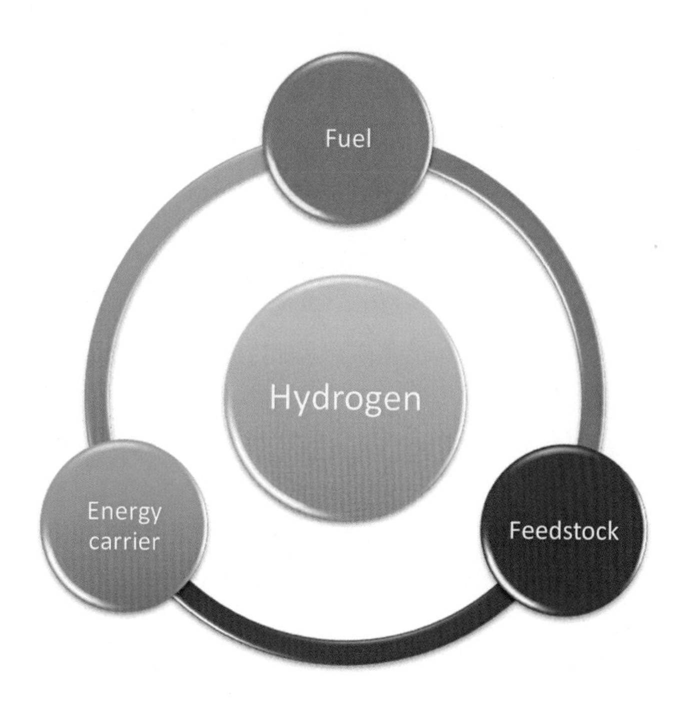

FIG. 1.7

Hydrogen utilization as fuel, feedstock and energy carrier.

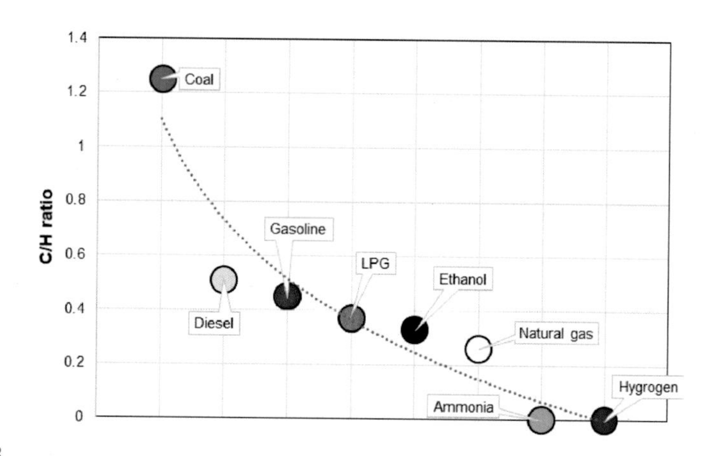

FIG. 1.8

Carbon–hydrogen ratio of common transportation fuels to illustrate the historical use of fuels.

Data from [24].

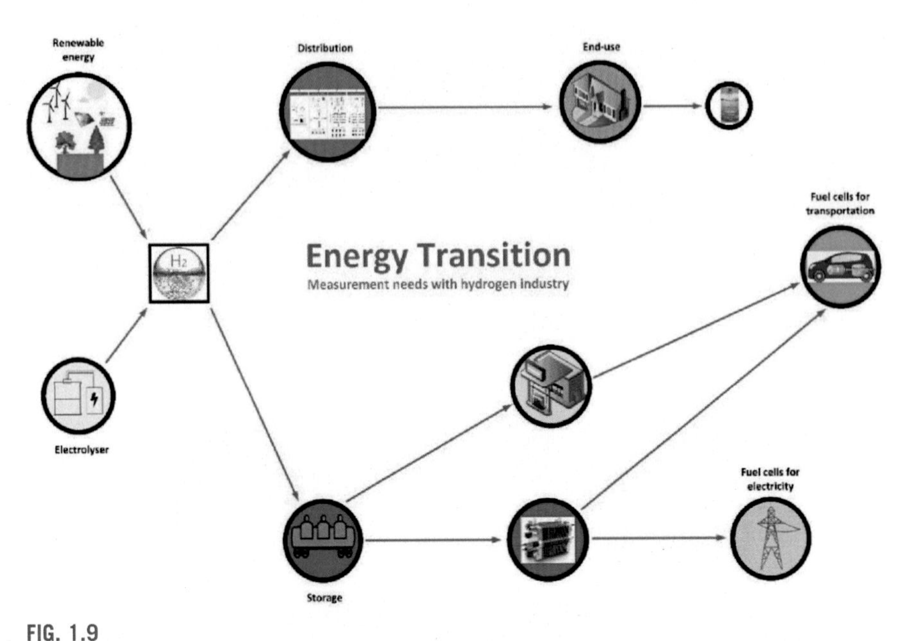

FIG. 1.9

Global energy transition for the hydrogen industry.

Modified from [22].

1.2 Hydrogen Properties and Sustainable Development

The global energy crisis since the 1970s impacted the hydrogen economy deliberation as a potential solution. A significant amount of hydrogen is produced through traditional natural gas reforming plants while environmental problems, greenhouse gases and carbon emission taxes are leading to the global transition toward 100% renewable energy. A large number of options of hydrogen usage is possible, where hydrogen is deployed for many fuels and chemicals as well as energy carrier as listed below:

- Ammonia synthesis
- Synthetic fuels
- Filling balloons
- Hydrogen containing chemicals
- Energy storage medium
- Hydrogenation of oils and fats
- Rocket fuel
- Welding
- Hydro-dealkylation and hydrocracking
- Hydrochloric acid
- Metallic ores reduction
- Cryogenics
- Combined heating and power
- Fuel for hydrogen fuel-cell vehicles

Hydrogen is the most abundant element in the universe and covers almost three quarters of all universal matter. Hydrogen is an odorless and colorless nonmetal and extremely combustible in the common form and tends to burst into a flame that makes it a dangerous and valuable resource. Robert Boyle, a British scientist, first discovered hydrogen in 1671. The significant properties of hydrogen are tabulated in Table 1.2.

The transition of conventional hydrogen production methods to renewable energy methods is a key requirement for sustainable energy systems, sustainable power generation, sustainable heat and power production, sustainable economy, and sustainable transportation. Hydrogen growing compatibility with renewable energy sources is extremely mandatory for improved energy security, sustainability and economic growth. The escaping from the depletion of the natural resources to uphold ecological balance is defined as sustainability, and three significant pillars of sustainability are environmental, social, and economic. The environmental step gets the greatest devotion as companies are concentrating on dropping the carbon footprints, waste packaging, and their impact on the environment. The social support bonds back to the concept of social license. The approval and support from the stakeholders, employees, and community are mandatory for sustainable business. A sustainable business must be profitable but the economic perspective cannot surpass the other two pillars. In fact, profit at any cost is not at all what the economic pillar is about. The economic pillar comprises of proper governance, compliance, and risk management. Fig. 1.10 displays the significant pillars of sustainability. The 5E approach covers energy, environment, economy, education, and ethics.

The variability of hydrogen production methods and fuel-cell technologies offer flexible options in numerous applications with increased efficiency and reduced environmental effects. Hydrogen offers decentralized power production using fuel cell and providing energy security. The hydrogen and fuel-cell technologies are more effective as compared to the conventional energy systems for sustainable power production and leading toward condensed resource consumption globally. Hydrogen storage is a challenging part as it possesses very low-density gas with a density of 40.8 g/m^3. A comprehensive sustainable hydrogen production using different renewable energy sources and hydrogen utilization in different applications is significant to establish sustainability and environmentally benign energy solutions. Fig. 1.11 exhibits the renewable energy-based sustainable hydrogen production methods. The renewable energy sources employed for hydrogen production are namely solar, wind, geothermal, hydro, ocean, and biomass gasification. The electricity generated by these renewable energy sources is fed to the electrolyzer for hydrogen production. The produced hydrogen may be employed to hydrogen fuel-cell electric vehicles, to aviation and locomotive, used for metal refining and

Table 1.2 Hydrogen Characteristics.

Properties	Value
Symbol	H
Atomic weight	1.008
Molecular weight	2.016
Electron configuration	$1s^1$
Category	Nonmetal
Color	Colorless
Odor	Odorless
Phase	Gas
Higher calorific value	141.9 MJ/kg
Density	0.083 kg/m^3
Lower calorific value	119.9 MJ/kg
Liquid density	70.8 kg/m^3
Boiling point	−252.87°C
Melting point	−259.14°C
Flame temperature	585°C
Critical points: • Temperature • Pressure • Density	 • 306°C • 12.84 bar • 31.40 kg/m^3
Diffusion coefficient	0.61 cm^2/s
Limit of air ignition	2045 vol. %
Specific heat	14.89 kJ/kg K
Standard chemical exergy	236.09 kJ/mol
Energy density	3 MJ/dm^3
Auto-ignition temperature	585°C
Laminar flame speed	230 cm/s
Minimum spark ignition energy	0.02 mJ
Flammability limits in air	Apr-75 vol. %
Ionization energy	13.5989 eV
Adiabatic flame temperature	2107°C
Flame velocity	2.65–3.25 m/s
Stoichiometric fuel/air mass ratio	0.029
Octane number	130
Limits of flammability (equivalence ratio)	0.1–7.1
Diffusive velocity in air	2 m/s

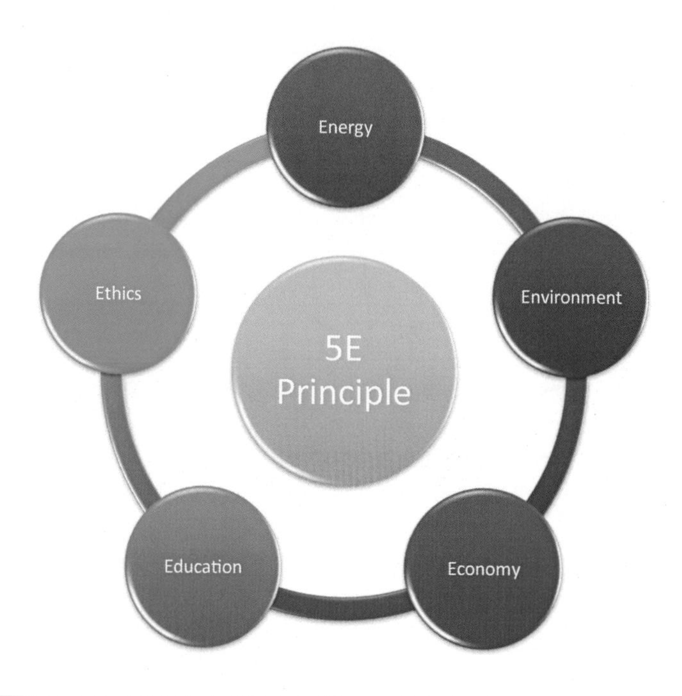

FIG. 1.10

Significant pillars of sustainability.

synthetic fuels and ammonia synthesis. The produced hydrogen can be employed as combustion fuels as well, which generates zero carbon emissions, and used for heating and power generation. The renewable energy sources are generally listed as follows:

- Solar
- Wind
- Geothermal
- Biomass
- Ocean
- Hydro

Fig. 1.12 exhibits the comparison of densities and volumetric and gravimetric densities of traditional and alternative fuels. Two hydrogen storage methods of cryogenic liquid and compressed hydrogen are indicated. The storage densities of the traditional and alternative fuels are presented on the primary *Y*-axis while the volumetric and gravimetric densities are presented on the secondary *Y*-axis. Unlike other traditional and alternative fuels, hydrogen possesses the maximum gravimetric density comparatively.

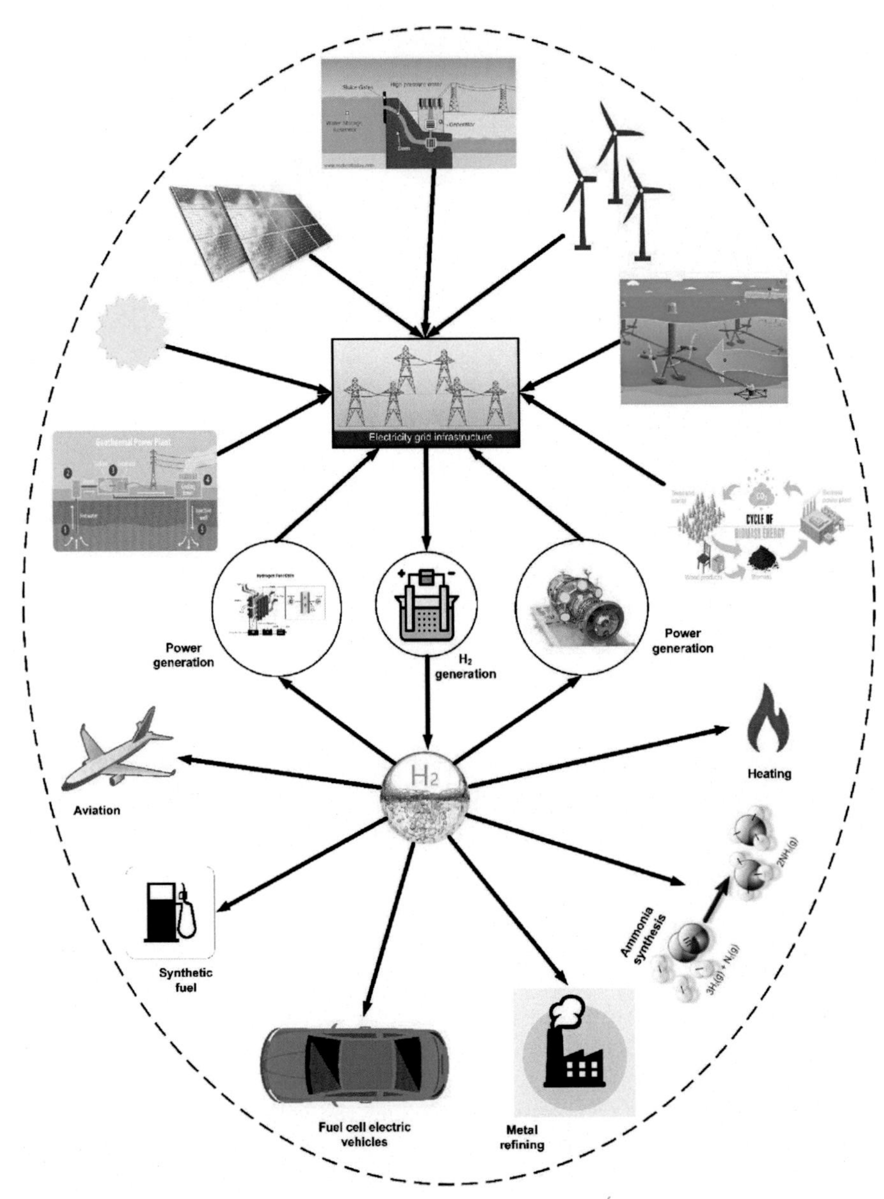

FIG. 1.11

An illustration of renewable energy-based sustainable hydrogen production methods.

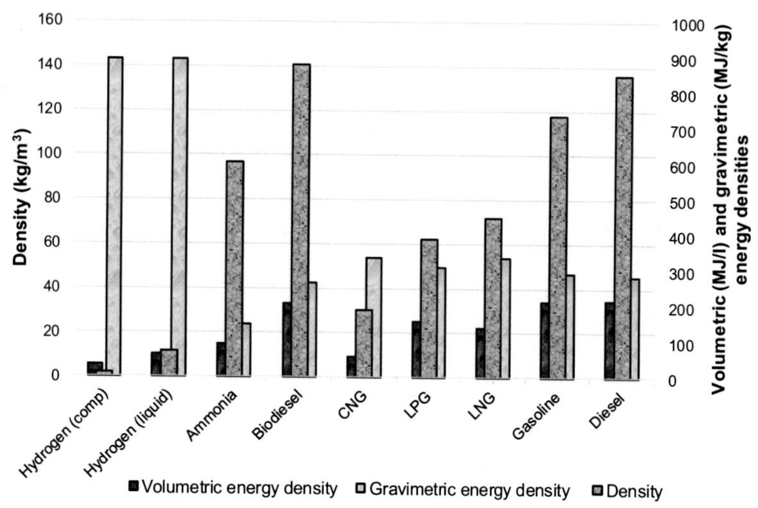

FIG. 1.12

Volumetric energy density, gravimetric energy density and mass storage densities of traditional and alternative fuels.

Modified from [24].

1.3 Hydrogen Storage

The hydrogen and fuel-cell systems can also be advantageous for residential applications. In the residential sector, the residence can be powered locally by employing hydrogen fuel cells. Additionally, hydrogen can directly be employed and combusted in the residence furnace to produce heat (thermal energy) for space heating and hot water on demand. The GHG reduction is one of the most significant advantages of using such a system. An additional benefit is reducing the peak loads from grids by promoting the capability of generating local power using fuel cells, and easy utilization of the delocalized renewable energy generation becomes easily applicable. Numerous environmental advantages are accompanied by hydrogen utilization and hydrogen-based fuel-cell systems and their applications in residences. A global transition is occurring in the transportation sector from traditional fuel combustion to hydrogen fuel-cell electric vehicles, and this transition is also occurring in aviation, boats, locomotives, and drones. In the fuel-cell electric vehicles, hydrogen tank is used instead of traditional fuel tanks, and stored hydrogen is fed to the hydrogen fuel cell to generate electric power. In the hydrogen fuel-cell hybrid electric vehicles, a battery storage system is also employed to store the additional electricity and is used to drive a vehicle when required. The hydrogen fuel cell is a well-developed and efficient technology used to generate electricity using hydrogen. A high-performance indicator is offered by the hydrogen fuel cells in terms of efficiencies as fuel cells do not follow the efficiency limits of the Carnot cycle. All over the world, a massive amount of H_2 fuel stations is projected to install in this year to facilitate the hydrogen fuel cell and fuel-cell hybrid vehicles. Hydrogen can be stored in multiple ways to be employed for different applications. Following are the three different methods of hydrogen storage:

- Renewable hydrogen storage for fuel-cell applications that can be employed for hydrogen fuel-cell electric vehicles and power generation
- Renewable hydrogen storage in the form of ammonia. As ammonia offers high energy density as compared to hydrogen, the produced hydrogen can be used to synthesize ammonia that can be cracked back using reformer into hydrogen using the following chemical reaction:

$$N_2 + 3H_2 \rightarrow 2NH_3$$

- Renewable hydrogen storage for different applications such as fuel for combustion, fuel for hydrogen fuel-cell vehicles, oil refining, synthetic fuels and ammonia synthesis, and heating and power production

Renewable hydrogen storage for fuel-cell applications that can be employed for hydrogen fuel-cell electric vehicles and power generation. Fig. 1.13 displays the representation of hydrogen storage methods starting from the renewable energy sources leading toward hydrogen that is stored to be employed for different applications such as fuel cells, ammonia and heating, cooling, and power. The different renewable energy sources that can be employed for clean hydrogen production are solar, wind, biomass, geothermal, hydro, and ocean. The electricity generated by these renewable energy sources is fed to the electrolyzer with water that produces hydrogen. The produced hydrogen is stored in the storage tank and employed to the fuel cell to generate power for different applications.

Renewable hydrogen is also stored in the form of ammonia as ammonia offers high energy density as compared to the hydrogen. The different renewable energy sources that can be employed for clean hydrogen production are solar, wind, biomass, geothermal, hydro, and OTEC cycle. The produced hydrogen is used to synthesize ammonia that can be cracked back into hydrogen on demand. The representation also shows the renewable hydrogen employed for ammonia that can be reformed back to hydrogen on demand. The ammonia reforming is a well-established Haber Bosch reversible process that is used for ammonia synthesis, and the same process is used to crack ammonia back to nitrogen and hydrogen. For ammonia synthesis, an air separation unit such as cryogenic air separation unit, pressure swing adsorption, or membrane separation is employed to separate nitrogen from the air. The unreacted gases are recycled to the reactor, and produced ammonia is stored to meet the hydrogen demand.

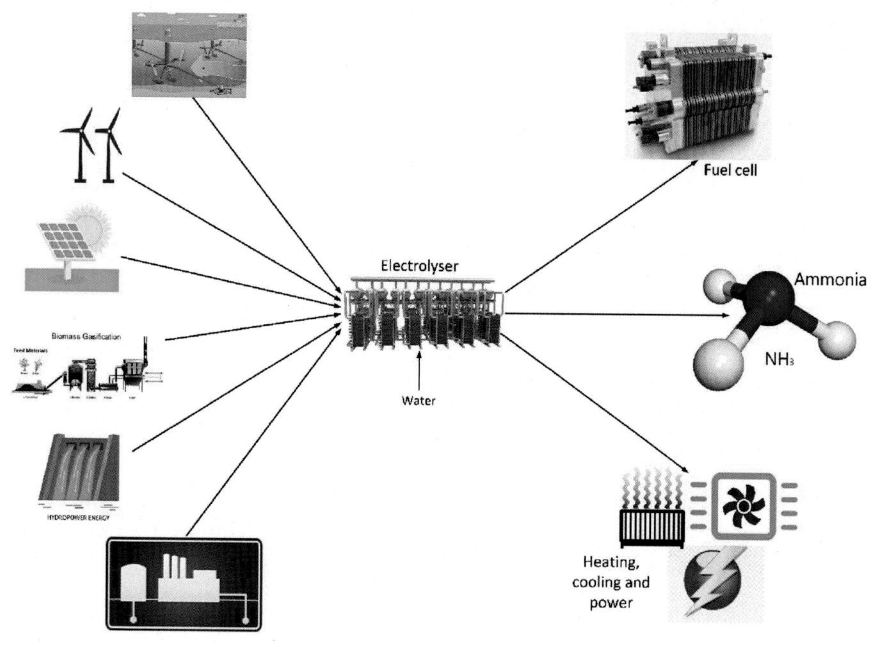

FIG. 1.13

Renewable hydrogen for various applications.

The storage of hydrogen and the utilization to meet high-demand peaks and other applications such as fuel, energy carrier and synthesis of different chemicals. The different renewable energy sources that can be employed for clean hydrogen production are solar, wind, biomass, geothermal, hydro, and OTEC cycle. The representation also shows the hydrogen applications for heating, cooling, and power. The stored hydrogen can be employed for multiple purposes depending upon the applications, such as it can be used as jet fuel, can be employed for ammonia and methanol synthesis, used for refining, can also be employed to hydrogen fuel-cell electric vehicles, and for also for sustainable energy systems.

A recent article published in the open literature[25] presents a comparative study of different hydrogen storage systems and their efficiencies employing solid-state materials for hydrogen storage. To evaluate the performances of different hydrogen storage, solid-state materials must contain characteristics of operating temperature and pressure, heat effects of hydrogen release and uptake, packing densities and reversible H_2 storage capacities. A performance assessment of systems for collecting 5 kg hydrogen in cylindrical containment full of solid hydrogen storage material containing, such as hydrides and the composites of reactive hydride as MgH_2 and AlH_3, was conducted. The performance assessment yielded volumetric and gravimetric hydrogen storage capacities as well as hydrogen storage efficiencies. They revealed that the weight efficiency of hydrogen was significantly influenced by the temperature-pressure conditions and packing density that evaluate the dimensions and type of containment. They recommended the usage of materials that undergo low heat effects and operate near reference conditions must be targeted to develop the new hydrogen stores that can offer the best operational efficiencies. Fig. 1.14 displays the classification of different hydrogen storage technologies (HST) including physical and chemical storage methods. An innovative way of synthesizing ammonia borane (hydrogen storage material) through copper (II)-ammonia complex oxidization in the liquid phase was published in a recent study.[26] Ammonia borane carries the extraordinary advantages of containing 19.6 wt.% hydrogen content that can control dehydrogenation and is regarded as a competitive material for hydrogen storage. They reported a unique synthesizing process to produce ammonia borane crystals employing Cu(II)$-NH_3$ complex as nitrogen and oxidizer source reactant. A recent study[27] conducted a thermodynamic analysis on the high-pressure compressed filling of gaseous H_2 storage tanks. Their thermodynamic analysis was based on energetic and exergetic approaches. Some sensitivity analyses were also conducted to explore the effects of the initial conditions on the filling process exergy destruction rate and exergetic efficiency. The transient analysis was also conducted to explore the filling process and determine the pressure and temperature pressure during filling inside the storage tank.

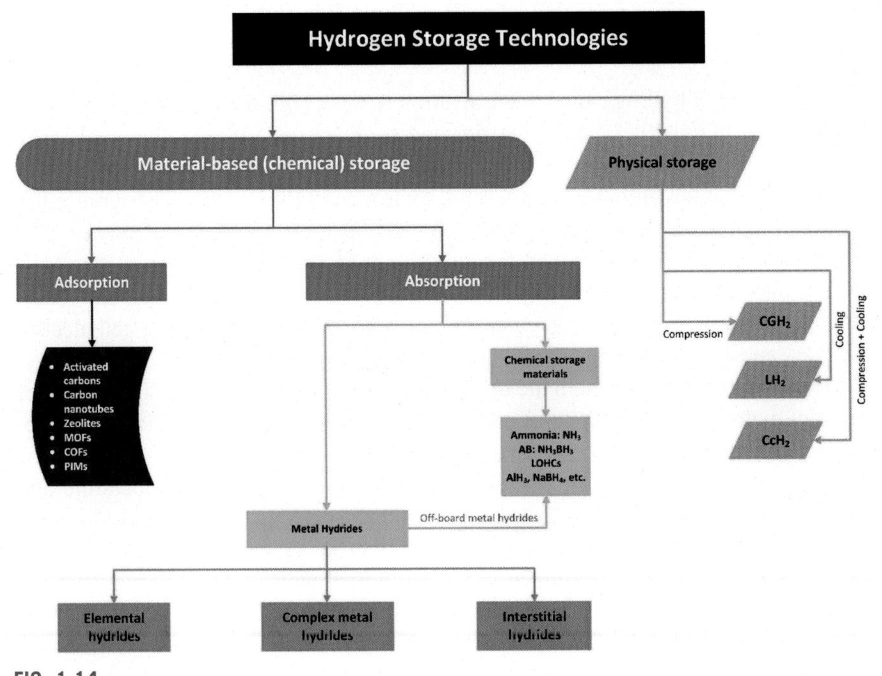

FIG. 1.14

Hydrogen storage technologies.

1.4 Hydrogen Infrastructure, Transportation, and Distribution

The hydrogen and fuel-cell systems can also be advantageous for residential applications. In the residential sector, the residence can be powered locally by employing hydrogen fuel cells. Additionally, hydrogen can directly be employed and combusted in the residence furnace to produce heat (thermal energy) for space heating and hot water on demand. For the hydrogen distribution purpose, either the existing natural gas distribution network is used or transported in reservoirs, loaded to the pressurized tanks and supplied to the residence. Fig. 1.15 displays the hydrogen transportation supply chain for potential future ramp-up. The hydrogen transportation supply chain for potential future ramp-up can be distributed in four different categories of on-site, semicentralized, centralized, and intercontinental modes of hydrogen transportation. Hydrogen can be fueled to the fuel stations directly in on-site supply chain category, hydrogen is fueled to the fuel stations indirectly in semicentralized supply chain category and tankers are used to cover tens of kilometers, hydrogen is fueled to the fuel stations indirectly in centralized supply chain category and tankers are used to cover hundreds of kilometers, and hydrogen is imported to the fuel stations in intercontinental classification and tankers are used to cover hundreds of kilometers.

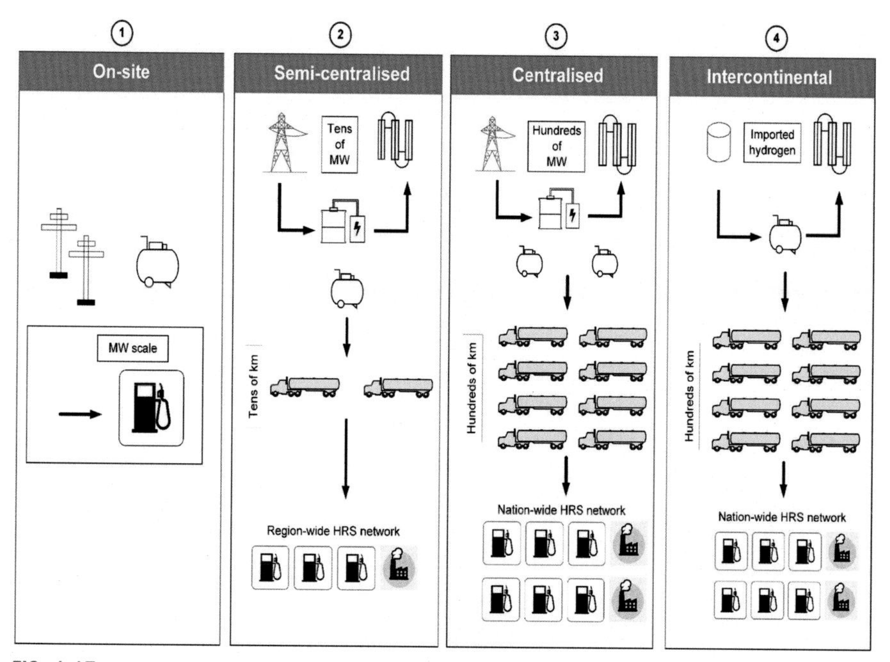

FIG. 1.15

Hydrogen transportation supply chain for potential future ramp up.

The demanding nature and geographic distribution are not the only factors that influence the supply chain structure, but also affected by the following factors:

- Firstly, the existing hydrogen sources accessibility or hydrogen production feedstock in the district, associated with the onsite production cost, as hydrogen production is the utmost capital-intensive supply chain part.
- Secondly, outside a convinced threshold consumption, on-site hydrogen production and delivery through devoted hydrogen pipeline that is feasible mainstream supply mode.
- Thirdly, according to a risk management standpoint, large-scale production capacity investment is conventionally made as if a large production proportion is sold to a single client.

In the last two stages, the facilities of new products can be leveraged to centralized or semicentralized hydrogen production sources. Fig. 1.16 displays the applications of hydrogen in the transportation sector. Hydrogen is also employed as jet fuel.

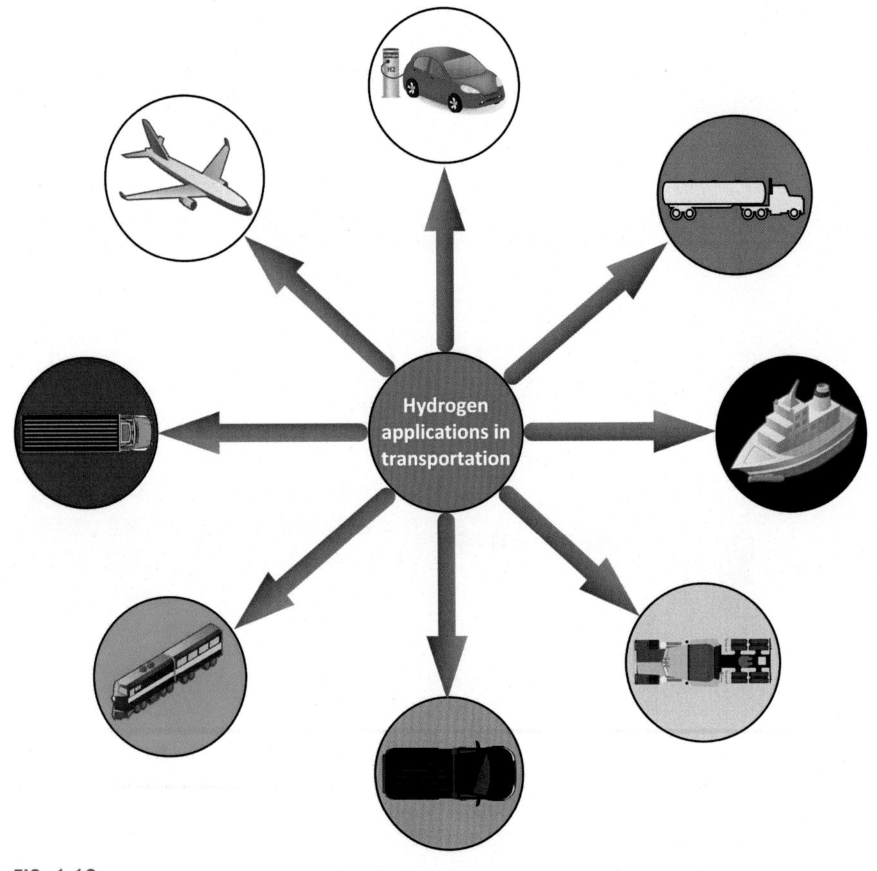

FIG. 1.16

Hydrogen applications in the transportation sector.

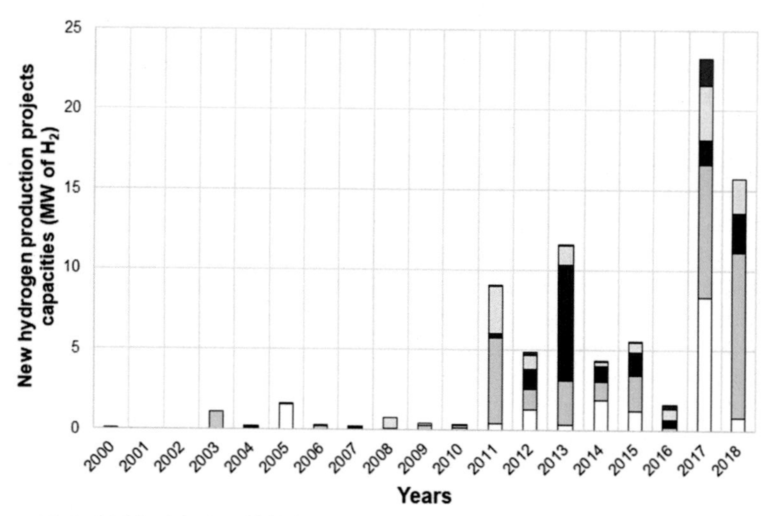

FIG. 1.17

New hydrogen production projects capacities.

Data from [28].

The hydrogen storage methods are already described in the previous section. Hydrogen can be used for multiple purposes such as it can be used as an energy carrier, as fuel, in chemical industries, in refining, in fuel cells, and as energy storage media. New hydrogen production project capacities are distributed in three significant sectors of industrial feedstocks, vehicles, gas grid injections, electricity storage, and heat, and hydrogen production project capacities of these sectors are shown in Fig. 1.17 starting from 2000 to 2018. The global hydrogen production and demand are increasing significantly as most of the developed countries are transitioning from conventional energy sources to renewable energy sources and hydrogen becomes the best option solution with unique characteristics described earlier in this chapter and being an environmentally benign solution. The continuous growth in the new hydrogen production project capacities is shown in Fig. 1.17, which is expected to rise further in the coming era.

1.5 Hydrogen Fuel-Cell Applications

Hydrogen storage is a challenging part as it possesses very low-density gas with a density of 40.8 g/m^3. Hydrogen is a secondary (storage) energy source and needs to be produced using primary energy source and reaction always undergoes losses through conversion processes; therefore, the hydrogen production cost is higher than the cost of energy employed to produce hydrogen. Natural gas reforming is the most significant source of global hydrogen production. To overcome various factors such as carbon footprints, environmental issues, greenhouse gas emissions, and carbon emission taxes, a global transition from conventional energy sources to renewable energy sources is desirable for environmentally benign energy solutions. Some technical and scientific hydrogen economy challenges for traditional energy sources are (1) lowering the hydrogen production cost, (2) to develop an environmentally benign and carbon-free clean hydrogen production system for mass production, (3) development of hydrogen delivery and distribution infrastructure, (4) hydrogen storage systems development for stationary and vehicular applications, and (5) drastic cost reduction and substantial development in the durability of fuel cells. Renewable energy-based hydrogen production is the solution to most of these hydrogen economy challenges. Hydrogen carries numerous benefits against fossil fuels as follows:

- Liquid hydrogen offers improved performance as a transportation fuel in comparison with other liquid fuels, namely gasoline, alcohols, and jet fuel
- Five different processes are used to convert hydrogen into useful energy forms such as mechanical, thermal, and electrical while only the combustion process is used to convert fossil fuels into useful energy
- Hydrogen offers high utilization efficiency in conversion to suitable energy forms, and hydrogen is 39% more efficient as compared to fossil fuels
- Hydrogen is the safest fuel in terms of toxicity and fire hazards

The significant fuel-cell application systems are directly linked with electrical power production and utilization for transportation vehicles. The fuel-cell applications are divided into four sections of power generation, propulsion for transportation, multigeneration, and special applications. The power generation sector can be used for further subsections such as automotive, portable, distribution, stationary, auxiliary, mobile, and backup power. The propulsion for the transportation sector can be further divided into auto-vehicles, buses, trains, boats, and aviation. The multigeneration sector can be used for further subsections such as district power, district heating, remote power, remote heating and water, industrial power, industrial heating, and oxygen production. The special applications sector can be further divided into medical, military, and aerospace, which is most common for hydrogen generation and oxygen production. Fig. 1.18 demonstrates the cataloging of applications for fuel cells. For the production of pure oxygen using fuel cells, an electrolyzer of smaller capacity is connected with a fuel cell of larger capacity that uses the pure water from the fuel cell to produce hydrogen and oxygen.

The autothermal reforming of different hydrocarbons and biodiesel was employed for fuel-cell applications in a recent study[29] conducted by the collaboration of Argonne National Laboratory with the University of Puerto Rico. Both of these institutes are working on reforming catalyst development program. The significant focus of this study was to investigate the new catalysts and their viability to convert glycerin, biodiesel, and methanol into hydrogen and also conducted a comparative study based on production potential, determining the coke accumulation conditions and identify the reactor temperature effect and S/C and O/C ratios. In initial autothermal reforming experiments, glycerol, methanol and biodiesel displayed a rise in producing hydrogen with increased reactor temperature and

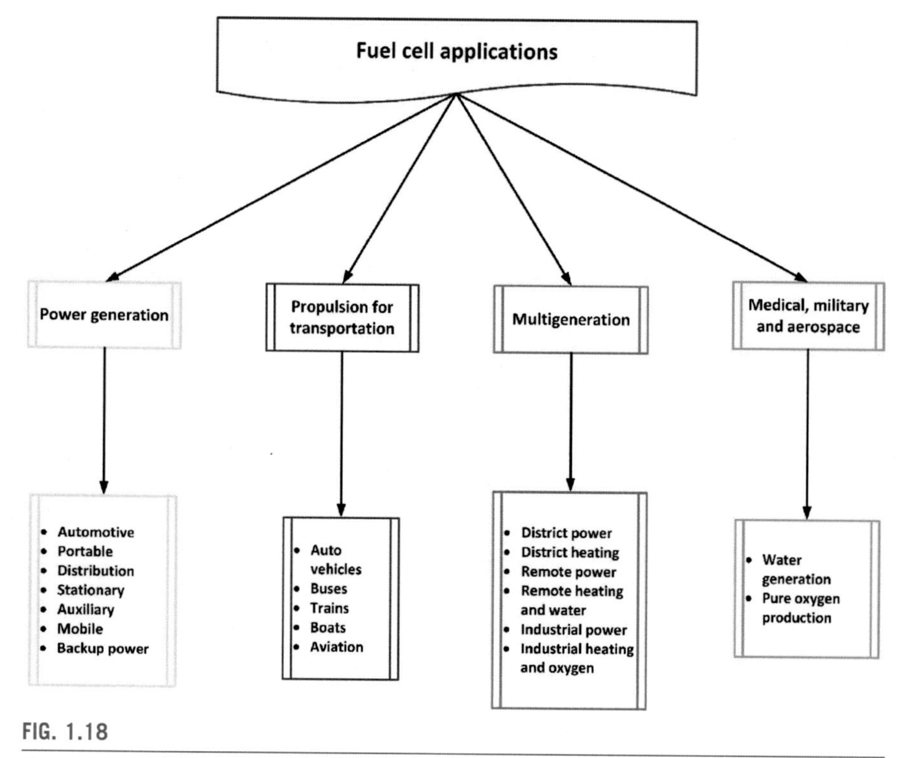

FIG. 1.18

Fuel-cell applications.

drop in O_2/C ratio. All glycerol and biodiesel experiments that were performed showed the formation of coke. A recent study[30] conducted a review study on the solar-hydrogen hybrid fuel-cell systems that can be employed for stationary applications. The most extensively employed solar energy-based hydrogen production technique is water electrolysis using solar energy and solar energy-based hydrogen production techniques were assessed in this study along with the solar-hydrogen hybrid fuel-cell systems and also conducted the preliminary analysis to evaluate the energetic and exergetic efficiencies analyses.

The hydrogen fuel-cell systems can be advantageous for residential applications as well. In the residential sector, the residence can be powered locally by employing hydrogen fuel cells. Additionally, hydrogen can directly be employed and combusted in the residence furnace to produce heat (thermal energy) for space heating and hot water on demand. In the propulsion for transportation and residential sectors, the portable devices for the utilization of hydrogen are the most suitable options such as proton exchange membrane (PEM) fuel cells, solid oxide (SO) fuel cells, phosphoric acid (PHA) fuel cells, and alkaline (ALK) fuel cells as these portable fuel cells demand tiny startup time. Some of the most significant applications of fuel cells are for fuel-cell vehicles and residential applications. Table 1.3 displays the descriptions and characteristics of different fuel-cell types.

Table 1.3 Descriptions and Characteristics of Different Fuel Cell Types [31].

Types of Fuel Cells	Electrolyte Material	Advantages	Disadvantages	Applications	Reactions A: Anode C: Cathode
PEM (polymer electrolyte membrane fuel cell)	Proton-conducting polymer	• Solid construction • Operate at relatively low temperatures (about 200°F).	• Sensitive to impurities • Precious metal catalysts • H_2 fuel	• Portable power • Transportation • Stationary power	A: $H_2 \rightarrow 2H^+ + 2e^-$ **C**: $\frac{1}{2} O_2 + 2H^+ + 2e^- \rightarrow H_2O$
MCFC (molten carbonate fuel cell)	Molten carbonate salt in a ceramic	• Used with many fuels • High efficiency	• Very corrosive electrolyte • High temperature (about 1200°F) • Precious metal catalysts	• Power plants	**A**: $H_2 + CO_3^{-2} \rightarrow H_2O + CO_2 + 2e^-$ $CO + CO_3^{-2} \rightarrow 2CO_2 + 2e^-$ **C**: $\frac{1}{2} O_2 + CO_2 + 2e^- \rightarrow CO_3^{-2}$
SOFC (solid oxide fuel cell)	Oxide ion-conducting ceramic	• Used with many fuels • Solid and rugged	• Very high temperature (1800°F) • Expensive materials	• Stationary power • Semi trucks	**A**: $H_2 + O^{-2} \rightarrow H_2O + 2e^-$ $CO + O^{-2} \rightarrow CO_2 + 2e^-$ $CH_4 + 4O^{-2} \rightarrow 2H_2O + CO_2 + 8e^-$ **C**: $\frac{1}{2} O_2 + 2e^- \rightarrow O^{-2}$
AFC (alkaline fuel cell)	Aqueous potassium hydroxide	• Efficiencies of up to 70%	• Sensitive to carbon dioxide • Very caustic medium • H_2 fuel	• Power and water • Space vehicles	**A**: $H_2 + 2(OH)^- \rightarrow 2H_2O + 2e^-$ **C**: $\frac{1}{2} O_2 + H_2O + 2e^- \rightarrow 2(OH)^-$
PAFC (phosphoric acid fuel cell)	Phosphoric acid in a matrix	• Most commercially developed fuel cell • Operating temperature range of PAFC is 400°F.	• Very corrosive electrolyte • Unstable at higher temperatures • H_2 fuel	• Stationary power • Power plants	**A**: $H_2 \rightarrow 2H^+ + 2e^-$ **C**: $\frac{1}{2} O_2 + 2H^+ + 2e^- \rightarrow H_2O$

1.5.1 Proton Exchange Membrane Fuel Cells

PEM fuel cells play a significant role in the fuel cell-based power generation for transportation and residential sectors because they offer reduced operating temperatures. Fig. 1.19 displays the basic illustration of a PEM fuel cell. The proton-conducting membrane (solid polymer electrolyte) is the particular characteristic of this fuel-cell type. The thin electrolyte allows proton conduction reaction using hydronium ions supported by the wet membrane. The anodic and cathodic reactions can be represented as follows:

Anodic reaction:

$$H_2\ (g) \rightarrow 2H^+\ (aq) + 2e^- \tag{1.3}$$

Cathodic reaction:

$$\frac{1}{2}O_2\ (g) + 2H^+ + 2e^- \rightarrow H_2O\ (aq) \tag{1.4}$$

A polytetrafluoroethylene is more suitable than polystyrene sulfonate polymer for electrolyte as it offers better conductivity. Water flow management in PEM fuel cells is a significant design subject as protons form hydronium ions and for that reason, membranes need to be hydrated to operate. For the reduced ohmic losses and better

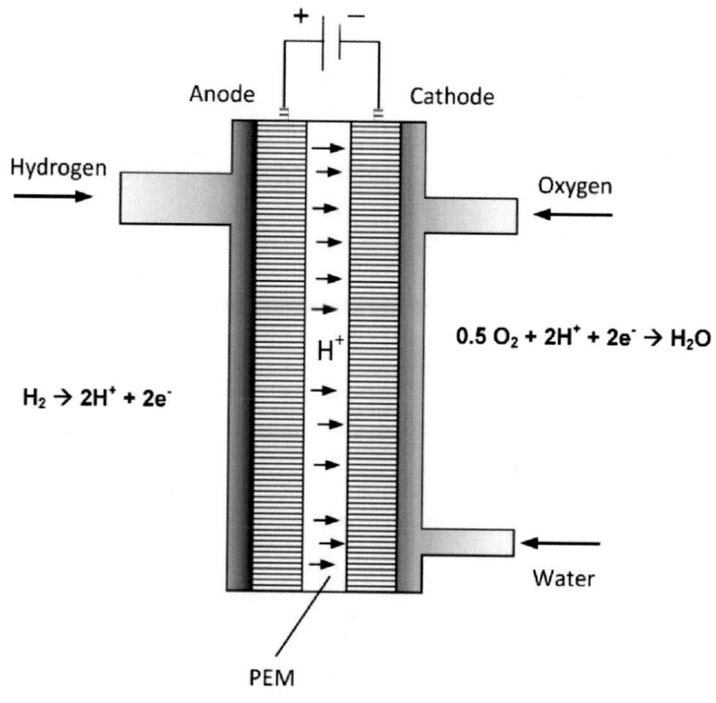

FIG. 1.19

Basic illustration of a PEM fuel cell.

performance, the membrane-electrode assembly is given significant importance in the PEM fuel-cell design. Noble metal such as platinum catalysts can be coated on the electrodes to recompense for slow kinetics, which is designed in a three-phase boundary. Also, the platinum catalyst is essential for anodic reaction to employ pure hydrogen, and platinum catalyst can be poisoned due to CO impurities present in hydrogen.

1.5.2 Phosphoric Acid Fuel Cells

The PHA fuel cells operate similarly to the electrode half-reactions in PEM fuel cells. Protons work as charge carriers in acidic electrolytes while water is formed at the cathode. The electrolyte creates the difference between PHA fuel cells and PEM fuel cells. In PEM fuel cells, the solid polymer electrolyte is used in which phosphoric acid fuel cells employ a liquid acidic electrolyte. In this type of fuel cell, approximately 100% pure and pressurized phosphoric acid electrolyte is used that carries low volatility under 175°C specific operating temperature. The PHA fuel cells are widely developed for commercial level up to 25 MW and tolerate the presence of CO_2 in the air stream.

Silicon carbide is used to stabilize the liquid electrolyte in a solid matrix and installation arrangement is the same as of PEM fuel cells displayed in Fig. 1.20. The stabilization in silicon carbide is achieved to minimize the loss of electrolytes caused by evaporation. The silicon carbide matrix comprises micrometer-sized particles that assure reduced ohmic losses.

1.5.3 Solid Oxide Fuel Cells

The SO fuel cell is an emerging technology with proton conduction. As being an acidic electrolyte, Eqs. (1.3) and (1.4) represent the half-reactions, though, the electrolyte is placed between two planar porous electrodes and formed of a proton-conducting metal oxide layer. Fig. 1.21 exhibits a basic illustration of a proton-conducting SO fuel cell. The SO fuel cells accompanied by proton conduction usually utilize barium oxide. The proton conduction phenomena in solid oxide membranes offer a significant benefit of allowing the protons to travel from anode (+) to cathode (−). As a result, the water is formed at the cathode. The complete utilization of hydrogen is consequently possible in SO fuel cells with straight implications to raise the system compactness and simplicity by eradicating the afterburner requirement. Furthermore, zero NO_x formation occurs as complete hydrogen reacts electrochemically at fuel-cell cathode, thus the emissions emitted by the fuel cell only consist of nitrogen and steam and that means it is clean. Enormous efforts are being devoted to developing proton-conducting membranes globally. The absence of the liquid phase in the SO fuel cells supports the design. In this regard, the materials of barium cerate ($BaCeO_3$) were recognized as outstanding solid oxide electrolytes as they offer high capability proton conduction over wide temperatures ranges from 300°C to 1000°C. The key problematic issue is the formation of a solid membrane due to barium cerate. Doping barium cerate with Samarium (Sm) appears

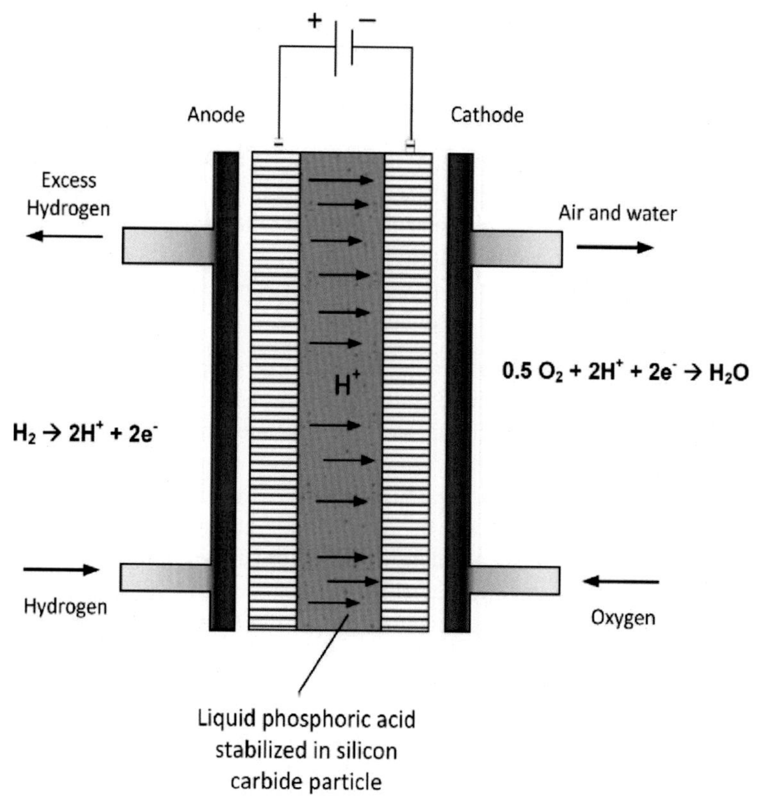

FIG. 1.20

Basic illustration of a phosphoric acid (PHA) fuel cell.

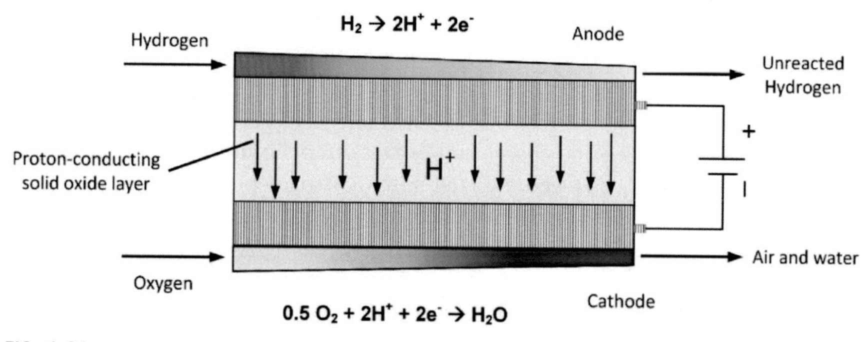

FIG. 1.21

Basic illustration of a solid oxide (SO) fuel cell.

to be a suitable option that allows for sintering membranes with low thickness and high-power densities. They carry yttria-stabilized zirconia electrolyte solid layer that operates under a high temperature of 1000°C and SO fuel cells offer the following significant advantages:

- As no noble metal catalysts are desirable, they are cost-friendly and offer a comparatively longer lifetime.
- Internal reforming fuels (e.g., methane, ammonia) to hydrogen is enabled, so a reduced fuel size tank can be employed.
- The high exergy associated with the exhaust gases can be used to produce low-temperature heating and supplementary power.

The absence of the liquid phase in the SO fuel cells supports the design with the existence of solid and gas (two-phase) solid–gas processes. Throughout the process, hydrogen is consumed at the anode and steam is generated and the partial pressure of hydrogen drops down because of this reason. As a result of decreased hydrogen partial pressure, the degradation in the reaction kinetics occurs and supplying excess hydrogen is the only way to resolve this problem and to provide the compensation for this effect and excess hydrogen is consumed, which can be completed in numerous ways. For instance, using the afterburner to combust surplus hydrogen and thermal energy is used to generate work through a gas turbine or releasing the recovered heat are the common methods. Therefore, certain NO_X amounts result during hydrogen combustion.

1.5.4 Alkaline Fuel Cells

Although ALK fuel cells require a complex electrolyte recirculation system, they are the favored space applications technology and are very striking for road vehicles, However, PEM fuel cell is the greatest technologically advanced option for vehicular application even though it entails a compromised driving range, subsequently with the hydrogen fuel. On the differing, high-temperature fuel cells, namely SO fuel cells are more suitable to the larger-power applications. Similarly, SO fuel cells can be employed to the vehicles as a supplementary power unit that can operate at a steady state. At this juncture, PEM fuel cells and ALK fuel cells perform a significant part because they offer reduced operating temperatures.

ALK fuel cells offer slightly higher efficiency as compared to the PEM fuel cells with more than 60% that can reach 70% potentially and take the hydroxyl benefit and utilize it as an outstanding charge carrier that offers faster oxygen reduction kinetics. These ALK fuel cells were utilized in NASA space programs, and excellent stability was demonstrated. The ALK fuel cells catalyst requirement is less stringent as compared to the PEM fuel cells as they use a cheap nickel-based catalyst. The 30%–40% by weight potassium hydroxide solution is used as the electrolyte. Following are the half-reactions at each electrode:

Anodic reaction:

$$H_2(g) + 2OH^-(aq) \rightarrow 2H_2O\ (l) + 2e^- \tag{1.5}$$

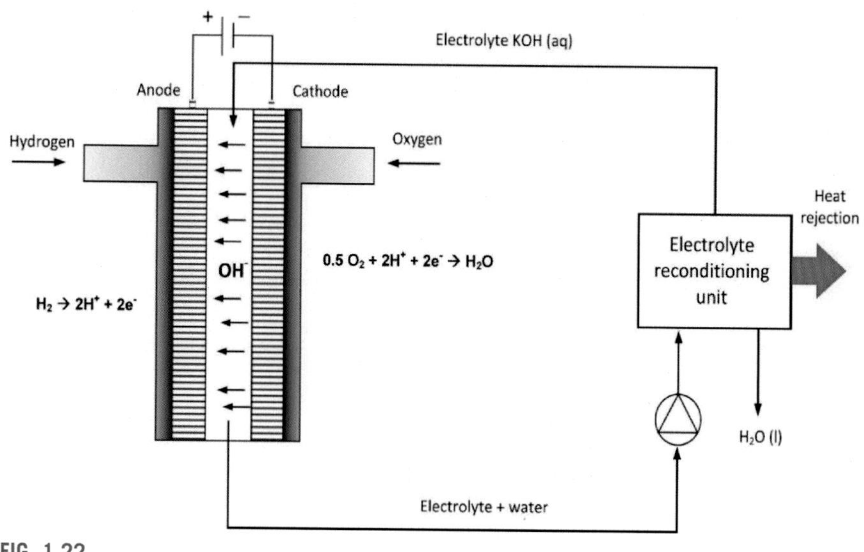

FIG. 1.22

Basic illustration of an alkaline (ALK) fuel cell.

Cathodic reaction:

$$\frac{1}{2}O_2\ (g) + 2\ H_2O + 2e^- \rightarrow 2OH^-(aq) \tag{1.6}$$

The spontaneous potassium hydroxide reaction with CO_2 in the air is a significant electrolyte problem. If CO_2 exists in the potassium hydroxide electrolyte, carbonate is formed in the aqueous CO_3^{-2} solution. As a consequence, the electrolyte is degraded. Fig. 1.22 exhibits the basic illustration of an ALK fuel-cell system configuration. Hydrogen enters via porous anode and generates an electrolyte–electrode interface. The reconditioning of electrolytes is done externally to eliminate the generated water, and fresh electrolyte is added if required. Subsequently, the electrolyte is entered back to the cell.

1.5.5 Ammonia Fuel Cells

The working principle of an ammonia fuel cell is similar to the hydrogen fuel cell involving electrode reactions in addition to membrane electrolytes. Additionally, hydrogen can be mixed with ammonia and fed to the fuel cell that improves the experimental performance and efficiencies of the ammonia fuel cell. Nevertheless, alkaline electrolyte-based ammonia fuel cell involves some differences in comparison with hydrogen-fueled PEM fuel cell. Fig. 1.23 displays a general schematic of the alkaline electrolyte-based direct ammonia fuel cell (DAFC) displaying the anodic and cathodic reactants and products.

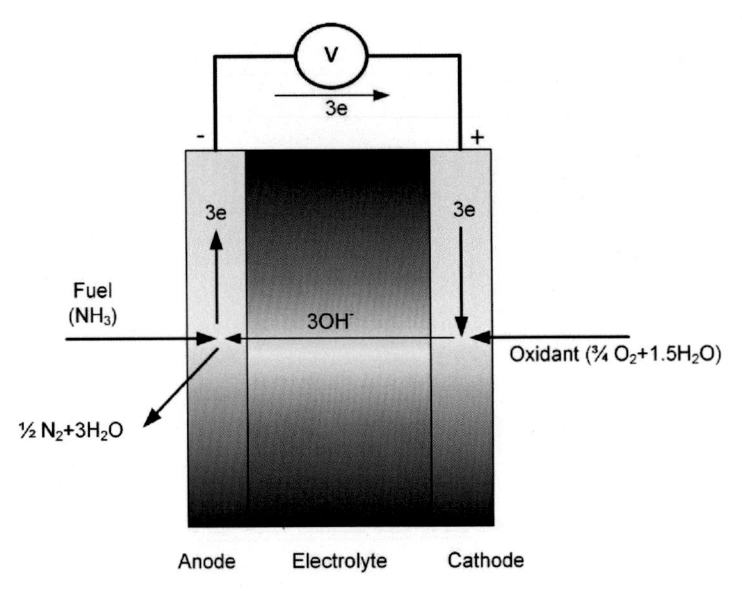

FIG. 1.23

Basic illustration of a direct ammonia fuel cell.

The input fuel feed in a DAFC contains direct ammonia (NH_3) inlet. In alkaline-based DAFC, this fuel input feed reacts at the anode electrochemically in the existence of catalyst through negatively charged hydroxyl ions. The anodic reaction can be expressed as

$$NH_3 + 3OH^- \rightarrow \frac{1}{2} N_2 + 3H_2O + 3e^- \tag{1.7}$$

The previously mentioned anodic electrochemical reaction releases electrons that generate electrons flow over any external load and permit the utilization of useful power. The cathodic electrochemical reaction that receives electrons and concludes the overall chemical reaction can be written as

$$\frac{3}{4} O_2 + 1.5H_2O + 3e^- \rightarrow 3OH^- \tag{1.8}$$

The mandatory cathodic reactants of water (H_2O) and oxygen (O_2) molecules are required to be made available at the cathode that also reacts in the existence of electrochemical catalyst to form the hydroxyl ions (OH^-). These ions that are shaped at cathode travel to the anodic lateral of fuel cell over an alkaline electrolyte that permits only negatively charged ions to pass through. Hence, the electrochemical reactions keep on taking place at either electrode with oxidant inputs and fuel and the valuable power can be consistently extracted from the cell. The overall reaction of an alkaline electrolyte-based DAFC can be expressed as

$$NH_3 + \frac{3}{4} O_2 \rightarrow \frac{1}{2} N_2 + 1.5H_2O \tag{1.9}$$

Even though the DAFC working principles are reasonably well established, and their performance is comparatively lower than hydrogen fuel cells. This is principally accredited to the electro-oxidation of unsatisfactory ammonia molecules. The catalyst activity practical in the hydrogen oxidation case for fuel cells is not established for ammonia oxidation case yet. This remains a key factor that prohibits the high-performance DAFC development.

1.6 Closing Remarks

The substantial growth in global economic development and population occurring is the reason for increased energy demand. Power generation plays a vital role in the industrial revolution of any country. The major portion of this energy demand is set to be covered by the traditional energy production methods employing fossils fuels, and a part of global energy demand is covered by renewable energy sources that are growing significantly. The utilization of traditional fossil fuels creates numerous environmental problems and GHG emissions. Thus, a transition toward 100% renewable energy is required to overcome the challenges of global warming, ozone layer depletion, carbon emissions taxes, and fossil fuel depletion. The intermittent nature of some renewable energy sources such as solar and wind increases the requirement of an energy storage media and hydrogen is the chief candidate that can be used for multiple purposes such as fuel, energy storage media, synthesis of methanol and ammonia, and energy carrier. A global revolution is occurring in the transportation sector with the transition of fuel combustion-based vehicles to hydrogen fuel-cell and hybrid hydrogen fuel-cell vehicles. The hydrogen fuel cell provides clean power using hydrogen with zero carbon emissions, and these applications are not only limited to the car, truck, tankers, and trackers but also include a locomotive, boating, and aviation vehicles. In the hydrogen fuel cell hybrid electric vehicles, a battery storage system is also employed to store the additional electricity and is used to drive a vehicle when required. The hydrogen fuel cell is a well-developed and efficient technology used to generate electricity using hydrogen. A high-performance indicator is offered by the hydrogen fuel cells in terms of efficiencies as fuel cells do not follow the efficiency limits of the Carnot cycle. Hydrogen can be employed for multiple applications including jet fuel, ammonia and methanol synthesis, oil refining, hydrogen fuel-cell electric vehicles and for sustainable energy systems.

CHAPTER

Hydrogen Production Methods

2

Although hydrogen is known as the most widely found element in almost everything on the earth, it is not really available alone. It is therefore important to disassociate it from various hydrogen containing sources, where extraction needs to be carried out in an environmentally-friendly manner to produce pure hydrogen. Undeniably, this extraction process requires energy; nevertheless, hydrogen extraction can be performed employing any primary energy source as hydrogen is an energy carrier but not an energy source. This indicates that a primary energy source among different candidates such as fossil fuels, natural gas, nuclear, solar, biomass, wind, hydro, or geothermal is used to produce hydrogen. Typically, hydrogen production can be performed employing miscellaneous energy resources namely fossil fuels for example natural gas and coal, nuclear energy, and different renewable energy sources, namely solar, wind, geothermal, hydro, ocean thermal energy conversion (OTEC), and biomass. The diversity and range of alternative potential energy sources also make hydrogen a promising energy carrier.

Even though steam methane reforming is the CO_2 intensive process that is being used globally for significant portion of hydrogen production today, renewable electricity can also be employed to produce green hydrogen undergoing zero CO_2 emissions as electrolysis is another conventional technique that splits water into oxygen and hydrogen employing electrical current and this electricity can be harvested from renewable energy sources to produce green hydrogen.

The hydrogen production cost is a significant subject. Steam reforming-based hydrogen production costs roughly three times as compared with the natural gas cost per unit of produced energy. Likewise, using electrolysis for hydrogen production with 5 cents/kWh of electricity will cost somewhat beneath two times of natural gas-based hydrogen production cost. According to the new report published by renewable world energy[32], the United States will be selling wind energy at the lowest recorded price of 2.5 cents/kWh, thus the electricity will cost somewhat less than four times of natural gas-based hydrogen production cost.

Hydrogen fuel is dragging the attention globally because of numerous reasons, and some of these significant reasons are listed as follows:

- Hydrogen production can be carried out using diverse energy resources.
- Hydrogen is suitable to meet all energy requirements and can be employed to hydrogen fuel-cell vehicles, used for residential applications, used as an energy carrier and can also be employed as a fuel for combined heating and power production systems

Renewable Hydrogen Production. https://doi.org/10.1016/B978-0-323-85176-3.00005-6

- Hydrogen is the slightest contaminating; subsequently, the hydrogen usage in fuel cells or combustion processes produces water
- Hydrogen is the flawless solar energy carrier, and it can also be used as an energy storage medium.

Table 2.1 arranges the comparative cost and performance evaluations of several hydrogen production processes. It incorporates the significant hydrogen production processes such as steam methane reforming, landfill gas dry reformation, coal gasification, H_2S methane reforming, naphtha reforming, methane/natural gas pyrolysis, steam-iron process, steam reforming of waste oil, partial oxidation of heavy oil and coal, grid electrolysis of water, high-temperature water electrolysis, chlor-alkali electrolysis, solar and photovoltaic (PV) water electrolysis, photolysis of water, biomass gasification, photobiological, thermochemical water splitting, photoelectrochemical water decomposition, and photocatalytic water decomposition and states the ideal and practical energy requirements and efficiencies and cost comparison.

A recent review article[33] conducted a comparative study and environmental impact assessment of nonrenewable and renewable sources based on hydrogen production approaches. They aimed at studying and comparing the hydrogen production methods performance and evaluate the social, economic, and environmental impacts. A number of methods were considered, namely coal gasification, natural gas reforming, solar and wind energy-based water electrolysis, biomass gasification, high-temperature electrolysis, S—I cycles, and thermochemical Cu—Cl cycles. Environmental impacts in terms of global warming potential (GWP), acidification potential and GWP, energetic and exergetic efficiencies, and production costs of mentioned methods are determined and compared. Moreover, the relationship between the capital cost of hydrogen production and plant capacity was considered. The results revealed that S—I cycles and thermochemical Cu—Cl cycles are considered environmentally benign as compared with other traditional methods. The solar, wind, and high-temperature electrolysis were also revealed to be environmentally attractive while electrolysis-based methods were least attractive comparing the production costs. Consequently, efficiencies enhancement and cost reduction of wind and solar energy-based hydrogen production methods were found to be potential options. Energetic and exergetic efficiencies comparison indicated the benefits of biomass gasification-based hydrogen production over other methods. Inclusive rankings indicated that S—I and thermochemical Cu—Cl cycles were found to be promising applicants to produce cost-effective and environmentally benign hydrogen.

Life cycle assessment was conducted in a recent study[34] for numerous hydrogen production approaches. Their study included five different methods of hydrogen production such as natural gas reforming, solar and wind energy-based water electrolysis, coal gasification, and thermochemical Cu—Cl cycle. Each described method was evaluated and compared considering the energy equivalents and CO_2 equivalent emissions, and they also conducted a case study for hydrogen fueling stations considering the geographical location of Toronto, Canada. It was revealed that the thermochemical Cu—Cl cycle becomes advantageous considering the CO_2 equivalent emissions followed by solar and wind electrolysis. The methods of natural gas reforming, thermochemical Cu—Cl cycle, and coal gasification were found to be beneficial over renewable energy approaches considering hydrogen production capacities.

Table 2.1 Comparative Cost and Performance Features of Hydrogen Production Processes [37].

Process	Energy Required (kWh N^{-1} m^{-3})		Status of Tech.	Efficiency [%]	Costs Relative To SMR
	Ideal	Practical			
Steam methane reforming (SMR)	0.78	2–2.5	Mature	70–80	1
H_2S methane reforming	1.5	–	R&D	50	<1
Methane/NG pyrolysis			R&D to mature	72–54	0.9
Landfill gas dry reformation			R&D	47–58	~1
Naphtha reforming			Mature		
Partial oxidation of heavy oil	0.94	4.9	Mature	70	1.8
Coal gasification (TEXACO)	1.01	8.6	Mature	60	1.4–2.6
Steam reforming of waste oil			R&D	75	<1
Steam-iron process			R&D	46	1.9
Partial oxidation of coal			Mature	55	
Grid electrolysis of water	3.54	4.9	R&D	27	10-Mar
Chlor-alkali electrolysis			Mature		By-product
High-temp. electrolysis of water			R&D	48	2.2
Solar and PV-electrolysis of water			R&D to mature	10	>3
Biomass gasification			R&D	45–50	2.0–2.4
Thermochemical water splitting			Early R&D	35–45	6
Photolysis of water			Early R&D	<10	
Photobiological			Early R&D	<1	
Photocatalytic decomp. of water			Early R&D		
Photoelectrochemical decomp. of water			Early R&D		

In the recently published book entitled accelerating the transition to a 100% renewable energy era[35], a chapter was included enlightening the hydrogen role in the global transition to potentially 100% renewable energy. This chapter enclosed all the renewable energy that can replace the traditional hydrogen production methods of natural gas reforming and coal gasification. Each renewable energy method was incorporated for producing hydrogen including wind, solar, hydro, geothermal, ocean thermal energy conversion, and biomass gasification. They proposed the transition toward renewable energy methods replacing the traditional energy methods for clean, sustainable, and environmentally benign hydrogen production. Another recent study[36] investigated hydrogen production methods according to green chemistry principles. They aimed this study to explore the hydrogen production approaches according to green chemistry principles. Every single method was analyzed and assessed using 12 principles to identify the requirement criteria. The methods of hydrogen production explored in this study were categorized into four different classifications of energy sources, namely electrical, hybrid, thermal, and biological. Investigations revealed that water electrolysis, photoelectrochemical method, biomass gasification and biophotolysis were found to produce green hydrogen among electrical, hybrid, thermal, and biological methods respectively.

Fig. 2.1a displays the global CO_2 emission distributions among different sectors of electricity, heat generation, transport, industry, buildings, and other sectors. The collective CO_2 emissions of the electricity heat generation sector and transport

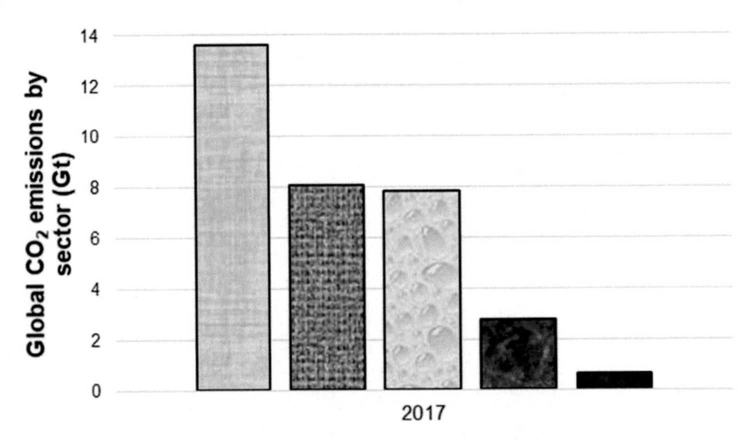

(a)

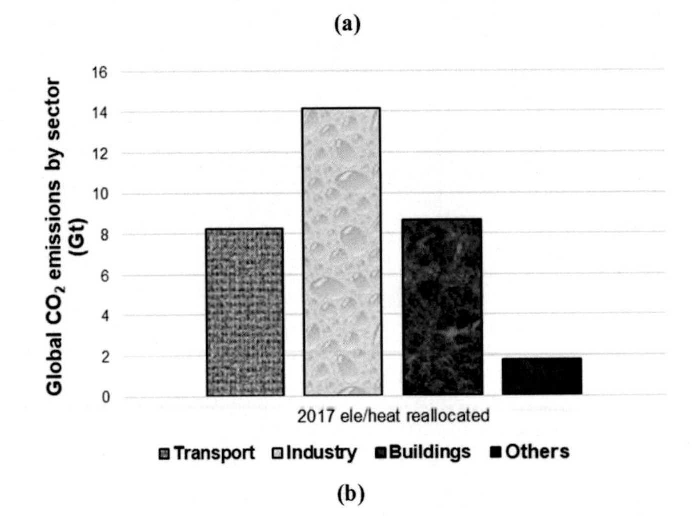

(b)

FIG. 2.1

(a) Global CO_2 emissions distribution among different sectors; (b) heat reallocation in different sectors.

sector constitute almost two-thirds of the total emissions, and these sectors were accountable to nearly complete global emissions growth since 2010 and an outstanding one-third is divided between buildings and industry. Fig. 2.1b displays the heat reallocation in the transport, industry, buildings, and other sectors. Figure displays the reallocation global CO_2 emissions from power generation. The industry sector is accountable for somewhat half of the total emissions while transport and building sectors are accountable for each quarter. The one-half of global electricity is consumed by the buildings sector and industry consumes almost the other half.

Dincer and Acar[38] firstly included a 3S approach originally introduced by Ibrahim Dincer to be employed for hydrogen production systems. Fig. 2.2 exhibits the steps of source, system, and service included in this approach. The sources included in this approach are the primary energy sources to be integrated with the hydrogen production system followed by the storage and distribution to encounter the supply and demand concerning the distribution employing the existing infrastructure and supply for the end-use amenities such as fuel cells for power production and hydrogen fuel-cell vehicles, internal combustion engines, manufacturing of different chemicals, namely ammonia, methanol, pharmaceuticals, and hydrochloric acid and to encounter the residential applications, namely heating, cooling, power, freshwater, combined heating

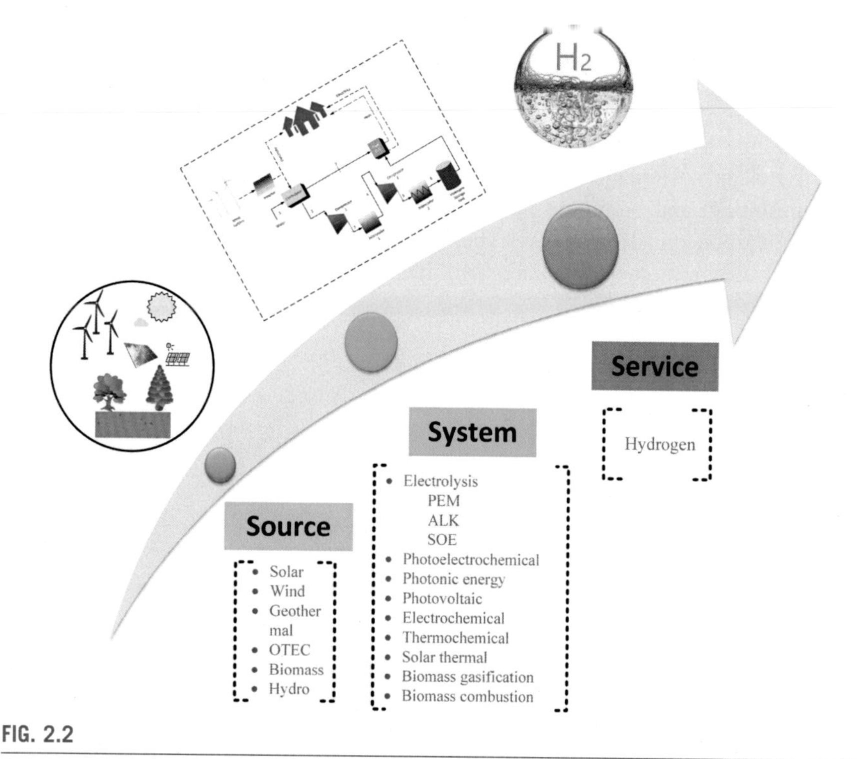

FIG. 2.2

3S approach for renewable energy-based hydrogen production systems.

and power production, and space heating and cooling need to be considered to deliver an environmentally benign and sustainable approach to eradicate the environmental problems. Intending to encounter the solutions of producing sustainable energy, environmental problems and greenhouse gas (GHG) emissions need to be eradicated.

The selection of the primary energy source for producing hydrogen is the first step and the source needs to be abundant, clean, plentiful, reliable, and affordable. Currently, the mainstream hydrogen (almost 95%) is produced using fossil fuels and more specifically natural gas reforming, methane partial oxidation and coal gasification. For some intermittent renewable energy sources, a consistent, reliable, trusted and principal storage system is required to be combined with the systems. According to the new report published by renewable world energy, the United States will be selling wind energy at the lowest recorded price of 2.5 cents/kWh, thus the electricity will cost somewhat less than four times of natural gas-based hydrogen production cost. To meet the optimum outcomes, the selection of a suitable system with a source is also substantial. The outcome of using this approach is environmentally benign, clean and sustainable hydrogen.

The mainstream hydrogen (almost 95%) is produced using fossil fuels and more specifically natural gas reforming and coal gasification. Additional significant methods of producing hydrogen consist of water electrolysis and biomass gasification. Hydrogen production methods are distributed between two categories of conventional and renewable energy methods. The significant conventional hydrogen production methods employing different sources are as follows:

- Natural gas reforming
- Coal gasification

Renewable hydrogen production methods are gaining much attention because they offer clean, sustainable, and environmentally benign energy and hydrogen production solutions and overcome the challenges of GHG emissions, fossil fuel depletion, and carbon emissions taxes. The significant hydrogen production methods employing different renewable energy sources are listed as follows:

- Solar
  - Thermochemical
  - Photochemical
  - Photoelectrochemical
  - Electrolysis
- Wind
- Hydro
- Ocean thermal energy conversion
- Tidal
- Geothermal
- Biomass

Fig. 2.3 exhibits the distribution of conventional and renewable hydrogen production methods. A brief description of each hydrogen production method subdivided into conventional and renewable methods is described in this chapter.

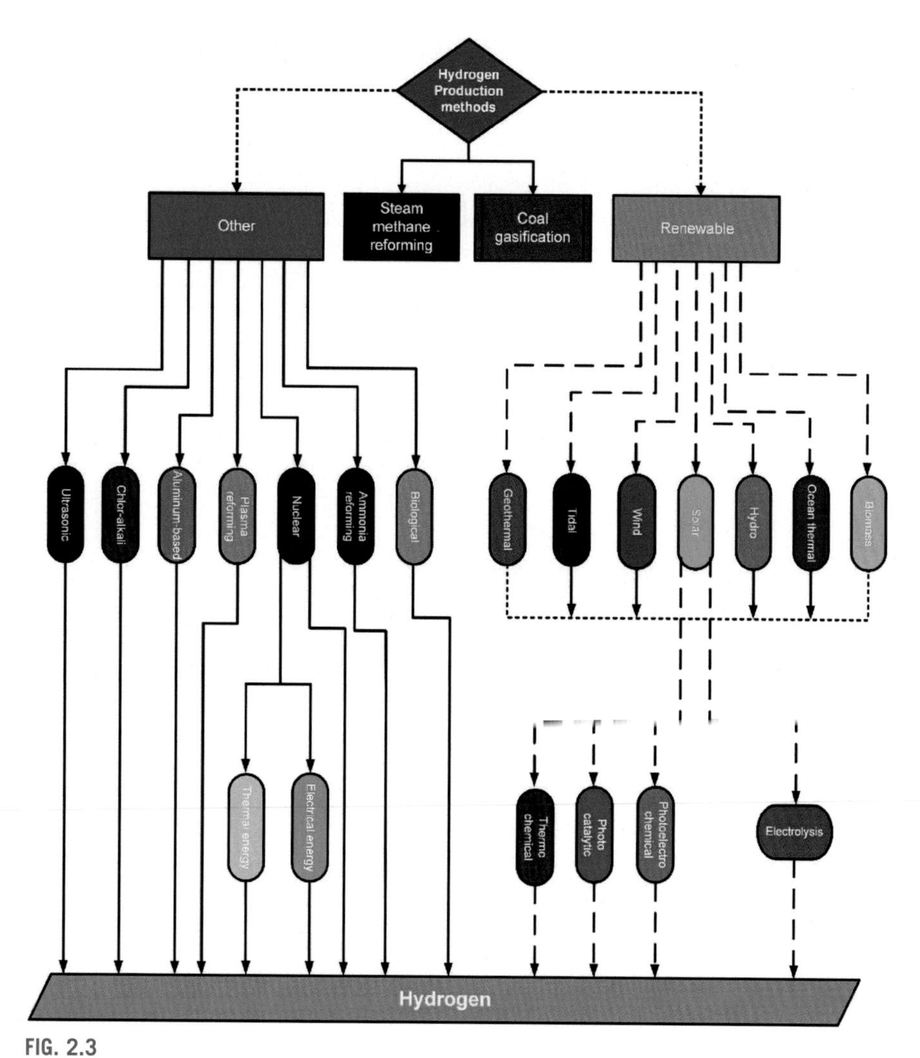

FIG. 2.3

Conventional and renewable hydrogen production methods.

2.1 Conventional Hydrogen Production Methods

This section encloses the different conventional hydrogen production methods, and a transition is required from these traditional energy sources to renewable energy sources that can produce environmentally benign hydrogen.

2.1.1 Natural Gas Reforming

A significant portion of the traditional hydrogen is produced using fossil fuels and more specifically natural gas reforming and methane partial oxidation. Fig. 2.4 displays the steps involved in the natural gas reforming for hydrogen production while Fig. 2.5 displays the schematic of the natural gas forming system.

A recent study[40] conducted the technoeconomic analysis and assessment of natural gas reforming-based hydrogen production. It is firm that hydrogen carries the potential that can play substantial involvement in global energy production. The economic and technological analysis was performed with intentions to overcome the

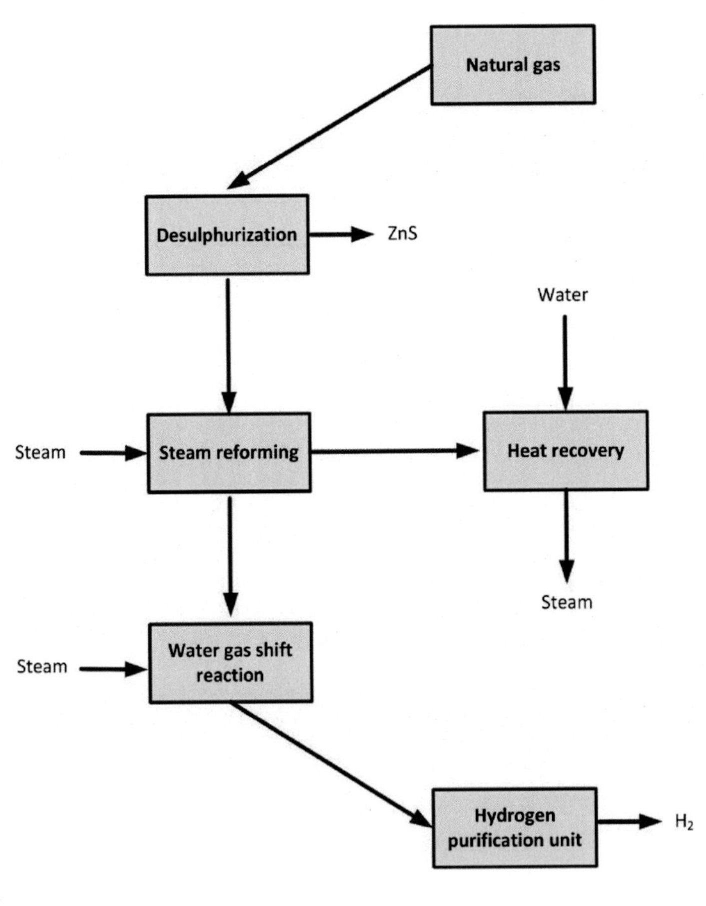

FIG. 2.4

Natural gas reforming system steps for hydrogen production.

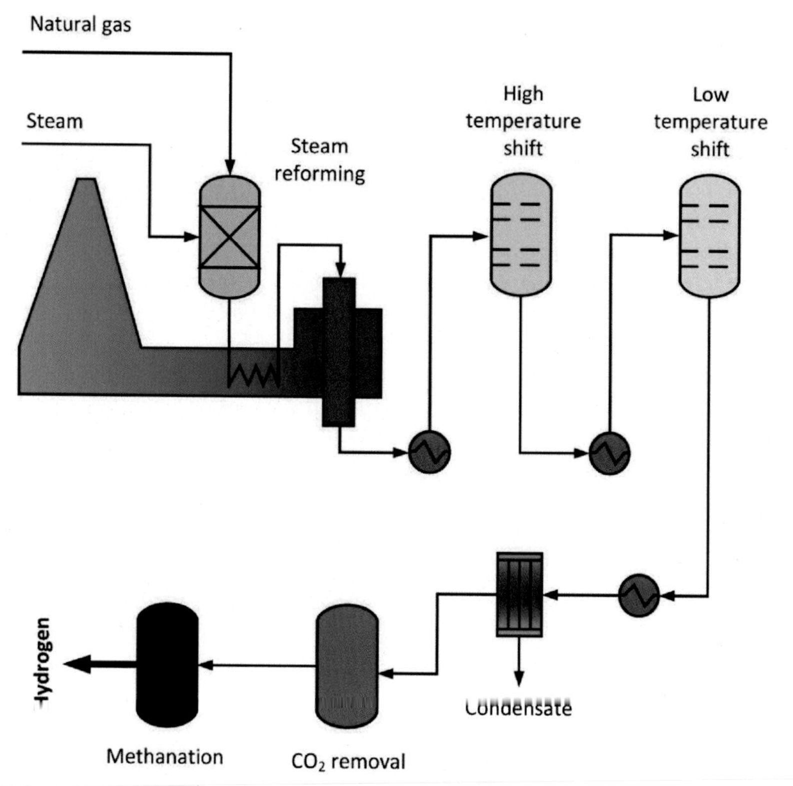

FIG. 2.5

Schematic of the natural gas reforming system for hydrogen production.

Modified from [39].

most probable pathway of producing hydrogen for the automotive industry. The two significant methods of centralized hydrogen production and ammonia thermal cracking were evaluated while natural gas reforming was found to be the most technically feasible options and mentioned processes were further assessed and estimated using comprehensive economic analysis. The outcomes revealed that the natural gas reforming offered less payback time as well as higher investment return. Numerous sensitivity analyses were also conducted to explore the effect of the influential variables on the hydrogen production cost such as capital cost, feedstock price, and operating capacity factor. This study revealed the natural gas reforming process as the most technologically and economically obtainable raw material for the time being before the transition to renewable energy.

Numerical model development for natural gas reforming integrated with furnace using steam reforming process (SRP) one-dimensional numerical model was established in a recent study.[41] The model was employed using the programming language C in the commercial Fluent software. The SRP model represented energy and mass balances along channels/tubes under uniform conditions excluding the temperature inside porous regions and also calculated the temperature profiles and concentrations of gas species along channels/tubes. They also validated the system through experimental results attained from a laboratory prototype tubular reactor. The designed model was formerly coupled with Fluent that enabled its employment in simulating the steam reformer operating under the capacity of 5 kg/day. A burner was employed in the middle of the reformer vessel that was bounded by channels through multiple passages. They employed the model for several mixtures and displayed their effect on the performance of the reformer.

A recent study[42] conducted the analysis, optimization, cost estimation, energy saving analysis and carbon footprints considering the carbon emissions taxes of a solar supported natural gas steam and autothermal reforming system for multigeneration of hydrogen, power, and ammonia. The CO_2 emissions generated through the steam methane reforming system were fed to autothermal reforming system. A cryogenic air separation method was incorporated with the system to supply oxygen to autothermal reforming system, and separated nitrogen was employed for ammonia synthesis. An aqueous ammonia-based carbon capture system was employed to capture CO_2 emissions emitted by autothermal reforming system. The power produced by the system covers the system requirement as well as produces power as a final output. The designed system offered a methodology to update the existing natural gas reforming system not only to eradicate CO_2 emissions but also to avoid carbon emissions taxes. The designed system also offered high exergetic and energetic efficiencies of 45.0% and 53.4%, respectively.

In the initial processing step, the desulfurization of natural gas is employed to separate the sulfur content from the natural gas and gaseous hydrogen sulfide (H_2S) is found in raw natural gas. The sulfur is recovered using the Claus process from gaseous H_2S. The produced gaseous hydrogen sulfide reacts with oxygen gas to separated sulfur in Claus plants, and the following is the reaction:

$$2H_2S + O_2 \rightarrow 2S + 2H_2O \tag{2.1}$$

The activated titanium or aluminum oxide catalyst is employed to the catalytic Claus process that boosts the conversion efficiency of sulfur. The sulfur dioxide (SO_2) that is formed through combustion reacts with gaseous hydrogen sulfide, and sulfur is separated from gaseous hydrogen sulfide by separating the elemental sulfur.

$$2H_2S + SO_2 \rightarrow 3S + 2H_2O \tag{2.2}$$

The step followed by the desulfurization of natural gas is steam methane reforming. In this step, steam reacts with natural gas at high temperatures and forms carbon monoxide and hydrogen gases. Following is the chemical reaction:

$$CH_4 + H_2O \rightarrow CO + 3H_2 \tag{2.3}$$

A steam-based autothermal reforming is another type of steam methane reforming that is also followed by a water gas shift reaction used to convert CO to CO_2. The reaction can be represented as follows:

$$2CH_4 + CO_2 + O_2 \rightarrow 3H_2 + H_2O + 3CO \tag{2.4}$$

$$3CO + 3H_2O \rightarrow 3H_2 + 3CO_2 \tag{2.5}$$

The carbon dioxide-based autothermal reforming is the second type of autothermal steam methane reforming that is also followed by a water gas shift reaction used to convert CO to CO_2. The reaction can be represented as follows:

$$4CH_4 + 2H_2O + O_2 \rightarrow 10H_2 + 4CO \tag{2.6}$$

$$4CO + 4H_2O \rightarrow 4H_2 + 4CO_2 \tag{2.7}$$

The steam methane reforming process is followed by the water gas shift reaction (WGSR). This step consists of two steps of high-temperature and low-temperature shift reactions. A water gas shift reactor uses steam to convert carbon monoxide to carbon dioxide and produces hydrogen gas. The reaction is as follows:

$$CO + H_2O \rightarrow CO_2 + H_2 \tag{2.8}$$

The aqueous ammonia-based carbon-capture can be used to capture CO_2 emissions released by the natural gas reforming process. The aqueous ammonia reacts as an absorbent for CO_2 capture. The water amount in the aqueous ammonia solution helps to determine the concentration and molarity of the absorbent. The absorbent decreases the CO_2 amount entailed in the input stream, and the following are the chemical reactions involved in the aqueous ammonia-based carbon dioxide-capturing system:

$$NH_3 + H_2O \leftrightarrow NH_4^+ + OH^- \tag{2.9}$$

$$NH_3 + CO_2 + H_2O \leftrightarrow NH_2COO^- + H_3O^+ \tag{2.10}$$

$$2H_2O \leftrightarrow H_3O^+ + OH^- \tag{2.11}$$

$$HCO_3 \leftrightarrow CO_2 + OH^- \tag{2.12}$$

$$CO_2 + 2NH_3 + H_2O \rightarrow (NH_4)_2CO_3 \tag{2.13}$$

$$HCO^{3-} + H_2O \leftrightarrow CO_3^{-2} + H_3O^+ \tag{2.14}$$

$$NH_4CO_{3(s)} + H_2O \leftrightarrow NH_4^+ + HCO_3^- + OH^- \tag{2.15}$$

$$CO_2 + OH \leftrightarrow HCO_3^- \tag{2.16}$$

2.1.2 Coal Gasification

Coal gasification is another significant method used for hydrogen production. Fig. 2.6 exhibits the coal gasification system steps for hydrogen production. The first step is the air separation that separates oxygen from air and supplies to the gasifier.

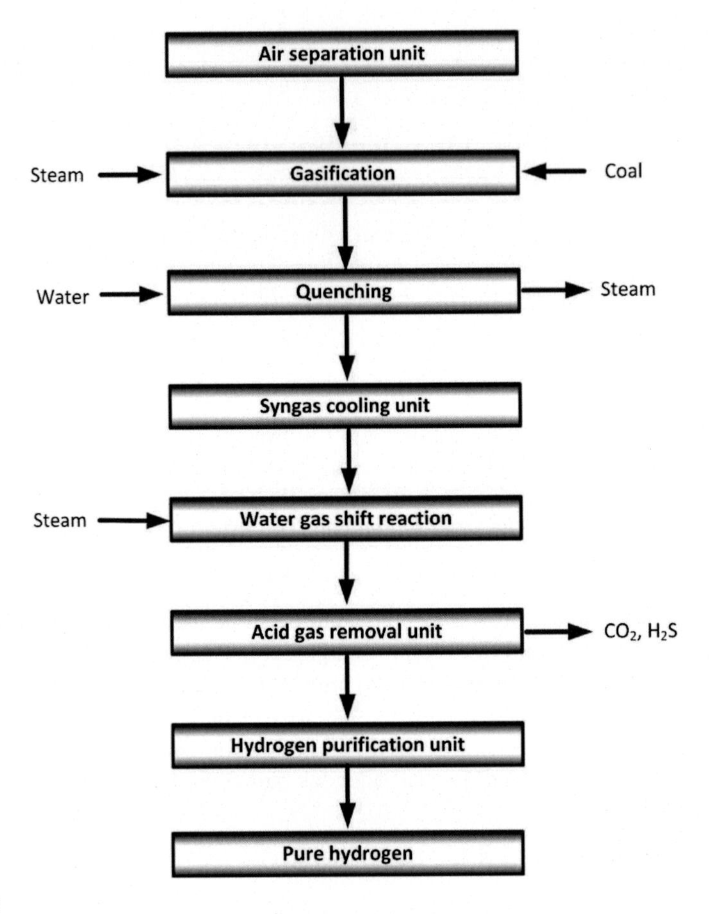

FIG. 2.6

Coal gasification system steps for hydrogen production.

One of the three well-established methods namely; membrane separation, cryogenic air separation, or pressure swing adsorption is used for air separation. The step followed by the air separation is the coal pyrolysis. The pyrolysis step requires two inputs of steam and oxygen with coal for gasification. In this step, coal is wrecked into volatile and char.

In a recent study[43], the development and analysis of coal gasification-based hydrogen production plants incorporating thermochemical Cu–Cl cycle with compression system for hydrogen, entrained flow gasifier, cryogenic air separation system, WGS membrane reactor, and hydrogen-driven combined power cycle was proposed. The gasifier yielded syngas that passed through WGS membrane reactor that also incorporated the hydrogen stripping by increasing the percentage of shift reaction and to capture more hydrogen. The outstanding syngas was fed to the Brayton cycle for combustion that yields power using a gas turbine. They also recovered the thermal energy from the turbine output to generate steam that is supplied to the thermochemical Cu–Cl cycle for hydrogen production. The produced power covered the system requirement and also extracted as a significant output. The formed hydrogen was compressed at 700 bar pressure for storage. The hydrogen production plant is developed and modeled in the Aspen Plus software package. The designed system offered high energetic and energetic efficiencies of 47.6% and 51.3%, respectively.

The supply-chain system for the technoeconomic analysis of biomass and coal gasification-based production of hydrogen was incorporated in a recent study.[44] They assessed the biomass cofiring effect on gasification for producing hydrogen considering the supply chain through CO_2 capture as of economical, technical, and environmental viewpoint. Numerous cases involving different feedstocks were employed to the gasification reactor to investigate the mixture of coal with wheat straw or sawdust. The energetic efficiency through performance, syngas composition as well as CO_2 were carried out. The overall operating costs and capital investment were also evaluated for the gasification plant, and a discrete model was also developed using results extraction using the Aspen Plus simulations. The results revealed that an increase in the quantity of biomass in feedstock decreased the production of hydrogen from 7% to 23% for wheat straw and 28% for sawdust.

The volatile matter contains moisture and tar, which is presented by C_6H_6. The pyrolysis reaction[43] is written as follows:

$$\text{Coal} \rightarrow \text{Char} + (C_6H_6 + H_2 + CO + H_2S + CO_2 + N_2 + CH_4 + H_2O) \tag{2.17}$$

Subsequent to the pyrolysis reaction, the combustion of the volatile matter can be expressed by the following reactions:

$$C_6H_6 + \frac{15}{2}O_2 \rightarrow 3H_2O + 6CO_2 \tag{2.18}$$

$$CO + \frac{1}{2}O_2 \rightarrow CO_2 \tag{2.19}$$

$$H_2 + \frac{1}{2}O_2 \rightarrow H_2O \tag{2.20}$$

$$CH_4 + 2O_2 \rightarrow 2H_2O + CO_2 \tag{2.21}$$

The following equations present the elemental composition of char:

$$\text{char} \rightarrow C + N_2 + O_2 + S + H_2 + \text{Ash} \tag{2.22}$$

$$C + 2H_2 \rightarrow CH_4 \tag{2.23}$$

$$C + H_2O \rightarrow CO + H_2 \tag{2.24}$$

$$C + CO_2 \rightarrow 2CO \tag{2.25}$$

$$C + O_2 \rightarrow CO_2 \tag{2.26}$$

$$CH_4 + 2H_2O \rightarrow CO_2 + 4H_2 \tag{2.27}$$

$$CO + H_2O \rightarrow CO_2 + H_2 \tag{2.28}$$

$$C + 0.5O_2 \rightarrow CO \tag{2.29}$$

$$S + H_2 \rightarrow H_2S \tag{2.30}$$

The gasification step is followed by the quenching process through which rapid cooling is applied. Syngas cooling unit comes next where heat is recovered for syngas cooling, and recovered heat can be used for multiple purposes such as power production, hot water, and space heating. The water-gas shift reaction is followed by the syngas cooling unit that converts CO to CO_2. The acid gas removal unit is followed by the water gas shift reaction where hydrogen sulfide (H_2S) and carbon dioxide (CO_2) are separated. The hydrogen purification unit is followed by the acid gas removal step, and the pressure swing adsorption process is usually used to separated hydrogen for the other gases. Gasification and water gas shift reaction are the processes that require steam input, and heat covered by the syngas cooling unit can be employed to convert water into steam at the desired temperature.

2.2 Renewable Hydrogen Production Methods

This section encloses the different renewable hydrogen production methods that can produce environmentally benign hydrogen. Fig. 2.7 exhibits the distribution of renewable hydrogen production methods. Three different commodities of electricity, thermal energy, or fuel can be obtained using renewable energy that can be employed to produce hydrogen. Electricity can be fed directly to the electrolyzer to produce hydrogen, thermal energy can be employed to the thermochemical cycles for hydrogen production, and fuels can be employed to produce hydrogen using biomass gasification. Wind energy source generates mechanical work that is converted to the electrical output using the generator, and electrical work is employed to the electrolyzer to produce hydrogen. High-temperature geothermal energy extraction methods generate electricity that is employed to the electrolyzer to produce hydrogen, and a geothermal heat pump generates thermal energy that can be employed to the thermochemical cycle to produce hydrogen. Hydro energy source rotates a turbine that generates mechanical work that is converted to the electrical output using the generator, and electrical work is employed to the electrolyzer to produce hydrogen. Solar energy extracted in terms of photic and photovoltaic energy produces electrical work that is used for hydrogen production through electrolysis.

The solar energy-based hydrogen produced is a promising technique that has the potential to play a significant role in contributing to sustainable energy supply. A recent study[45] published a comparative investigation of thermochemical, photoelectrolytic, electrolytic, and photochemical technologies to produce hydrogen using the solar-to-hydrogen technique. A recently published study[46] investigated the hydrogen

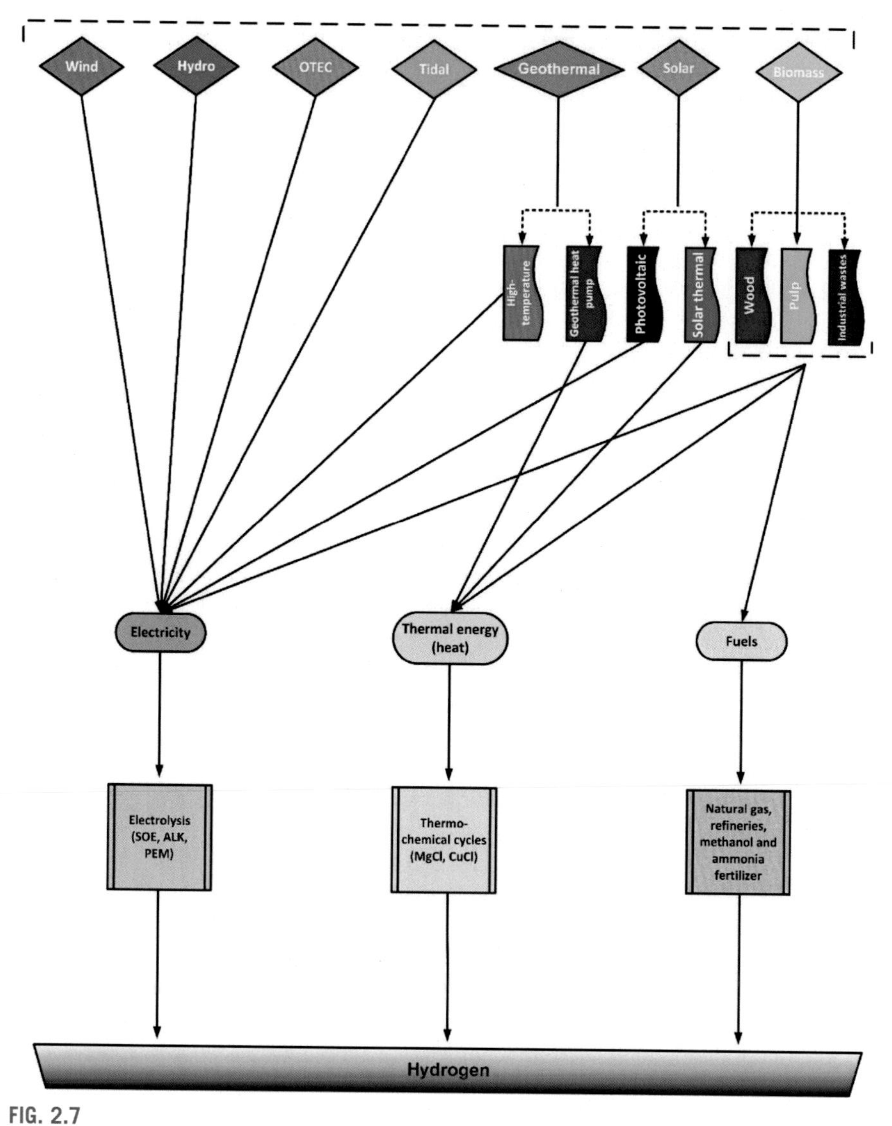

FIG. 2.7

Distribution of the renewable methods of hydrogen production.

Modified from [35].

production using wind energy sources integrated with the hydrogen storage system. They modeled a curtailed wind energy integrated with hydrogen production and storage system and recognized the design system based on real-time data for a whole year concerning a 10 min average. Two different scenarios of hydrogen production using electrolysis employing wind energy were offered, and the effect was applied on electrolyzer power payback period. The outcomes of this study stressed the hydrogen storage technologies and their significance to reveal some economical regularities.

A review study[47] was published recently discussing the production of hydrogen using geothermal energy. To achieve complete renewable energy utilization, the production of hydrogen is measured as an exceptional choice. A summary of advancements in hydrogen production technologies and the current status is followed by inclusive review research linked to the production of hydrogen employing geothermal energy, which is followed by geothermal energy-assisted hydrogen production systems and their process descriptions accompanied by the economic, technical, and environmental characteristics.

Lastly, a comparative analysis was conducted in terms of environmental aspects and costs of the hydrogen production methods employing dissimilar energy sources was established. This study revealed the cost-effectiveness of hydrogen production employing geothermal energy in comparison with solar and wind energy sources. A recent review study[49] offered an inclusive review of the processes of hydrogen production. This study enclosed the economic and technical features of 14 dissimilar methods of hydrogen production including both traditional and alternate resources of energy namely; coal, natural gas, nuclear, solar, biomass, and wind. They conducted a detailed comparative study of the considered renewable and conventional hydrogen production methods. The thermochemical gasification and pyrolysis were found to be economically feasible methods that offer high potential to turn out to be competitive on a large-scale in near future.

Solar thermal energy is extracted using a solar collector that is employed in the thermochemical cycle for hydrogen production. Ocean thermal energy conversion source generates mechanical work that is converted to the electrical output using the generator and electrical work is employed to the electrolyzer to produce hydrogen. Biomass energy source is distributed among numerous categories that are used to generate electricity through pyrolysis and gasification using proximate and ultimate analysis, and produced electricity is employed to the electrolyzer for hydrogen production.

The consumption of renewables is increasing significantly around the globe. Fig. 2.8 displays the global consumption of renewables in a different region of the world (Mt). Continuous and significant growth of renewable energy sources can be depicted in all regions of the world including North America, South and Central America, Europe, Commonwealth Independent States, Middle East, Africa, and the Asia Pacific from 2008 to 2018. The renewable consumption in the North America increased from 34.2 to 118.8 Mt between 2008 and 2018, increased from 7.8 to 35.4 Mt in South and Central America, increased from 54.1 to 172.2 Mt in Europe, increased from 0.1 to 0.6 Mt in the Commonwealth Independent States, increased from 0.1 to 1.7 Mt in the Middle East, increased from 1 to 7.2 Mt in Africa, and increased from 26.6 to 225.4 Mt in the Asia Pacific.

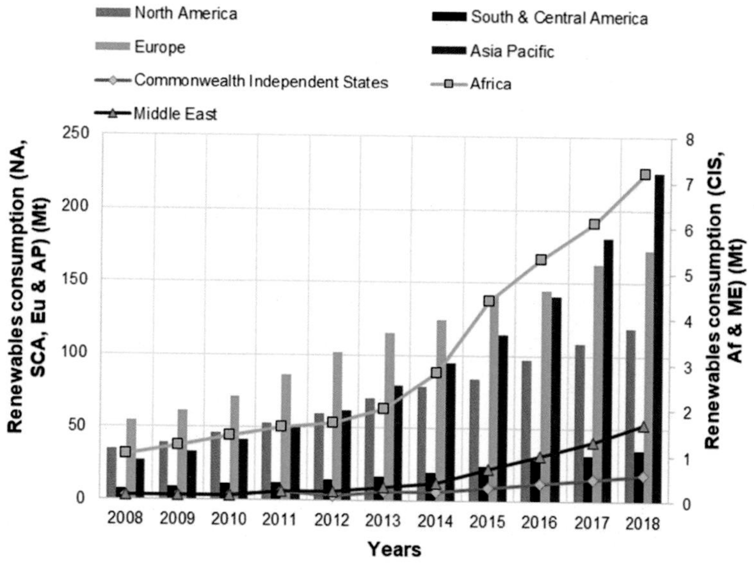

FIG. 2.8

Global consumption renewables in a different region of the world (Mt).

Modified from [48].

2.2.1 Solar Energy

Solar energy is one of the significant primary sources that can be utilized for clean hydrogen production. Fig. 2.9 displays the distribution of solar energy sources that can be employed for hydrogen production.

The solar energy source can be distributed among four following categories that can be used for hydrogen production:

- Concentrated solar thermal energy
- Photovoltaic
- Photoelectrolysis
- Biophotolysis

The concentrated solar thermal energy source can be used to produce hydrogen using multiple routes namely; solar thermolysis, solar thermochemical cycle, mechanical energy to electrical energy, solar gasification, solar cracking, and electrolysis. Photoelectrolysis and biophotolysis are capable of producing hydrogen directly. The photovoltaic source generates electrical power which is employed to the electrolysis for hydrogen production. The thermal energy extracted from the concentrated solar thermal energy source is also employed to hydrogen production routes of solar gasification and solar ammonia reforming.

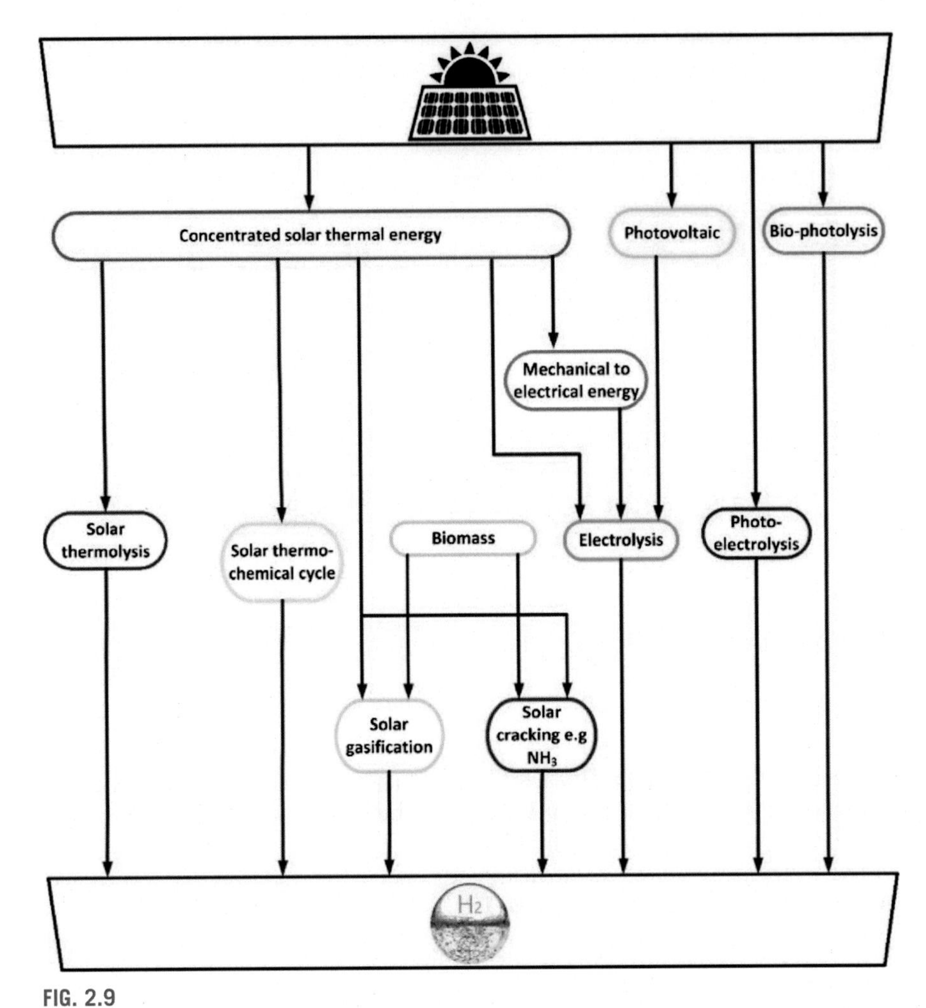

FIG. 2.9

Solar energy-based hydrogen production.

Modified from [50].

2.2.2 Wind Energy

Wind power defines the process of generating electricity using wind energy. Wind turbines are employed to extract wind energy. The wind rotates the turbines that convert kinetic energy into mechanical energy, and the generator converts mechanical energy into electrical energy. The electrical energy extracted from the wind energy source is AC; thus, AC/DC converter is employed to feed the DC electrical power to the electrolyzer for hydrogen production. Fig. 2.10 displays the wind

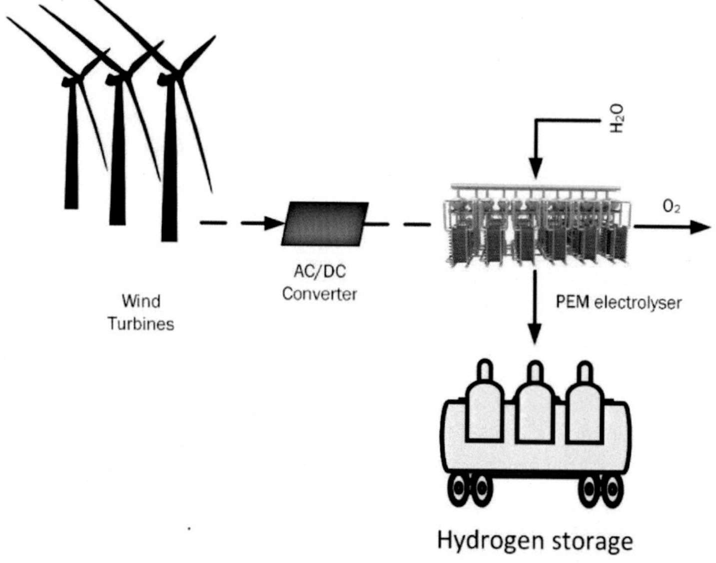

FIG. 2.10

Wind energy-based hydrogen production.

energy based hydrogen production system. The electrical energy extracted from the wind energy source is converted from AC to DC and employed to the electrolyzer that split water into oxygen and hydrogen. The wind energy source for electricity generation carries numerous benefits that can be listed as follows:

- Environment-friendly and clean fuel source that offers environmentally benign energy solutions undergoing zero emissions.
- A renewable and sustainable energy source that uses the sun-based atmospheric heated wind and this source can never run out.
- Wind power is completely free and cost-effective.
- Domestic and industrial installation can be employed on existing farms. Wind turbines utilize a land fraction for power production that can be employed to the electrolyzer for hydrogen production.
- Jobs are being created for the wind turbine manufacture, installation, and maintenance as well as in wind energy consulting.

Wind sources-based electricity can be employed to the electrolyzer for hydrogen production and produced hydrogen can be employed to the fuel-cell vehicles for electricity generation and hydrogen can also be stored to meet the load during the low wind speed intervals. The technology of producing hydrogen using wind energy is an efficient, clean, and sustainable energy production mode. For the maximum theoretical efficiency of the wind turbine, the Betz limit is followed, which states that the maximum wind turbine theoretical efficiency can be 59.3%. Most of the turbines generate around 50% wind energy passing over the rotor area. The wind turbine capacity factor is the average output power divided by the maximum power capability.

2.2.3 Geothermal Energy

Geothermal energy can be defined as the heat extracted from the earth's subsurface. The geothermal energy is transmitted to the earth's surface using steam or hot water. The geothermal energy can be utilized for multiple purposes depending on nature such as cooling, heating, or producing electricity. The international geothermal association reported that 10,715 MW of geothermal power was produced by 24 countries in 2010 that displayed 20% raise since 2005 and 18,500 MW of geothermal power was reported in 2015 that displayed almost a 60% raise since 2010 in an online capacity.

The complete or partial amount of electricity generated by the geothermal plant can be used to produce hydrogen by employing electrolyzer. A geothermal source with a high temperature is required for the lower cost of producing hydrogen and liquidating process. Geothermal power plants extract heat from deep inside of earth crust to produce steam that is employed to the turbine to generate electricity. Geothermal heat pumps are used for water heating or to provide heating to the building to the surface of the earth.

Fig. 2.11a exhibits the geothermal energy-based hydrogen production with reinjection, and Fig. 2.11b displays the geothermal energy-based hydrogen production without reinjection. Flash steam, dry steam and binary cycle are three geothermal plant types. Dry steam is the oldest geothermal power production technology that extracts steam from the ground that is employed to the turbine. The deep hot high-pressure water is pulled into cold low-pressure water using flash plants. The three significant geothermal energy usages include direct usage in a residential heating system that employs reservoir or spring hot water, electricity generation and ground source cooling, and cooling.

2.2.4 Hydro Energy

Hydropower denotes the conversion of water-flowing energy into electricity. Hydropower is measured as a renewable energy source since the cycle of water is continually renewed. In history, hydropower was first employed for mechanical milling. Hydropower covers around 17.5% of global electricity employing the water potential energy. Excluding a few countries that are abundant in this source, hydropower is generally used as a backup to meet the peak-load demand as it can be voluntarily started and stopped.

Hydropower is among the oldest sources of planet power to produce power using the flow of water to drive a turbine or wheel. This power source was used by ancient Greece farmers for mechanical work such as grinding grain. Hydropower is measured as a renewable energy source and undergoes zero toxic byproducts.

Hydro energy is accessible in numerous forms such as potential energy that can be retained from high water dam heads, kinetic energy from rivers water flow and tidal bombardments, and kinetic energy from waves movement on comparatively stationary water masses. Numerous resourceful methods have been established to harness energy nevertheless employing water to the turbine to produce electricity. Those do not typically contain employing the water movement to operate other pneumatic or hydraulic mechanism forms to accomplish a similar task.

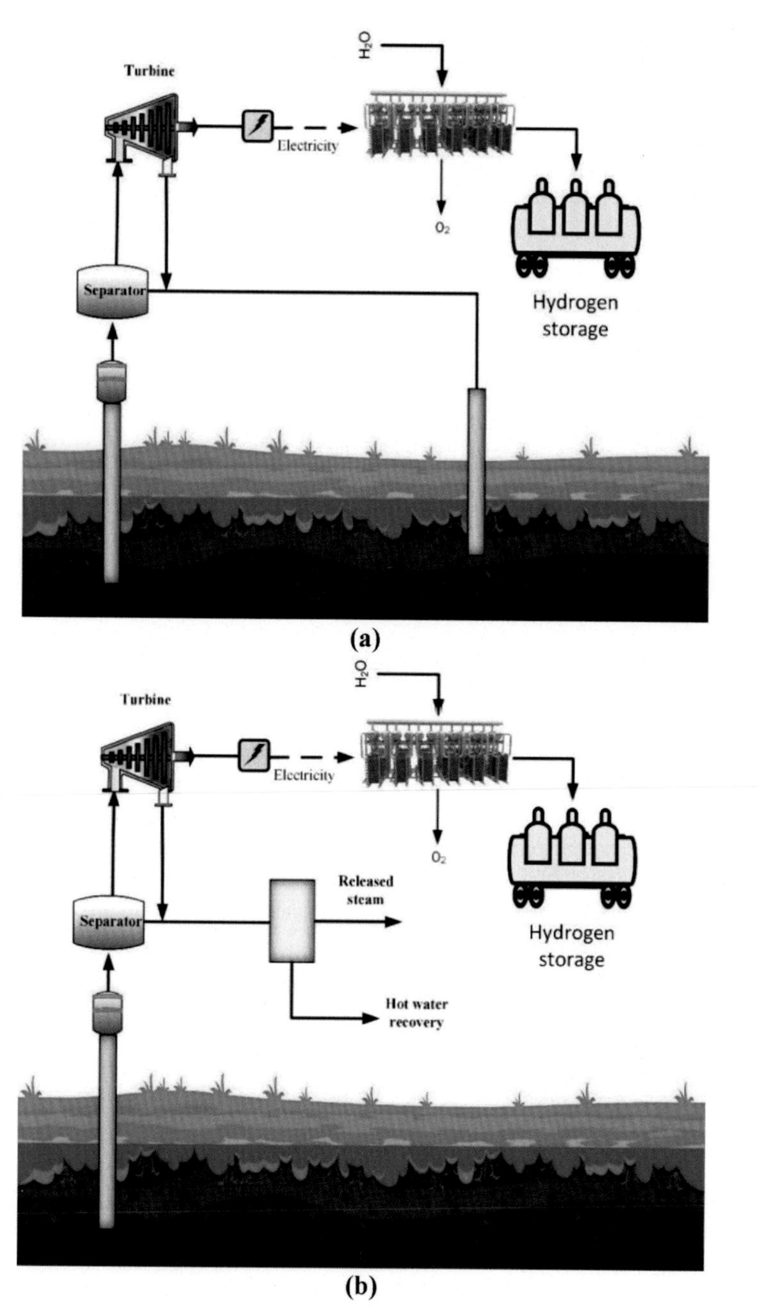

FIG. 2.11

Geothermal energy-based hydrogen production: (a) with reinjection; (b) without reinjection.

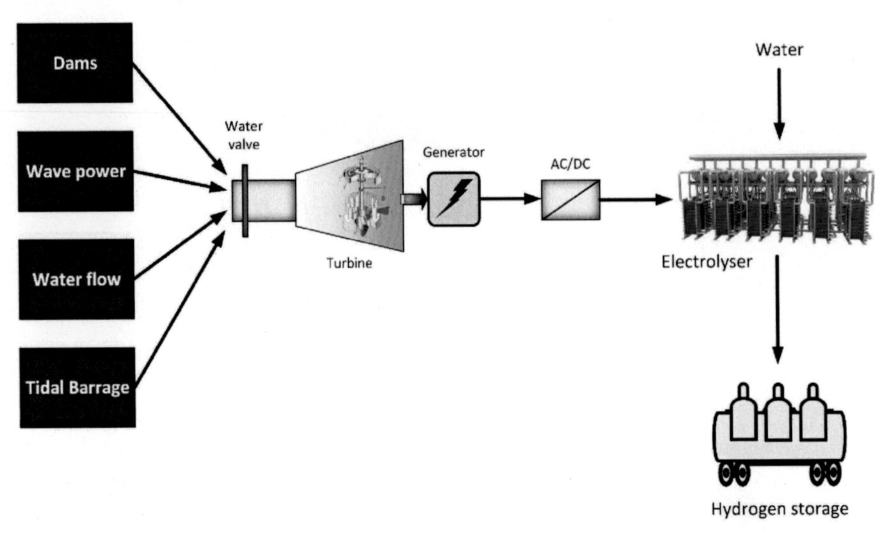

FIG. 2.12

Routes of hydro energy generation.

Fig. 2.12 shows the different routes of hydro energy generation. Similar to the steam turbines, water turbines depend on the working fluid impulse impact on the turbine blades or else reaction between the blades that turn the turbine and operate the generator and working fluid. Numerous dissimilar types of turbines have also been established to achieve optimized performance under particular conditions of water supply.

Hydropower generation is certainly the greatest effective method for large-scale electricity generation. Hydropower energy flows can be concentrated and controlled. The energy conversion process extracts kinetic energy that is converted to electrical energy directly and it contains no ineffective intermediate chemical or thermodynamic processes undergoing heat losses. The process can never be completely efficient undergoing 100% efficiency nevertheless extracting 100% kinetic energy of the flowing water represents the flow must stop.

The hydroelectric power plant conversion efficiency is subjected to the water turbine type that is employed, and this efficiency can reach 95% for large-scale installations while conversion efficiencies of 80%–85% are displayed by the smaller power plants undergoing electrical power less than 5 MW. However, it is hard to generate power using low flow rates.

Hydroelectric power capacity is a special power plant category that employs the water-flowing to produce electricity. This is done by driving water through a series of turbines that convert kinetic and potential energy into turbine rotational motion. The turbine is further integrated with the generator, and blade rotation generates electricity.

2.2.5 Ocean Thermal Energy Conversion

OTEC process generates electricity employing temperature differences in cold ocean deep water and tropical surface warm water. OTEC power plants employ large cold deep and surface seawater quantities to drive a turbine that operates the power cycle and generates electricity. OTEC technology is known as a marine renewable energy system that extracts the ocean water energy absorbed from solar energy to produce electric power. OTEC system employs the warm surface ocean water with a temperature of approximately 25°C that is used for the vaporization of the working fluid such as ammonia that carrier low-boiling point. Fig. 2.13 exhibits the OTEC-based hydrogen production system.

The OTEC cycle works under the principle of using warm water for working fluid evaporation that has a low boiling point. The pressurized vapor rotates the turbines that generate mechanical work, and the generator produces electricity. The deep cold seawater is employed for working fluid vapor condensation. The Carnot thermal efficiency, which is the maximum possible efficiency of an OTEC cycle, is approximately 7% as the temperature difference in surface and deep ocean water is somewhat over 20°C in tropical waters.

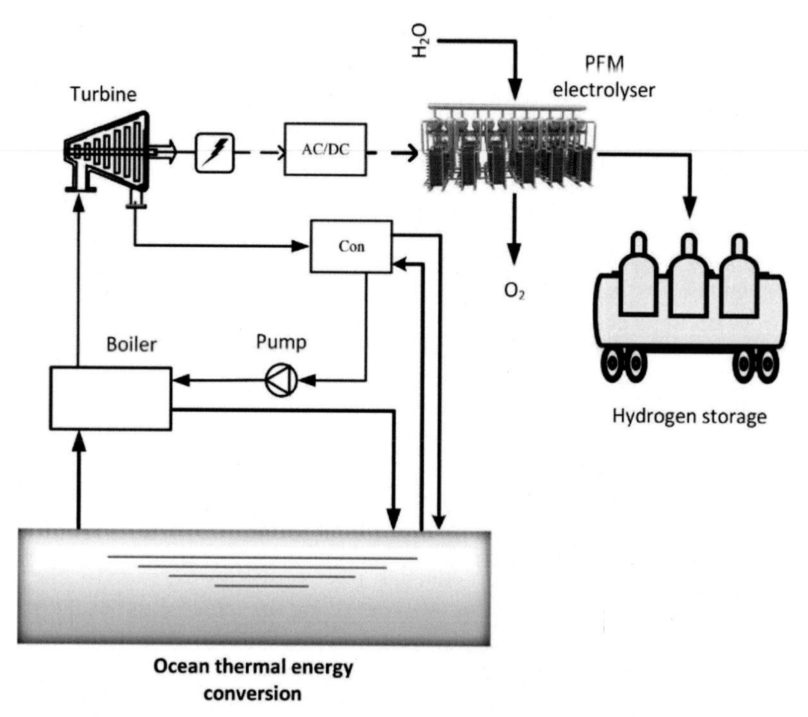

FIG. 2.13

Ocean thermal energy conversion-based hydrogen production.

The OTEC plants carry the significant advantage of producing environmentally benign renewable energy that is dissimilar to the solar and wind energy that have intermittent nature and can generate electricity under explicit conditions. Many well-known parties approximate 3–5 TW of baseload power production that can be produced without disturbing the ocean temperature and environment. That has about twice the global electricity demand. The OTEC cycle is also subjected to hydrogen production by employing the generated power to the electrolyzer. Hydrogen is also used as a storage medium using liquefaction under certain conditions.

2.2.6 Biomass Gasification

Gasification is known as a process that converts organic materials or carbonaceous fossil fuels into carbon monoxide, carbon dioxide, and hydrogen, and the resulting mixture is known as producer gas or syngas. The biomass gasification process is used to convert solid biomass fossil fuels into combustible gaseous producer gas employing several categorized thermochemical reactions. The producer gas is low-heating value fuel whose calorific value is in the range of 1000–1200 kcal/Nm^3.

Biomass gasification encompasses biomass burning under a limited air supply to offer carbon dioxide, hydrogen, methane, carbon monoxide, nitrogen, steam accompanied by chemicals such as char, tars, and ash particles containing combustible gas. Gasification is a considerably cleaner process of conversion as compared with combustion as gasification produces fuel in the form of syngas rather than burning it, which prevents many pollutants to be emitted such as NO_X, SO_X, and other particulates that occur at higher temperatures than typical gasification temperature. Fig. 2.14 displays the hydrogen production schematic using biomass gasification.

Gasification allows the fuel cleaning opportunity former to combustion. Biomass treatment prior to the burning in the boiler offers an exceptional opportunity to eliminate contaminants and other particulates from the gas stream before being employed. Biomass energy offers some advantages, which are listed as follows:

- Biomass is an abundant and extensively accessible source of renewable energy.
- Biomass is a carbon-neutral fuel and releases identical carbon amounts to the air as was captivated by plants in their life cycle.
- Biomass decreases fossil fuel overreliance.
- Biomass technology is less expensive as compared with fossil fuels.
- Solid waste burning reduces 60%–90% of the garbage dumped in landfills.

The two significant hydrogen production routes from biomass are biochemical and thermochemical. The thermochemical route comprises three significant methods of pyrolysis, supercritical water gasification, and gasification while biological conversions consist of fermentative and photosynthesis-based hydrogen production, biological gasification, and supercritical water gasification. The pretreatment of biomass is followed by the biomass gasification step that produces the syngas that passes through the syngas cooling unit that is used for heat recovery, which is

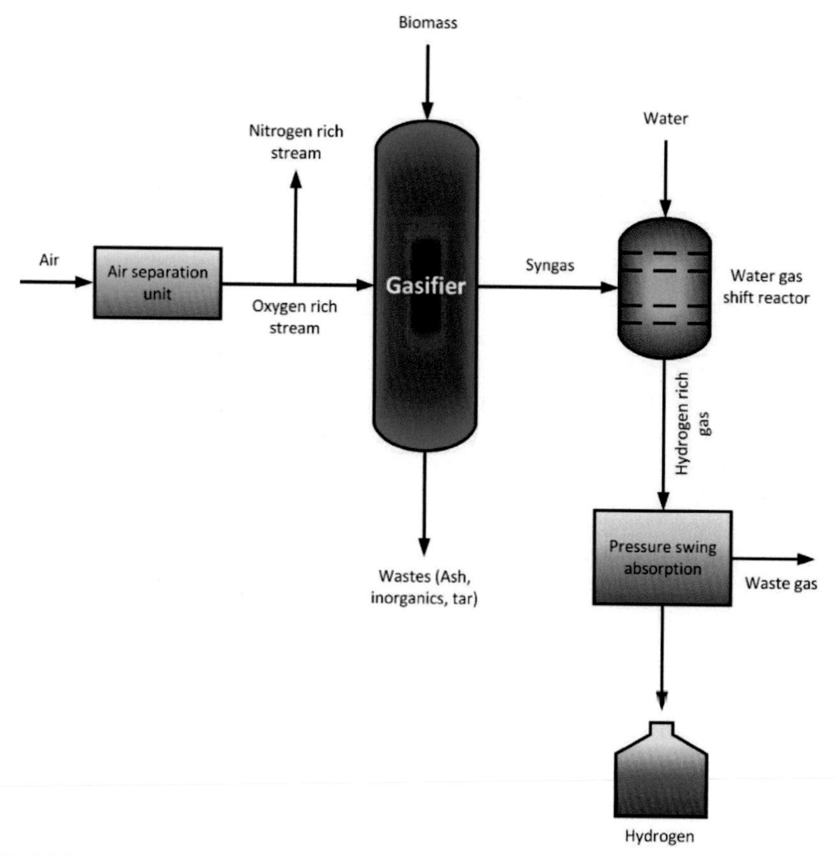

FIG. 2.14

Hydrogen production schematic using biomass gasification.

followed by the water gas shift reaction. The acid gas removal takes place after WGSR, which is followed by a carbon dioxide-capturing unit that captures CO_2 and hydrogen is purified in the last stage.

2.3 Other Hydrogen Production Methods

A significant amount of hydrogen is also produced using other hydrogen production methods that are listed as follows:

- Nuclear
  - Thermal energy (process/waste heat)
  - Electrical energy

- Aluminum-based
- Plasma reforming
- Ammonia reforming
- Ultrasonic
- Chlor-alkali
- Biological

2.3.1 Nuclear Energy-Based Hydrogen Production

Nuclear energy is originated from uranium atoms splitting using the process named fission. Nuclear energy produces heat to generate steam, and the generated steam is employed to the turbines to produce electricity. Nuclear plants do no undergo fuel burning, and they do not generate GHG emissions. In the reactors employed to the nuclear power plants, fission reaction is employed as it can be controlled. The reason for not using fusion reaction to generate power is the difficulty level to control the fusion reaction, and providing the conditions required by the fusion reaction is costly as well.

A recent study[51] enclosed the Canadian advancements in hydrogen production using nuclear energy and the thermochemical Cu—Cl cycle. They presented the recent development in hydrogen production employing nuclear energy through thermochemical Cu—Cl cycle and electrolysis. They included not only the specific process and reactor design in the Cu—Cl cycle but also the thermochemical characteristics, safety, controls, advanced materials, electrolyzer economic analysis, reliability, and integration of Canadian nuclear power plants with hydrogen plants.

A recent book[52] was conducted to provide a comprehensive route of producing hydrogen using nuclear energy. With the renaissance of global nuclear power and increasing hydrogen significance as a carrier of clean energy, the nuclear energy utilization for large-scale production of hydrogen can play a significant role in sustainable energy future. The cogeneration of hydrogen and electricity using nuclear plants can be attractive. The significant focus was directed to thermochemical Cu—Cl and sulfur-based cycles. They presented and deliberated the cycle configurations, modeling, equipment design, and implementation problems. The book offers an overview of significant qualifying technologies in the direction of industrialization and the design of hydrogen plants integrated with nuclear plants.

A solitary nuclear power plant with a 1000-MW capacity is capable of 200,000 tonnes or more of hydrogen production capacity every year. Fig. 2.15 displays the nuclear energy-based hydrogen production for fuel-cell applications. The thermal energy obtained from the nuclear power plant can be used for hydrogen production using high- and medium-temperature thermochemical water splitting, high- and medium-temperature steam reforming, or high- and medium-temperature steam electrolysis and can also be employed to generate power that can produce hydrogen upon demand.

In a nuclear power plant, the fission reaction is employed to the nuclear reactor, which is a heat source of thermal energy. Similar to the characteristic stations generating thermal power, steam is generated using heat produced by the nuclear reactor

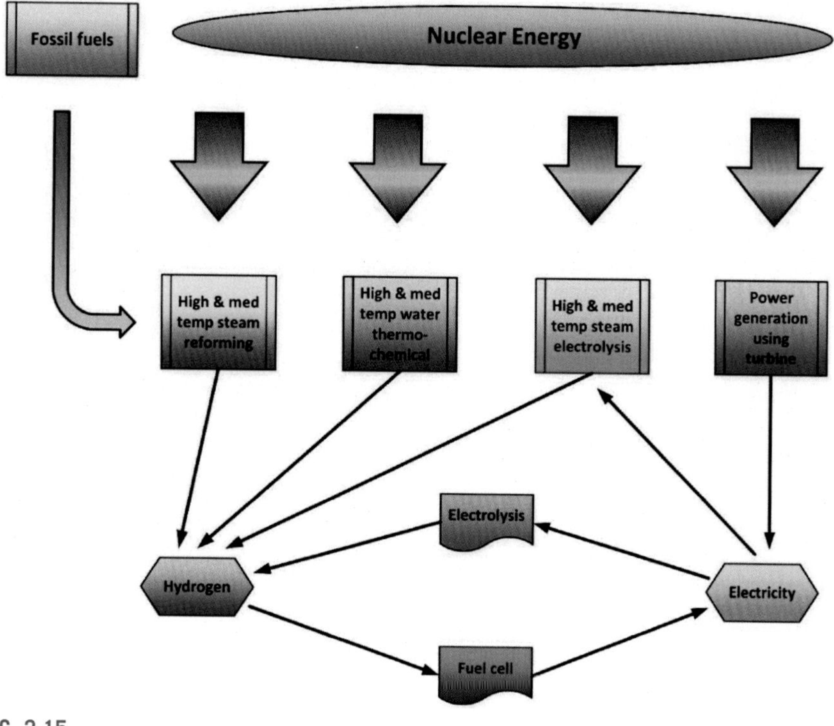

FIG. 2.15

Nuclear energy-based hydrogen production for fuel-cell applications.

and this steam is employed to the turbine and generates electricity using a generator. Nuclear energy is generated using atoms splitting in the nuclear reactor that converts water into steam, employed to the turbine and electricity is generated. In 29 states, 95 nuclear plants produce approximately 20% of national electricity using uranium. Fig. 2.16 shows the nuclear energy extraction for different applications including hydrogen production.

The nuclear reactor is used to undergo a fission reaction and generate thermal energy. This thermal energy generated by the nuclear power plant can be used for multiple purposes.[53] The most common method of utilizing this thermal energy is to produce power using a turbine that can later be employed to the fuel cells for hydrogen production. Some other methods are to employ this thermal energy to the different processes such as thermochemical water-splitting cycles, steam reforming methods, or water electrolysis for hydrogen production.

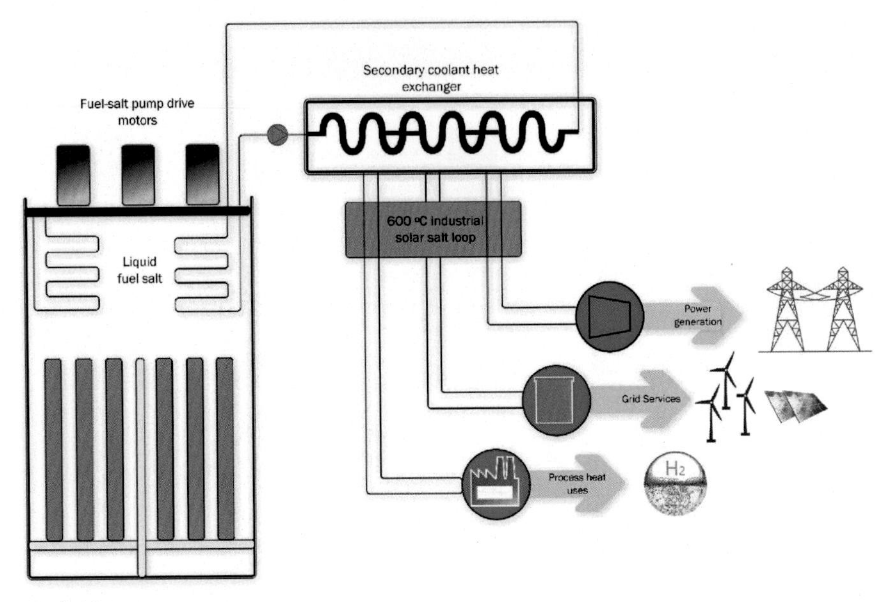

FIG. 2.16

Nuclear energy extraction for different applications including hydrogen production.

2.3.2 Aluminum-Based Hydrogen Production

Aluminum has the potential of producing hydrogen gas by reacting with water. This kind of hydrogen production technique can be employed for multiple purposes such as feeding hydrogen fuel-cell vehicles, Even though the hydrogen production idea using the concept of water reaction with aluminum metal is not new, some recent studies published recently[54] claimed the hydrogen production using aluminum reaction with water can be made more efficient and hydrogen can be employed directly to the fuel-cell devices that can generate power for the portable applications, namely laptop computers, chargers, and emergency generators, and this method may also be employed for the residential applications and also for the transportation industry to feed fuel-cell-powered vehicles. Fig. 2.17 displays the aluminum-based hydrogen production for fuel-cell applications.

A recent study[55] conducted experimental investigations on solar-driven aluminum hydrogen production using water. Sustainable hydrogen production cost at low and high capacities is a significant challenge for hydrogen to be a future alternative potential fuel. They designed and assessed a new system for producing hydrogen employing solar energy and aluminum. They performed numerous experiments to quantify the rate of hydrogen production. Furthermore, expressions were developed between the volume and time of hydrogen production as well as conversion ratio and energy efficiency to be evaluated. Likewise, a methodology was established to attain the stable and optimum rates of hydrogen production. It was detected that the volume of hydrogen production and system efficiency were increased at high molarity of sodium hydroxide under low temperatures.

At 3.7 wt.% hydrogen weight fraction, the reaction of water with aluminum metal produces 1 kg hydrogen by consuming 9 kg of aluminum (considering

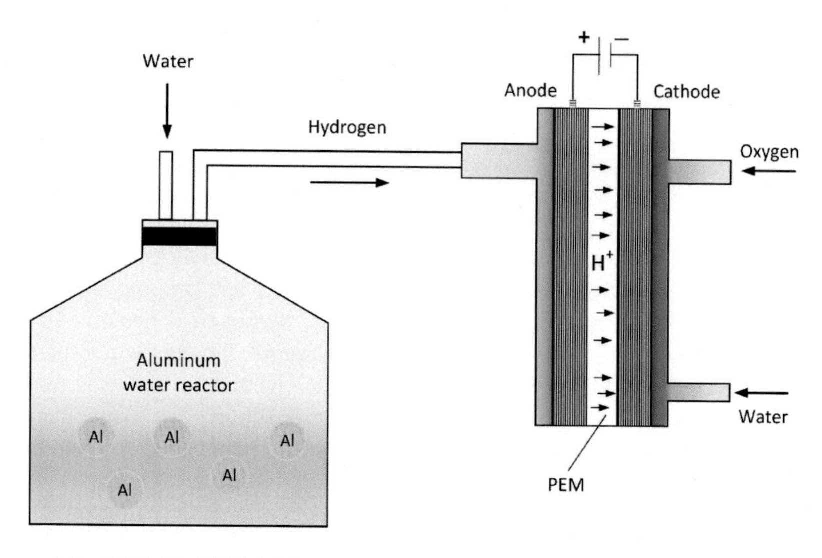

FIG. 2.17

Aluminum-based hydrogen production for fuel-cell applications.

100% yield). Consuming 15.5 kWh/kg of aluminum, a total of 140 kWh energy will be required to yield 1 kg of hydrogen. In the ambient temperature conditions, the reaction occurs between water and aluminum metal to form hydrogen gas along with aluminum hydroxide ($Al(OH)_3$). The volumetric capacity of hydrogen followed by the reaction is 46 g H_2 per liter, and the gravimetric capacity of hydrogen is 3.7 wt.%. The possible reactions of aluminum and water process are written as follows:

$$2Al + 6H_2O \rightarrow 2Al(OH)_3 + 3H_2 \quad (2.31)$$

$$2Al + 4H_2O \rightarrow 2AlO(OH) + 3H_2 \quad (2.32)$$

$$2Al + 3H_2O \rightarrow Al_2O_3 + 3H_2 \quad (2.33)$$

2.3.3 Plasma Reactor-Based Hydrogen Production

The Kvaerner carbon black process is a plasma reforming method that was developed by a Norwegian company in the 1980s for hydrogen production, and liquid hydrocarbons were used for carbon black. In the total available feed energy, hydrogen contains around 48%, activated carbon contains around 40%, and superheated steam contains the remaining 10%. A distinction of this defined process was offered in 2009 employing plasma arc technology for hydrogen and production.

Mizeraczyk and Jasinski[57] published a research article on plasma processing-based hydrogen production. In their research article, the plasma-based hydrogen production method was the key focus and hydrogen production efficiency was given much significance. Their article also answered the queries if small-scale plasma-based hydrogen production methods can be used for high production rates, low investment, and high reliability.

Chehade et al.[56] proposed a novel microwave-based plasma generation method to dissociate water into hydrogen gas. In the experimental setup, steam was flowing through direct discharge inside the developed reactor placed in 900 W microwave. The highly energetic electrons formed through the electric field provided by the 2.45 GHz microwave strike the molecules of water vapor and become ionized and dissociate into oxygen and hydrogen radicals. The hydrogen concentration was found in the range of 20,000–228,000 ppm. Furthermore, the energetic efficiency ranged from 11.6% to 53.7%. Fig. 2.18 exhibits the schematic of the experimental setup demonstrating microwave-based hydrogen production. The figure demonstrates a unique water plasmolysis hydrogen production method for numerous fuel-cell and portable applications.

Water vapor direct decomposition is a natural method considered for producing hydrogen. They determine the water vapor decomposition energetic efficiency by plasma, and production of hydrogen was evaluated by exergetic loss in reverse and forward reactions. Using ceriated tungsten, 50% highest energetic efficiency and 44% highest exergetic efficiencies were achieved; however, 36% lowest

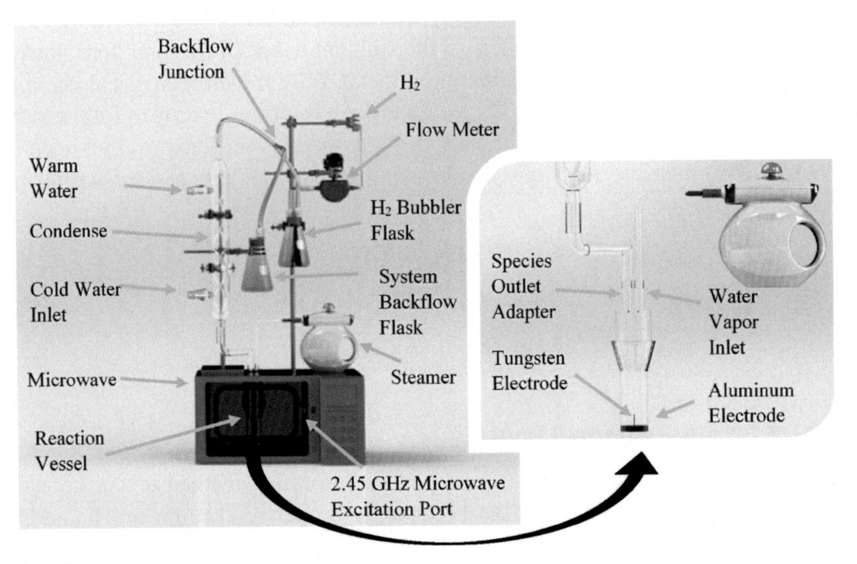

FIG. 2.18

Schematic of the experimental setup demonstrating microwave-based hydrogen production.

Modified from [56].

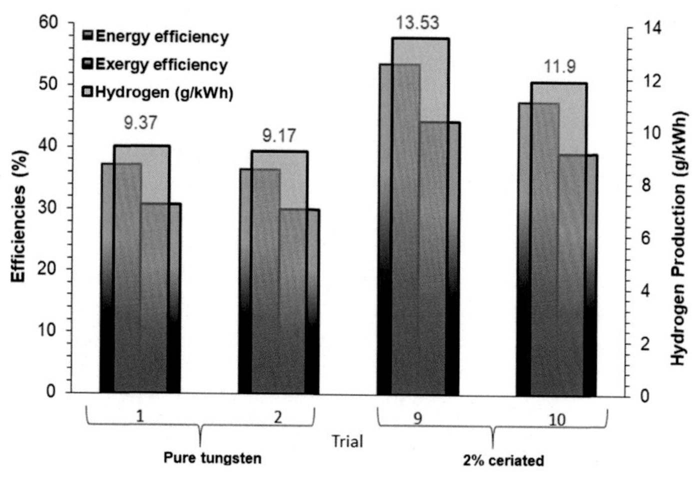

FIG. 2.19

System efficiencies and hydrogen production rates using pure ceriated tungsten.

Modified from [56].

energetic efficiency and 30% lowest exergetic efficiencies were accomplished as shown in Fig. 2.19. The theoretical water vapor decomposition efficiency to produce plasma-based hydrogen is ranged from 50% to 70%.[58] The range of experimental efficiency and theoretical work demonstrates that the decomposition of water vapors entails high ionization degree channeled by both electron and vibrational excitation.

2.3.4 Ammonia Cracking for Hydrogen Production

Ammonia can be cracked to produce hydrogen over a catalyst accompanied by nitrogen nontoxic gas. Each mole of ammonia comprises one and a half mole of hydrogen and a half mole of nitrogen. Hydrogen can be produced from ammonia cracking. First, ammonia decomposes into hydrogen and nitrogen molecules using a catalyst. After decomposition, the mixture of ammonia cracking molecules pass through the hydrogen membrane that blocks all gases and allows only hydrogen gas to pass. The 17.65% of ammonia mass is constituted by hydrogen.

A detailed technoeconomic analysis is conducted on the distribution of natural gas-based produced hydrogen in a previous study.[40] It is well recognized that hydrogen can play a substantial contribution to global energy production. The mainstream natural gas reforming plants in the United States are centralized for producing hydrogen. They aimed to justify the most reasonable hydrogen production route for the automotive sector by performing economic and technological analysis and assessment. From the assessment, some processes namely; ammonia thermal cracking and central hydrogen production providing substantial delivery are eradicated through on-site natural gas and methanol steam reforming are recognized as the most technically feasible options. The economic analysis conducted on the mentioned processes revealed that natural gas reforming offers high investment returns and short payback time. Sensitivity analysis was also conducted to estimate the effect of different variables such as the price of natural gas feedstock, operating capacity factor, and capital investment on overall hydrogen production cost. The natural gas reforming was revealed to be suitable technologically and economically for short-term production of hydrogen, and the raw material is readily available. Thus, this process is suitable to be employed before the transition to the solar, wind, and biomass renewable energy sources.

Cracking of ammonia is recognized as a basic process and cost-effective hydrogen production method; nevertheless, it is suitable to employ this cracked hydrogen in applications where nitrogen presence is adequate otherwise a mixture of nitrogen and hydrogen is processed through a hydrogen membrane that blocks all gases and allows only hydrogen gas to pass. The significant ammonia cracking applications, namely steel annealing furnace for constant heat treatment. The greatest effective catalysts for ammonia cracking are rare metals based, namely cobalt and ruthenium. Although iron can crack ammonia efficiently at the temperature of 600°C, developing and discovering the inexpensive catalysts that operate at lower temperatures ranging from 350 to 500°C offer required efficiencies are desirable for ammonia reforming. Fig. 2.20 displays the ammonia reforming process for hydrogen production. The ammonia reforming process can be displayed as follows:

$$NH_3 \rightarrow \frac{1}{2}N_2 + \frac{3}{2}H_2 \tag{2.34}$$

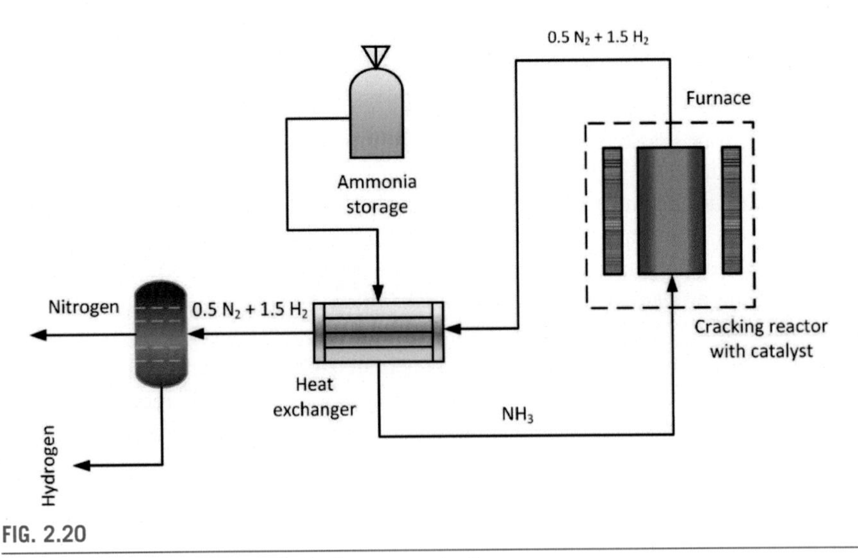

FIG. 2.20

Ammonia reforming process for hydrogen production.

2.3.5 Ultrasonic-Based Hydrogen Production

Ultrasonic-based hydrogen production is among the emerging technology and the process undergoes hydrogen production employing ultrasound power in a liquid water medium. The minute the ultrasound waves pass through liquid water with high frequency, this process generates the mechanical vibration in liquid water resulting in the production of acoustic cavitation bubbles.

The penetrating and powerful ultrasonic waves undergoing the range of frequency from 20 to 1100 kHz passing through liquid trigger the minor acoustic cavities that oscillate, grow, and enlarge generating an incredible amount of heat.[59]

The oscillation of the bubbles creates shear forces and high-velocity jets on the surrounding of bulk water. The collapse generates high-amplitude shock waves that go around 10,000 atm, and the shock wave intensity typically depends on dissimilar aspects counting the acoustic power, ultrasound frequency, and bulk temperature. Subsequent to the collapse, vast pressures and temperature are produced because of the absorption of the sound waves, the pressure reaches 2000 atm, and the temperature reaches 5000 K within the bubbles providing an infrequent chemical environment. In the collapse, every bubble behaves as a microreactor undergoing typical flame reaction. These microbubbles act as microcombustor where reaction or combustion chemistry takes place. At that point, the collapse leads to the hydrogen molecules formation as shown in Fig. 2.21. This figure displays the ultrasonic source and its effect on the formation of bubbles and the dissociation hydrogen production

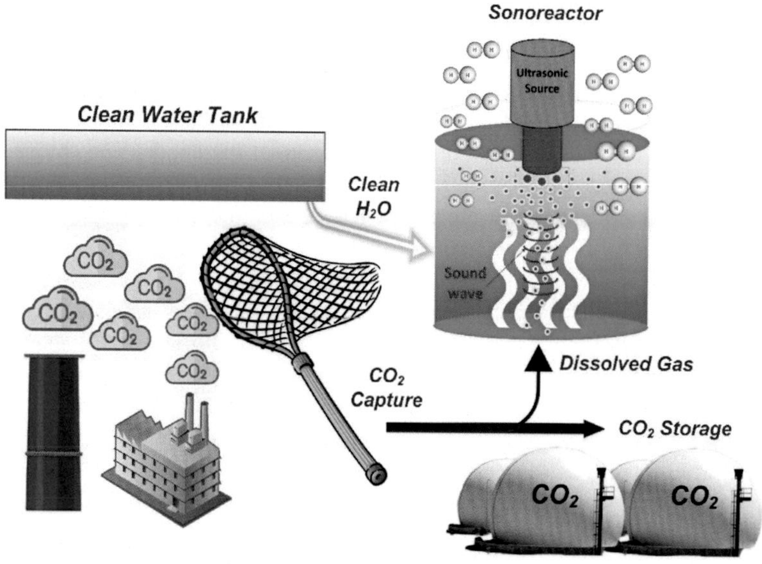

FIG. 2.21

Schematic representation of ultrasonic-based hydrogen production.

Modified from [60].

mechanism. The process consists of three successive stages; ultrasonic source immersed in the water container releases sound waves by frequency ranged from 20 to 40 kHz, acoustic cavitation bubbles produced at the ultrasonic source bottom side. The dynamics of the bubble undergo a 4-step mechanism, counting formation of bubbles, gradual growth, uneven phase, and collapse. The excessive temperature and pressure reache 2000 atm and 5000 K that is contingent to the acoustic operating conditions and the water vapor dissociation mechanism into hydroxyl.

The bubble is acting as a microcombustion reactor at which a series of chemical reactions are taking place.

$$H_2O + CO_2 \leftrightarrow OH^* + H^* + CO_2$$

$$OH^* + M \leftrightarrow H^* + O + M$$

$$H^* + O_2 \leftrightarrow HO_2$$

$$HO_2 + H^* \leftrightarrow H_2 + O_2$$

$$H_2 + O_2 \leftrightarrow OH^* + OH^* + CO_2$$

2.3.6 Chlor-Alkali Electrochemical Process

Chlor-alkali is among the leading electrochemical processes that produce hydrogen and chlorine. The chlor-alkali process includes the brine electrolysis that yields sodium hydroxide at cathode and chlorine gas at the anode. The general chemical reaction of sodium chloride and water can be expressed as follows:

$$2NaCl(aq) + 2H_2O \rightarrow 2NaOH(aq) + Cl_2(g) + H_2(g) \quad (2.35)$$

A chlor-alkali method is carried out by employing three different approaches of diaphragm cell, mercury cell, and membrane cell. In all of these methods, hydrogen is formed at the cathode while chlorine is produced at the anode. The chlor-alkali diaphragm cell is constructed as shown in Fig. 2.22. The cathode is made of asbestos deposition metal screens while the anode is made of the copper plate. Asbestos avoids the reaction of chlorine with caustic soda. The brine solution is added to the anode side.

The chlor-alkali electrochemical process diagram is shown in Fig. 2.27. At the anode, sodium chloride solution is employed, and chlorine ions are converted to chlorine gas. At the cathode, sodium ions pass through the membrane, and hydroxyl ions are formed by water reduction. The hydroxyl ions combine with migrated sodium ions. Subsequently, hydrogen gas is produced at the cathode. The process undergoes through the following reactions:

Cathodic reaction:

$$2H_2O(l) + 2e^- \rightarrow H_2(g) + OH^-(aq) \quad (2.36)$$

Anodic reaction:

$$2Cl^-(aq) \rightarrow Cl_2(g) + 2e^- \quad (2.37)$$

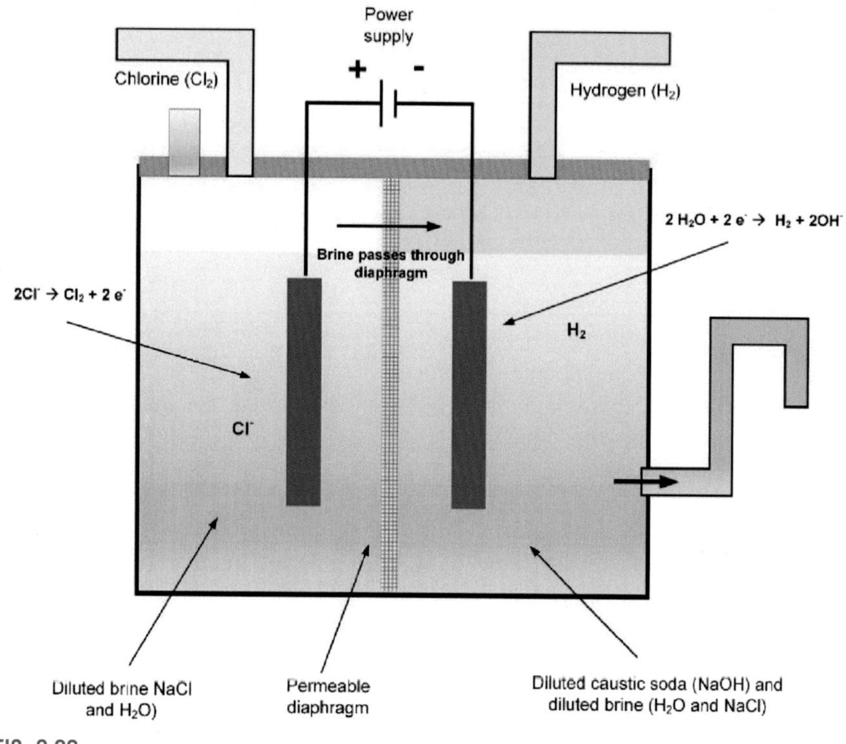

FIG. 2.22

Chlor-alkali electrochemical process.

Modified from [61].

This chlor-alkali technology produces a sodium chloride stream polluted weak caustic having a 30% concentration. On the contrary, the process of mercury cell is separated into two units of the electrolyzer and decomposer underdoing electrochemical reactions. In the mercury cell process, a 50% concentrated strong caustic solution is produced while in the electrolyzer, sodium amalgam is formed at the cathode and chlorine gas is formed at the anode. The reactions undergoing caustic soda formation in catholyte and sodium ions migration from anolyte through sodium chloride dissociation can be expressed as follows:

$$Na^+(aq) + OH^-(aq) \rightarrow NaOH(aq) \tag{2.38}$$

$$NaCl(aq) \rightarrow Na^+(aq) + Cl^-(aq) \tag{2.39}$$

Table 2.2 provides a comparison of three key chlor-alkali technologies presented earlier. The table shows that membrane cell technology entails less input energy as compared with diaphragm and mercury cells. Moreover, the highest purity of chlorine, hydrogen, and sodium hydroxide is produced by membrane technology. Table 2.2 arranges cathodic and anodic reversible potentials for characteristic

Table 2.2 Reversible Half-Cell Potentials in Chlor-Alkali Cell.

Reductant	Reaction	$E^0(V)^a$
Cathode	Hydrogen production: $2H_2O(l) + 2e^- \rightarrow H_2(g) + OH^-(aq)$	−0.99
Anode	Chlorine production: $Cl_2(g) + 2e^- \rightarrow 2Cl^-(aq)$	1.23
Overall	$2NaCl(aq) + 2H_2O \rightarrow 2NaOH(aq) + Cl_2(g) + H_2(g)$	−2.23

operating conditions of 90°C temperature, 1 bar pressure, 10 M of NaOH, and 3.5 M of NaCl in chlor-alkali membrane cell process.

The chlorine ion oxidation potential helps in determining the anodic reversible potential using the following equation:

$$E_a^o = E_{Cl^-/Cl_2}^o + \frac{RT}{2F}\ln\left(\frac{P_{Cl_2}}{(a_{Cl^-})^2}\right) \tag{2.40}$$

here E_{Cl^-/Cl_2}^o is achieved from Table 2.2, a_{Cl^-} is chlorine ion activity, and P_{Cl_2} is chlorine partial pressure. The Nernst equation helps in determining the water reduction potential:

$$E_c^o = E_{H_2O/H_2}^o + 2.303\frac{RT}{2F}\ln\left(\frac{P_{H_2}}{(a_{OH^-})^2}\right) \tag{2.41}$$

Here, E_{H_2O/H_2}^o is achieved from Table 2.2, a_{OH^-} is hydroxyl ion activity, and P_{H_2} is hydrogen partial pressure.

The overall reversible cell potential was correlated by Chandran and Chin[62] using partial pressure, cell temperature, and appropriate chemical activities. Table 2.3 arranges the energy requirement comparison of three chlor-alkali techniques. Following is the correlation for reversible cell potential:

$$E_{cell}^o = -2.18 + 0.000427T + 2.303\frac{8.314\ln\beta}{96500} \tag{2.42}$$

Here, β is partial pressures (H_2 and Cl_2) and activity coefficients of sodium hydroxide and sodium chloride dependent and T is in K, as follows:

$$\beta = \frac{a_{NaCl}\sqrt{P_{Cl_2}}\sqrt{P_{H_2}}}{a_{NaOH}} \tag{2.43}$$

The correlation for activation overpotential can be expressed as follows:

$$\Delta E_{act,i} = 0.0277\log\left(\frac{J}{J_0}\right) \tag{2.44}$$

Here, $J_0 = 0.0125\ A/cm^2$ for anode and $J_0 = 0.0656\ A/cm^2$ for cathode.

Table 2.3 Energy Requirement Comparison of Three Chlor-Alkali Techniques.

Reductant	Diaphragm Cell	Mercury Cell	Membrane Cell
Current density (A)	0.2	1.0	0.4
Cell voltage (V)	−3.45	−4.4	−2.95
Current efficiency for Cl_2 (%)	96	97	98.5
H_2 purity (%)	99.9	99.9	99.2
Cl_2 purity (%)	98	99.2	99.3
Cl^{-1} is 50% NaOH solution (%)	1.120	0.003	0.005
Land area for plant of 10^5 tons NaOH per year (m^2)	5300	3000	2700
Production rate per single cell, tons NaOH per year	1000	5000	2700
Energy consumption kWh per ton of NaOH			
(a) Electrolysis plus evaporation to 50% NaOH	3260	3150	2520
(b) Electrolysis only	2550	3150	2400

2.3.7 Biological Hydrogen Production

Biological hydrogen is named biohydrogen as it is produced using biological methods. Biological hydrogen production technology attracts researchers because of the production of clean hydrogen that can be utilized for numerous applications that can be formed using many biomass types. The classification of different biological hydrogen production methods that is shown in Fig. 2.23 is as follows:

- Biophotolysis
  - Direct biophotolysis
  - Indirect biophotolysis
- Bioelectrochemical system
  - Microbial electrolysis cell
- Fermentation
  - Dark fermentation
  - Photo fermentation

The challenges associated with the biological hydrogen production methods are as follows:

- Poising of the hydrogen producing organisms by oxygen
- Low hydrogen yield
- Economic feasibility
- Hydrogen separation required from CO_2 and H_2 mixture
- High BOD level in the effluent

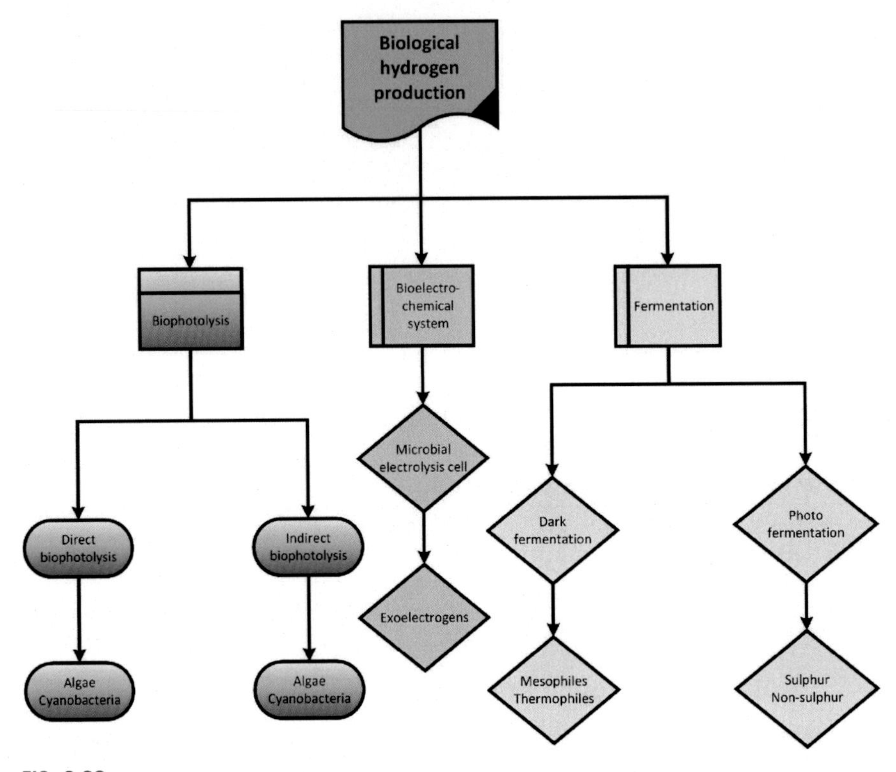

FIG. 2.23

Classification of biological hydrogen production methods.

- Requirement of an external light source
- Low hydrogen production rate
- Low light conversion efficiency requirement
- Customized photobioreactor requirement

The reactions occurring during the different processes of biological hydrogen production are as follows:

Mesophiles and thermophiles are the major organisms in the dark fermentation and the reaction is as follows:

$$C_6H_{12}O_6 + 2H_2O \rightarrow 2CH_3COOH + 2CO_2 + 4H_2$$

Sulfur and nonsulfur are the major organisms in the photo fermentation and the reaction is as follows:

$$C_6H_{12}O_6 + 6H_2O \rightarrow 6CO_2 + 12H_2$$

Algae and cyanobacteria are the major organisms in indirect biophotolysis, and the reaction is as follows:

$$2H_2O \rightarrow 2H_2 + O_2$$

Algae and cyanobacteria are the major organisms in indirect biophotolysis, and the reaction is as follows:

$$C_6H_{12}O_6 + 12H_2O \rightarrow 6CO_2 + 12H_2$$

Exoelectrogens are the major organisms in the microbial electrolysis cell, and the reaction is as follows:

$$C_6H_{12}O_6 + 2H_2O \rightarrow 2CH_3COOH + 2CO_2 + 4H_2$$

2.4 Thermochemical Cycles

The thermochemical cycle combines heat source with chemical reactions merely toward water splitting into its components. The terminology contains cycle word that represents that chemical constituents are uninterruptedly recycled while only water is used. If the thermochemical cycle is supplied with partial work with heat, the subsequent thermochemical cycle is called a hybrid thermochemical cycle. A system undergoing thermochemical water-splitting reaction follows the following reaction:

$$H_2O(l)\ H_2(g) + \frac{1}{2}O_2(g) \tag{2.45}$$

The thermodynamic reactor can undergo two fractions under strict imposed thermodynamics conditions. Work must be supplied through one fraction named Gibbs free reaction energy change. Heat must be provided through the other fraction by raising the species thermal agitation and can be represented as follows:

$$\Delta H = \Delta G + T\Delta S \tag{2.46}$$

At least two different temperature heat sources are mandatory for cyclical operation. This is insignificant in the thermolysis case, an inverse reaction consumes fuel. Therefore, a single thermolysis temperature that follows the maximum fuel-cell work recovery is equivalent to the reverse of water-splitting reaction Gibbs free energy at the same temperature. The endothermic reaction offers positive entropy change with the aim of a favored reaction with a rise in temperature and reverses in exothermic reactions. Fig. 2.24 represents the Carnot representation of thermochemical cycles and thermolysis-based engines. Following equations are prerequisites for multiple reactions water splitting for enthalpy, entropy, and work:

$$\sum_i \Delta H_i^0 = \Delta H^0 \tag{2.47}$$

$$\sum_i \Delta S_i^0 = \Delta S^0 \tag{2.48}$$

$$\sum_i \Delta G_i^0 = \Delta G^0 \tag{2.49}$$

$$\sum_i \Delta H_i^0 = \Delta H^0 \tag{2.50}$$

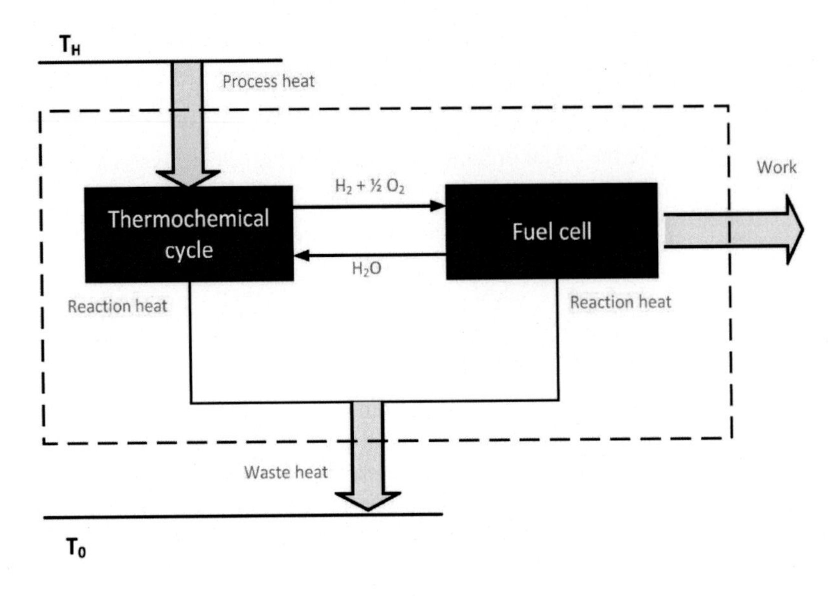

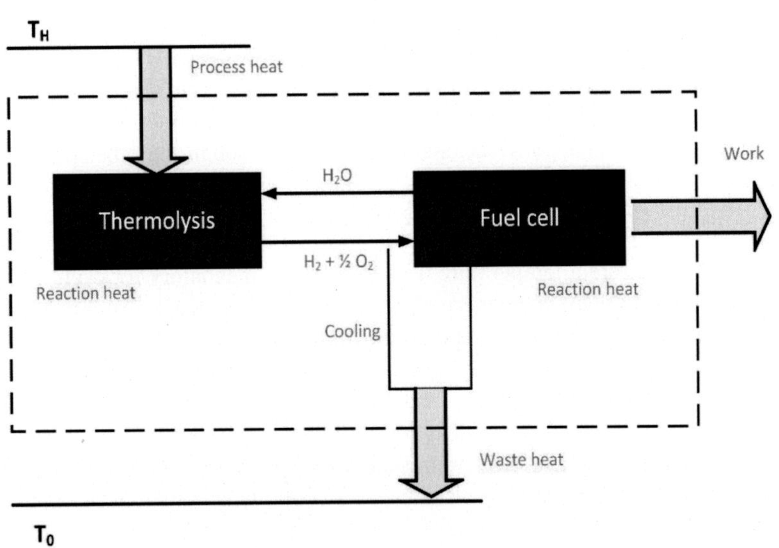

FIG. 2.24

Carnot representation of thermochemical cycles and thermolysis-based engines including fuel cell.

Finally, the relationship can be deducted for work requirement:

$$\sum_{p} \Delta S_i^0 \geq \frac{\Delta G^0}{(T_H - T^0)} \tag{2.51}$$

A Carnot approach representation of thermochemical cycles and thermolysis-based engines is displayed in Fig. 2.24. The heat is supplied to the thermochemical cycle from a heat source temperature TH and the thermochemical cycle splits water into hydrogen and oxygen. The produced hydrogen is supplied to the fuel cell which is used to generate power. The power generated by the fuel cells can be employed for numerous applications such as fuel-cell vehicles, portable and residential applications and combined heating and power system.

Fig. 2.25 exhibits the basic schematic of the 2-step water-splitting thermochemical cycle. In the schematic representation, XO represents metal-based redox material. The metal-based redox material utilization concept in the 2-step thermochemical water-splitting cycle was primarily established in the late 1970s.[63]

Note that XO signifying metal-based redox material can be either oxidized (XO_{ox}) or reduced (XO_{red}). The primary step is solar-driven metal oxide dissociation that undergoes an endothermic reaction of either lower-valence metal-oxide or elemental metal and the reaction is named thermal reduction while the water splitting takes place in the hydrolysis reaction that produces the hydrogen by material reduction.

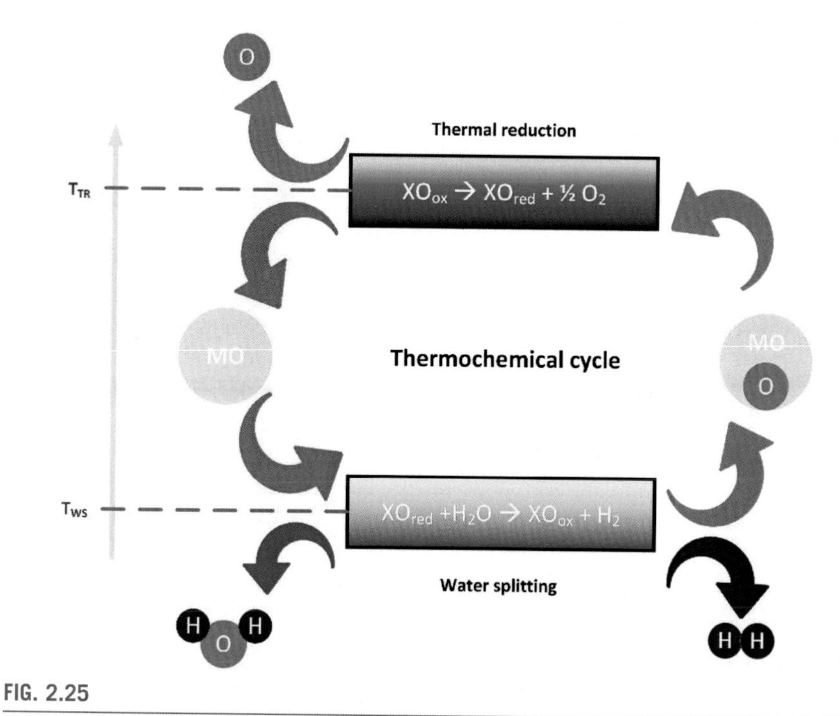

FIG. 2.25

Basic schematic of 2-step water-splitting thermochemical cycle.

Modified from [64].

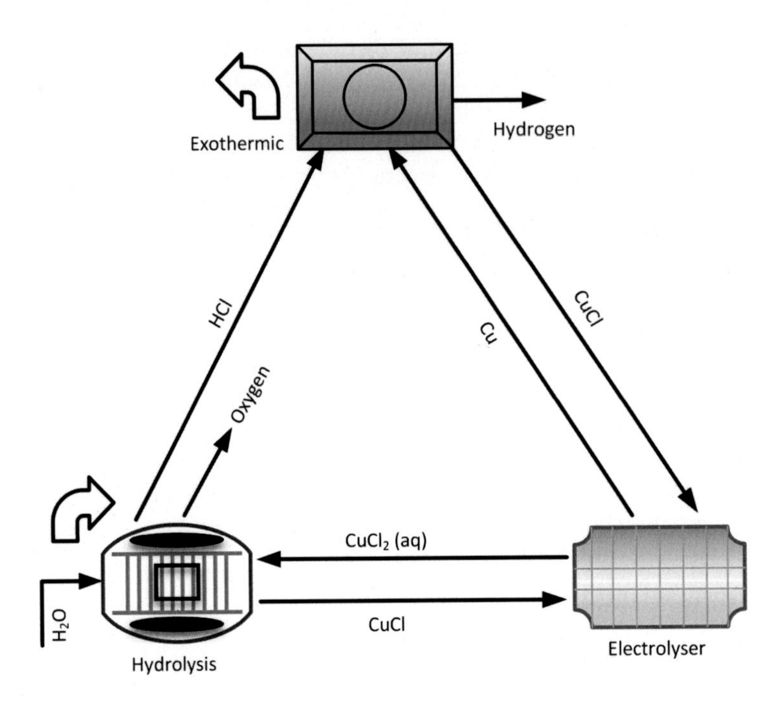

FIG. 2.26

General schematic of 3-step thermochemical Cu—Cl cycle.

A lab-scale prototype of the copper—chlorine (Cu—Cl) thermochemical cycle is established by a clean energy research laboratory at Ontario Tech University. This experimental setup can undergo three different configurations of 3-step, 4-step, and 5-step and offer high energetic and exergetic efficiencies. Fig. 2.26 represents a 3-step Cu—Cl cycle, Fig. 2.27 exhibits a 4-step Cu—Cl cycle and Fig. 2.28 displays a 5-step Cu—Cl cycle. Ishaq and Dincer[65] presented the simulation and comparative assessment of the three different configurations of Cu—Cl cycles. Each configuration consists of a different number of steps and undergoes different energetic and exergetic efficiencies. The significant steps of each configuration are followed by the general schematic displaying each component. The chemical reactions undergoing the three significant steps followed by the 3-step Cu—Cl cycle are as follows:

Electrolysis:

In this step, copper reacts with hydrogen chloride gas at the temperature of 450°C to form cuprous chloride and hydrogen gas.

$$2\mathrm{Cu}(s) + 2\mathrm{HCl}(g) \xrightarrow{450^{\circ}\mathrm{C}} 2\mathrm{CuCl} + \mathrm{H_2}(g) \tag{2.52}$$

Cupric chloride formation:

In the cupric chloride formation step, cuprous chloride split into copper and cupric chloride at the temperature of 30°C.

$$4\mathrm{CuCl}(s) \xrightarrow{30^{\circ}\mathrm{C}} 2\mathrm{Cu}(s) + 2\mathrm{CuCl_2}(aq) \tag{2.53}$$

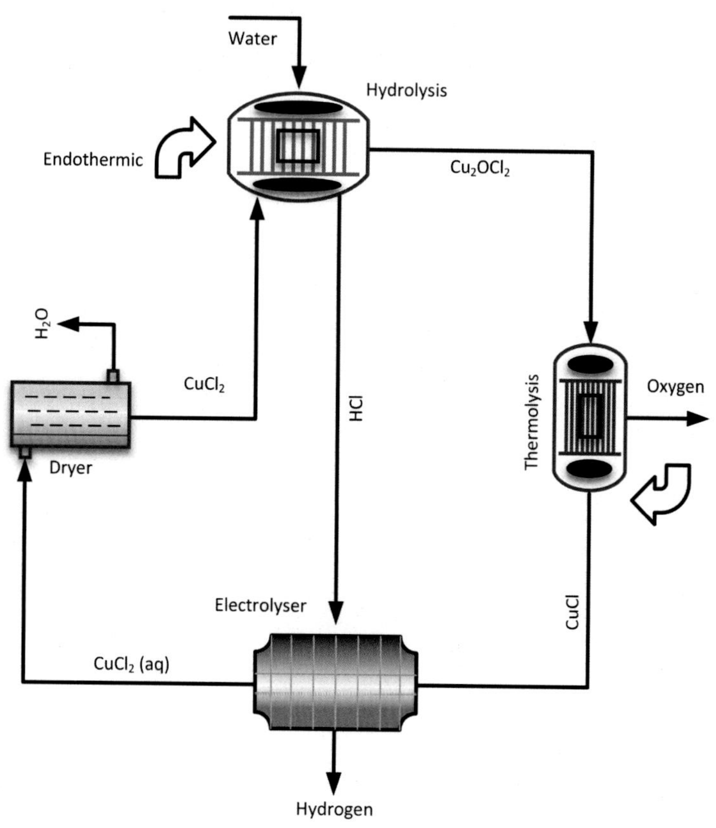

FIG. 2.27

General schematic of 4-step thermochemical Cu—Cl cycle.

Hydrolysis:

In the hydrolysis step, cupric chloride reacts with steam at the temperature of 530°C to form cupric chloride, hydrogen chloride, and oxygen gas.

$$CuCl_2(aq) + H_2O(l) \xrightarrow{530^\circ C} 2CuCl + 2HCl(g) + \frac{1}{2}O_2(g) \tag{2.54}$$

The chemical reactions undergoing the four significant steps followed by the 4-step Cu—Cl cycle are as follows:

Electrolysis:

In the electrolysis step, cuprous chloride reacts with hydrogen chloride gas at the temperature of 25°C to form cupric chloride and hydrogen gas.

$$2CuCl(aq) + 2HCl(aq) \xrightarrow{25^\circ C} H_2(g) + 2CuCl_2(aq) \tag{2.55}$$

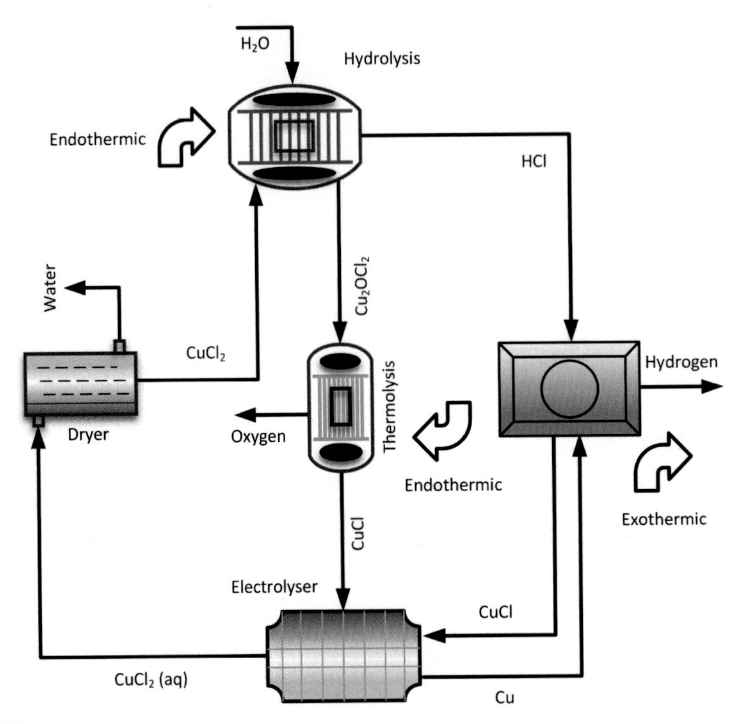

FIG. 2.28

General schematic of 5-step thermochemical Cu—Cl cycle.

Drying:

The aqueous cuprous chloride is passed through the drying process to separate the water from the aqueous cuprous chloride.

$$CuCl_2(aq) \xrightarrow{80^\circ C} CuCl_2(s) \tag{2.56}$$

Hydrolysis:

In the hydrolysis step, cupric chloride reacts with steam at the temperature of 400°C to form copper oxychloride and hydrogen chloride gas.

$$2CuCl_2(s) + H_2O(g) \xrightarrow{400^\circ C} Cu_2OCl_2(s) + 2HCl(g) \tag{2.57}$$

Thermolysis:

In the thermolysis step, copper oxychloride splits into cuprous chloride and oxygen gas at the temperature of 500°C.

$$Cu_2OCl_2(s) \xrightarrow{500^\circ C} 0.5\ O_2(g) + 2CuCl(l) \tag{2.58}$$

The chemical reactions undergoing the five significant steps followed by the 5-step Cu—Cl cycle are as follows:

Hydrolysis:

In the hydrolysis step, cupric chloride reacts with steam at the temperature of 530 °C to form cupric chloride, hydrogen chloride, and oxygen gas.

$$2CuCl_2(s) + H_2O(g) \xrightarrow{400^\circ C} Cu_2OCl_2(s) + 2HCl(g) \tag{2.59}$$

Thermolysis:

In the thermolysis step, copper oxychloride splits into cuprous chloride and oxygen gas at the temperature of 500°C.

$$Cu_2OCl_2(s) \xrightarrow{500^\circ C} 0.5\ O_2(g) + 2CuCl(l) \tag{2.60}$$

Cu production:

In the copper production step, cuprous chloride splits at the temperature of 25°C into copper and cupric chloride.

$$4CuCl(aq) \xrightarrow{25^\circ C} 2Cu(s) + 2CuCl_2(aq) \tag{2.61}$$

Electrolysis:

In the electrolysis step, copper reacts with hydrogen chloride gas at the temperature of 430°C to form cuprous chloride and hydrogen gas.

$$2Cu(s) + 2HCl(g) \xrightarrow{430^\circ C} H_2(g) + 2CuCl(l) \tag{2.62}$$

Drying:

The aqueous cuprous chloride is passed through the drying process to separate the water from the aqueous cuprous chloride.

$$CuCl_2(aq) \xrightarrow{80^\circ C} CuCl_2(s) \tag{2.63}$$

2.5 Electrolysis

Water electrolysis is used to split water into its components of oxygen and hydrogen using electricity. The significant electrolyzer types are proton exchange membrane (PEM) electrolyzer, solid oxide (SO) electrolyzer, and alkaline (ALK) electrolyzer. This section encloses all significant electrolyzer types with basic illustrations. Table 2.4 displays the standard electrode potentials for some general reactions.

2.5.1 Proton Exchange Membrane Electrolyzer

A PEM splits water into its constituents of oxygen and hydrogen employing acidic electrolytes. In the acidic electrolytes, the positive ions are transported that are known as cations. In water electrolysis, the protons are transferred from anode to cathode in the ionic transport. The overall electrochemical reactions are listed as follows:

Table 2.4 Standard Electrode Potentials for Some General Reactions.

Reductant	Reaction	$E^0(V)^a$
$NH_3(aq)$	$3N_2(g) + 2H^+ \leftrightarrow 2NH_3(aq)$	−3.09
H^-	$H_2 + 2e^- \leftrightarrow 2H^-$	−2.23
Ca	$Ca^+ + e^- \leftrightarrow Ca$	−3.8
$Al(s) + 3OH^-(aq)$	$Al(OH)_3(s) + 3e^- \leftrightarrow Al(s) + 3OH^-(aq)$	−2.31
$Ca(s)$	$Ca^{2+} + 2e^- \leftrightarrow Ca(s)$	−2.868
$Na(s)$	$Na^+ + e^- \leftrightarrow Na(s)$	−2.71
$H_2(g) + 2OH^-(aq)$	$2H_2O + 2e^- \leftrightarrow H_2(g) + 2OH^-(aq)$	−0.828
$3Fe(s) + 4H_2O(aq)$	$Fe_2O_3(s) + 8H^+ + 8e^- \leftrightarrow 3Fe(s) + 4H_2O(aq)$	+0.085
$SO_2(aq) + 2H_2O$	$HSO_4^- + 3H^+ + 2e^- \leftrightarrow SO_2(aq) + 2H_2O$	+0.160
Cl^-	$Cl_2(g) + 2e^- \leftrightarrow 2Cl^-$	+1.360
$Ag(s)$	$Ag^+ + e^- \leftrightarrow Ag(s)$	+0.780
$Au(s)$	$Au^+ + e^- \leftrightarrow Au(s)$	+1.830
$SO_2(aq) + 2H_2O$	$SO_4^{2-} + 4H^+ + 2e^- \leftrightarrow SO_2(aq) + 2H_2O$	+0.170
$Cu(s)$	$Cu^+ + e^- \leftrightarrow Cu(s)$	+0.520

Anodic reaction:

$$H_2O \rightarrow 2H^+ + \frac{1}{2}O_2\ (g) + 2e^- \tag{2.64}$$

Cathodic reaction:

$$2H^+ + 2e^- \rightarrow H_2\ (aq) \tag{2.65}$$

The operating temperature of the PEM electrolyzer is ranged 25−80°C. The characteristic consumption of electricity for a commercial PEM electrolyzer is ranged from 540 to 580 MJ/kg or 23−26 MJ/Nm^3 of hydrogen production. The PEM electrolyzer energy efficiency drops with current density quasi-exponentially and approaches to 54% at 10 kA/m^2 undergoing 80%−85% conversion and produced hydrogen is more than 99.999% pure. The industrial PEM electrolyzer capacity ranged from 0.2 to 60 Nm^3/h or 0.01 to 2.5 kg/h. The electricity consumption of a commercial PEM electrolyzer can reach 400 kW with water consumption of approximately 25 L/h.

Fig. 2.29 depicts the general illustration of a proton exchange membrane electrolyzer. It includes both-sided channeled bipolar plates with two endplates. Each plate is attached to a porous, planar, and electrically conductive layer serving as anode or cathode, and a proton exchange membrane is positioned between electrodes.

A polytetrafluoroethylene is more suitable than polystyrene sulfonate polymer for electrolyte as it offers better conductivity. Water flow management in PEM electrolyzer is a significant design subject as protons form hydronium ions and for that reason, membranes need to be hydrated to operate. For the reduced ohmic losses and better performance, the membrane-electrode assembly is given significant importance in the PEM electrolyzer design. Noble metal such as platinum catalysts can

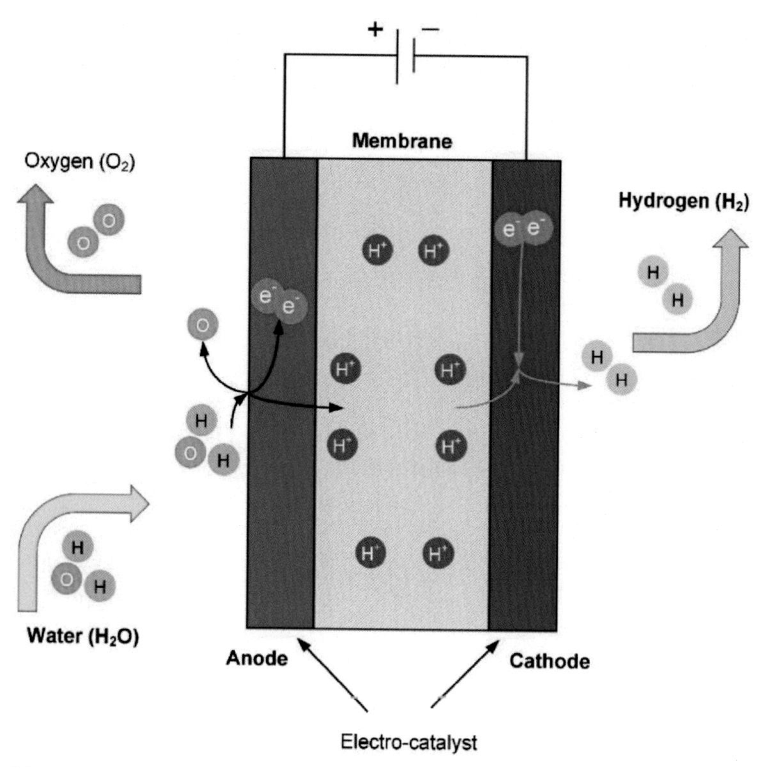

FIG. 2.29

Basic illustration of a proton exchange membrane electrolyzer.

be coated on the electrodes to recompense for slow kinetics that is designed in a three-phase boundary.

The electrical potential gradient is the most substantial driving force used to transport charge in the membrane cell process that is formed due to anions depletion on anode surface while cations depletion takes place on the cathode surface. Consequently, the cations are transported from anode to cathode partition. The conductivity shows the capability of a material to conduct electric current. To relate the resistance of a conductor with conductivity, the following correlation is used:

$$R_c = \frac{l}{\sigma A} \tag{2.66}$$

where R_c signifies the conductor resistance, σ denotes the conductivity, A indicates area, and l represents the conductor length. Ohmic losses calculations use the specific membrane electrical conductivity equations that is determined using the following correlation presented by Ni el al.[66]:

$$\sigma(c, T) = (0.5139c - 0.326)\exp\left(1268\left(\frac{1}{303} - \frac{1}{T}\right)\right) \tag{2.67}$$

Here, T signifies membrane temperature, and c denotes water molar concentration inside the membrane. Molar concentration changes across membrane thickness from high anodic concentration to lower cathodic concentration. The assumption of the linear change in molar water content, across the membrane is reasonable.

$$c(x) = c_c + \left(\frac{c_a - c_c}{\delta}\right)x \tag{2.68}$$

The electrical conductivity is well defined as $\sigma = \frac{dx}{dR}$ where R represents electrical resistance and differential equation can be represented as $dR = \sigma^{-1}dx$:

$$R_{\Omega,\mathrm{PEM}} = \int_0^{\delta} \sigma^{-1}(c(x), T)\, dx \tag{2.69}$$

The PEM electrolyzer concentration overpotentials and water vapor concentration variation across the membrane are essential to be taken into consideration. The correlation for the concentration overpotential can be expressed as follows:

$$\Delta E_{\mathrm{conc}} = J^2\left(\beta\left(\frac{J}{J_{\mathrm{lim}}}\right)^2\right) \tag{2.70}$$

The factor β can be defined as follows:

$$\beta = (7.16\times10^{-4}T - 0.622)P + (-1.45\times10^{-3}\,T + 1.68) \quad \text{if } P < 2 \text{ atm} \tag{2.71}$$

$$\beta = (8.66\times10^{-5}T - 0.068)P + (-1.6\times10^{-4}\,T + 0.54) \quad \text{if } P \geq 2 \text{ atm} \tag{2.72}$$

where $P = \frac{P_i}{0.1173} + P_{\mathrm{sat}}$ is local cathodic or anodic pressure, P_i denotes cathodic or anodic partial pressure, and P_{sat} signifies water saturation pressure and index i is denoted as c for cathode and a for the anode.

The cathodic and anodic activation potential in the PEM electrolyzer undergoes electron transfer at the membrane-electrode assembly. Ni et al.[66] research article helps to determine the correlation for activation potential as follows:

$$\Delta E_{\mathrm{act},i} = \frac{\mathrm{RT}}{F}\ln\left(\frac{J}{J_{0,i}} + \sqrt{\left(0.5\frac{J}{J_{0,i}}\right)^2 + 1}\right) \tag{2.73}$$

where $J_{0,i}$ is cathodic and anodic exchange current density.

Exchange current density is a substantial constraint that helps to calculate activation overpotential. The capability of the electrode is symbolized in the electrochemical reaction. High exchange current density indicates high electrode reactivity that outcomes lower overpotential. The Arrhenius equation can be used to express electrolysis exchange current density:

$$J_{0,i} = J_{\mathrm{ref},i}\, \exp\left(-\frac{\Delta E_{\mathrm{act},i}}{\mathrm{RT}}\right) \tag{2.74}$$

where $J_{\mathrm{ref},i}$ represents preexponential factor.

2.5.2 Solid Oxide Electrolyzer

SO electrolyzer emerged as an oxygen ion conduction promising hydrogen production technique by splitting water at high temperatures. The SO electrolyzer uses a nonporous solid metal oxide electrolyte that is oxygen ion (O^{2-}) conductive. The overall electrochemical reactions are as follows:

Cathodic reaction:

$$H_2O(g) + 2e^- \rightarrow H_2 + O^{2-} \tag{2.75}$$

Anodic reaction:

$$2O^{2-} \rightarrow O_2\,(g) + 4e^- \tag{2.76}$$

Fig. 2.30 exhibits the basic illustration of a SO electrolyzer. The electrolyte assembly entails a thin packed SO layer placed between porous anode and cathode.

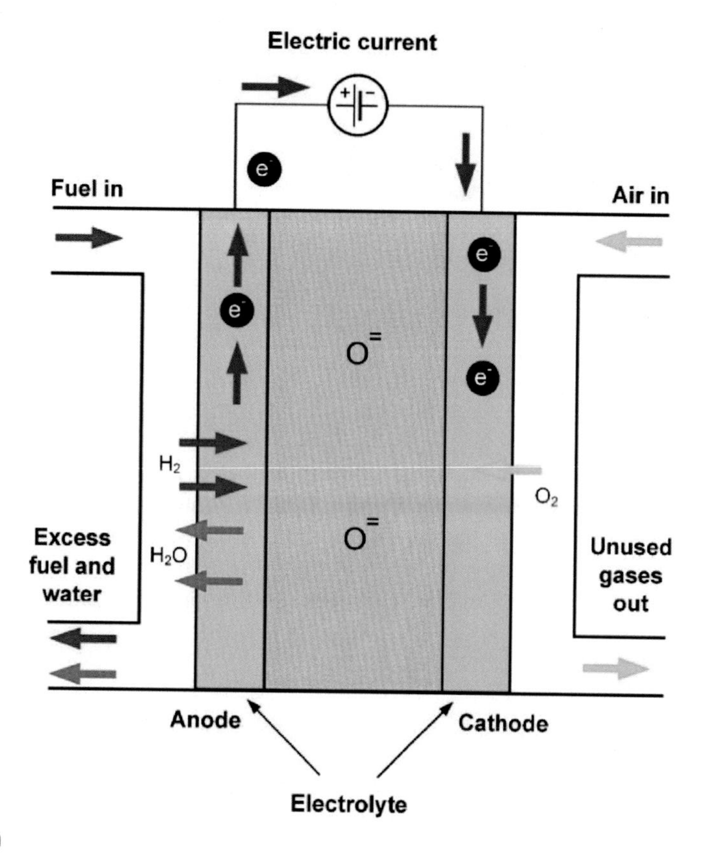

FIG. 2.30

Basic illustration of solid oxide electrolyzer.

The cathode is permeable to steam and hydrogen. The SO electrolyzer is an emerging technology with proton conduction. As being an acidic electrolyte, Eqs. (2.64) and (2.65) represent the half-reactions, though, the electrolyte is placed between two planar porous electrodes and formed of a proton-conducting metal oxide layer. The complete utilization of water is consequently possible in SO electrolyzer with straight implications to raise the system compactness and simplicity.

The absence of the liquid phase in the SO electrolyzer supports the design with the existence of solid and gas (two-phase) solid–gas processes. The cathode contains a composite structure containing a yttrium-stabilized zirconium-made porous matrix joint with cerium and/or nickel. The nickel atoms play the electrocatalyst role that is spread over a porous matrix. Hydrogen is evolved in the reaction that occurs in cathode porous structure in the vicinity of active centers.

Enormous efforts are being devoted to developing proton-conducting membranes globally. The absence of the liquid phase in the SO electrolyzer supports the design. In this regard, the materials of barium cerate ($BaCeO_3$) were recognized as outstanding SO electrolyte as they offer high capability proton conduction over wide temperatures ranges from 300 to 1000°C. The key problematic issue is the formation of a solid membrane due to barium cerate. Doping barium cerate with Samarium (Sm) appears to be a suitable option that allows for sintering membranes with low thickness and high-power densities. They carry yttria-stabilized zirconia electrolyte solid layer that operates under a high temperature of 1000°C, and SO electrolyzer offers several significant advantages. As no noble metal catalysts are desirable, they are cost-friendly and offer a comparatively longer lifetime.

The electrical conductivity of SO electrolyte changes exponentially with temperature as shown by Ni et al.[67]:

$$\sigma = 33000 \exp\left(-\frac{10300}{T}\right) \tag{2.77}$$

The cathodic and anodic polarization overpotential and reversible half-cell potentials calculation consider that reaction occurs in the gaseous phase. Consequently, the molar concentrations and fugacity with partial pressure can be substituted with activity while calculating the reversible half-cell potentials.

The associated activation overpotentials can be formulated as follows:

$$\Delta E_{\mathrm{act},i} = \frac{\mathrm{RT}}{zF}\sinh^{-1}\left(\frac{J}{2J_{0,i}}\right) \tag{2.78}$$

where i represents cathode and anode and cathodic exchange current density of 2000 A/m^2 while anodic exchange current density of 5000 A/m^2.

The Arrhenius equation can be used to express electrolysis exchange current density:

$$J_{0,i} = J_{\mathrm{ref},i} \exp\left(-\frac{\Delta E_{\mathrm{act},i}}{\mathrm{RT}} \right) \tag{2.79}$$

For SO electrolyzer, the activation energy magnitude order is approximately 100 kJ/mol, and anodic preexponential factors are around 5000 s^{-1} and cathodic preexponential factors 50,000 s^{-1}.

For SO electrolyzer, two different cathodic and anodic concentration overpotentials equations can be represented as follows:

$$\Delta E_{\mathrm{act},a} = \frac{\mathrm{RT}}{4F} \ln\left(\frac{\sqrt{\left(P_{\mathrm{O_2}}^{\mathrm{in}}\right)^2 + \frac{\mathrm{JRT}\mu\delta_a}{2\mathrm{FB}_g}}}{P_{\mathrm{O_2}}^{\mathrm{in}}} \right) \tag{2.80}$$

Here, μ signifies dynamic viscosity for molecular oxygen, $P_{O_2}^{in}$ symbolizes oxygen partial pressure in the gas stream at inlet anodic port, δ_a represents anode thickness of the anode, and B_g denotes flow permeability that is calculated using Carman–Kozeny equation:

$$B_g = \frac{2r^2\varepsilon^3}{72\xi(1-\varepsilon)^2} \tag{2.81}$$

where r signifies electrode pore radius, ε denotes electrode porosity, and ξ symbolizes electrode tortuosity.

$$\overline{\lambda} = 0.5(\lambda_{\mathrm{H_2}} + \lambda_{\mathrm{H_2O}}) \tag{2.82}$$

Collision-diffusion integral is determined based on mean collision time.

$$\tau = \frac{k_b T}{\overline{\varepsilon}} \tag{2.83}$$

where ε is binary system characteristic Lenard-Jones potential, and $k_b = 1.38066 \times 10^{-23} \frac{J}{K}$ is Boltzmann constant.

If P_i^{TPB} representing partial pressure can be obtained, then the following correlation is used to represent anodic concentration overpotential:

$$\Delta E_{\mathrm{conc},a} = \frac{\mathrm{RT}}{2F} \ln\left(\frac{P_{\mathrm{H_2}}^{\mathrm{TPB}} P_{\mathrm{H_2O}}}{P_{\mathrm{H_2}} P_{\mathrm{H_2O}}^{\mathrm{TPB}}} \right) \tag{2.84}$$

Likewise, the cathodic concentration overpotential can be represented as follows:

$$\Delta E_{\text{conc},c} = \frac{\text{RT}}{4F} \ln\left(\frac{P_{O_2}}{P_{O_2}^{\text{TPB}}}\right) \tag{2.85}$$

2.5.3 Alkaline Electrolyzer

For large-scale production, alkaline electrolysis is the main candidate for clean hydrogen production. Manabe et al.[68] presented a review study on the potential of alkaline electrolysis technology, and they revealed that a zero-gap technology was recently developed. ALK electrolyzer uses an alkaline electrolyte with pH > 7. In alkaline electrolytes, anions are mobile ionic species of the hydroxyl group. Typical electrolytes used in ALK electrolyzer are liquid bases solutions such as NaOH or KOH.

ALK electrolyzer offers slightly higher efficiency as compared to the PEM electrolyzer with more than 60% that can reach to 70% potentially and takes the hydroxyl benefit and utilizes it as an outstanding charge carrier that offers faster oxygen reduction kinetics. The ALK electrolyzer catalyst requirement is less stringent as compared to the PEM electrolyzer as they use a cheap nickel-based catalyst. Fig. 2.31 exhibits the basic illustration of an ALK electrolyzer system configuration.

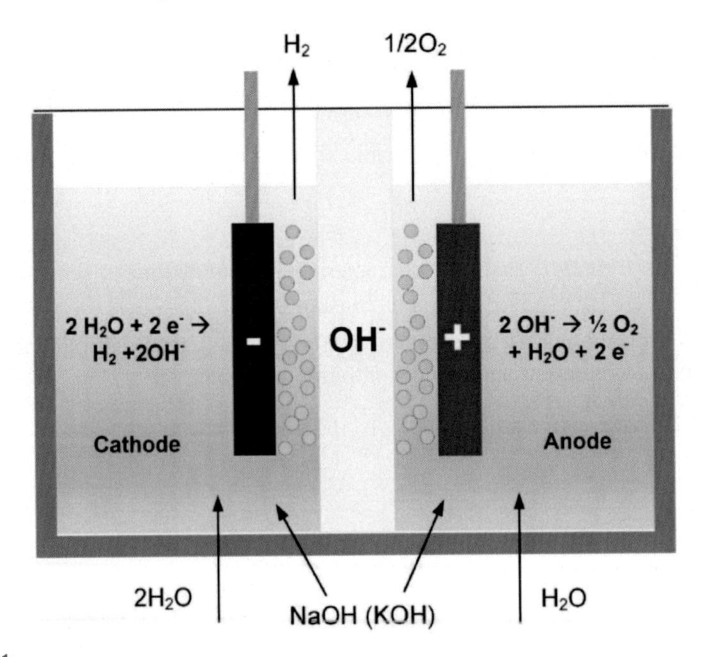

FIG. 2.31

Basic illustration of an alkaline electrolyzer.

The 30%–40% by weight potassium hydroxide solution is used as the electrolyte. Following are the half-reactions at each electrode:

Cathodic reaction:

$$2H_2O(l) + 2e^- \rightarrow H_2(g) + 2OH^- \tag{2.86}$$

Anodic reaction:

$$2OH^- \rightarrow H_2O + \frac{1}{2}O_2(g) + 2e^- \tag{2.87}$$

A commercial alkaline electrolyzer may offer an efficiency of 60%–70% through advanced systems that are currently under development can exceed 90% efficiency. The operating temperature of the commercial alkaline electrolyzer is 80–200°C and production capacities are ranged 500–30,000 Nm^3/h or 0.5–40 MW per unit concerning the hydrogen produced lower heating value. The electrolyte concentration should be sufficiently high, and such high concentration guarantees good mobility of KOH ions that is 30%. The electricity-to-hydrogen technique is used to calculate energy efficiency, and commercial electrolyzer efficiency ranges from 55% to 90%. While operating at temperatures higher than 120°C, a partial energy amount needed for water splitting can be transmitted through heat transfer.

The alkaline electrolyzer is built-in bipolar and monopolar configurations. In the monopolar configuration, numerous cells of electrolysis are packed closely such that the net voltage is the summation of the applied voltage of each cell.

$$E_{ALK} = \sum E_{cell} \cong (N_{cell} - 1)E_{cell} \tag{2.88}$$

For constructive reasons, the Ohmic losses are higher in monopolar electrolyzer as compared with bipolar configuration. In a bipolar configuration, cells are parallelly connected so that voltage remains the same, and the net current can be calculated as follows:

$$I_{ALK} = \sum I_{cell} \cong N_{cell} I_{cell} \tag{2.89}$$

In 1905, Julius Tafel published an equation that is well known for the relation between polarized electrode activation overpotential and current density. The Tafel equation can be written as follows:

$$\Delta E_{act} = a + b\ \ln(J) \tag{2.90}$$

The Tafel equation exhibits the polarization potential value required for establishing unit current density. The parameter b represents Tafel's slope, and it ranges from 0.04 to 0.150 V at a reference temperature.

$$a = b\ln(J_o) \tag{2.91}$$

$$b = \frac{RT}{\alpha F} \tag{2.92}$$

In the alkaline electrolyzer, the diffusion layer thickness practical range is 50–500 μm while limiting current density ranges from 8000 to 10,000 A/m^2. If limiting current is identified, the bulk concentrations and surface concentrations ratio can be attained by oxidation half-reaction and reduction half-reaction.

$$\left(\frac{c_{x=0}}{c_0}\right)_O = 1 + \frac{J}{J_{\lim,O}} \tag{2.93}$$

The correlation used for concentration overpotential is as follows:

$$\Delta E_{\text{conc},R} = \frac{\text{RT}}{zF}\ln\left(\frac{c_{x=0}}{c_0}\right) \tag{2.94}$$

Eq. (2.82) is effective for both anodic and cathodic concentration overpotential. Moreover, Eqs. (2.81) and (2.82) can be used to represent the cathodic concentration overpotential as follows:

$$\Delta E_{\text{conc},R} = \frac{\text{RT}}{zF}\ln\left(1 - \frac{J}{J_{\lim,R}}\right) \tag{2.95}$$

In the same way, Eqs. (2.81) and (2.82) are employed together to express anodic concentration overpotential as follows:

$$\Delta E_{\text{conc},O} = \frac{\text{RT}}{zF}\ln\left(1 - \frac{J}{J_{\lim,O}}\right) \tag{2.96}$$

The net concentration overpotential and cell overpotential because of species concentration are expressed as follows:

$$\Delta E_{\text{conc}} = \left\{\frac{\text{RT}}{zF}\ln\left(\frac{1 - \frac{J}{J_{\lim,R}}}{1 - \frac{J}{J_{\lim,O}}}\right)\right\} \tag{2.97}$$

Bagotski[70] used the correlation that defines combined kinetic diffusional polarization process as follows:

$$\frac{J}{J_o} = \left\{\frac{\exp\left(\frac{\alpha F}{\text{RT}}\Delta E_{\text{pol}}\right) - \exp\left(-(1-\alpha)\frac{F}{\text{RT}}\Delta E_{\text{pol}}\right)}{\left(1 + \exp\left(\frac{\alpha F}{\text{RT}}\Delta E_{\text{pol}}\right)\frac{J_0}{J_{\lim,R}}\right) + \left(\exp\left(-(1-\alpha)\frac{F}{\text{RT}}\Delta E_{\text{pol}}\right)\frac{J_0}{J_{\lim,O}}\right)}\right\} \tag{2.98}$$

where E_{pol} is electrode polarization overpotential owing to the mutual effects of concentration gradients and activation energy. Table 2.5 represents the development in the technoeconomic characteristics of PEM and ALK electrolyzers from 2017 to 2025.

Table 2.5 Technoeconomic Characteristics of PEM and ALK Electrolyzers (2017–25).

Technology		PEM		ALK	
	Unit	2017	2025	2017	2025
Efficiency	kWh electricity per kg of H_2	58	52	51	49
System lifetime	Years	20		20	
Efficiency (LHV)	%	57	64	65	68
Typical output pressure	Bar	30	60	Atmospheric	15
Stack lifetime	Operating hours	40,000 h	50,000 h	80,000 h	90,000 h
OPEX	Initial CAPEX % per year	2%	2%	2%	2%
Load range		0%–160% nominal load		15%–100% nominal load	
Ramp-up/ ramp-down		100% per second		0.2%–20% per second	
Start-up (warm-cold)		1 s to 5 min		1–10 min	
Shutdown		Seconds		1–10 min	

Data from IRENA. Hydrogen From Renewable Power: Technology Outlook for the Energy Transition; *2018. Available from: www.irena.org.*

2.6 Closing Remarks

The mainstream hydrogen (almost 95%) is produced using fossil fuels and more specifically natural gas reforming, methane partial oxidation, and coal gasification. Additional significant methods of producing hydrogen consist of water electrolysis and biomass gasification. The conventional hydrogen production method of natural gas reforming undergoes greenhouse gas emissions that cause environmental problems. Fossil fuels do not only possess the disadvantage of greenhouse gas emissions but accompany the faster depletion of fossil fuels. Renewable hydrogen production methods are gaining much attention because they offer clean, sustainable, and environmentally benign energy and hydrogen production solutions and overcome numerous challenges, namely GHG emissions, fossil fuel depletion, and carbon emissions taxes. Renewable energy is extracted using naturally driven sources, namely solar, wind, hydro, geothermal, ocean thermal energy conversion, and biomass gasification or pyrolysis.

Renewable energy is used to extract three different commodities, namely electricity, thermal energy, or fuel, and each of these commodities can be employed to produce hydrogen. Electricity can be employed directly to the electrolyzer to produce hydrogen, thermal energy can be employed to the thermochemical cycles for hydrogen production, and fuels can be employed to produce hydrogen using biomass gasification or pyrolysis. Electricity can be extracted directly using renewable energy sources such as solar PV, wind energy, hydropower, geothermal energy, and ocean thermal energy conversion while thermal energy is extracted using different sources such as solar thermal collectors, solar heliostats, and biomass gasification and pyrolysis. Mechanical work generated by energy sources such as wind, hydro, tidal, and ocean thermal energy conversion is converted to electrical power using the generator, and electrical work is employed to the electrolyzer to produce hydrogen. High-temperature geothermal energy extraction methods generate electricity that is employed to the electrolyzer to produce hydrogen, and a geothermal heat pump generates thermal energy that can be employed to the thermochemical cycle to produce hydrogen. Solar energy extracted using photic and photovoltaic energy produces electrical work that is employed for hydrogen production through electrolysis. Some renewable energy sources are combined for multigeneration A global transition is taking place in the transportation sector with the transition of fuel combustion-based vehicles to the hydrogen fuel-cell and hybrid hydrogen fuel-cell vehicles. To provide clean, environmentally benign and sustainable energy solutions, a global transition is required to replace the traditional energy sources with renewable energy sources where hydrogen is produced as a green one and used widely for systems and applications.

CHAPTER

Solar Energy-Based Hydrogen Production

3

Solar energy is recognized as a primary renewable energy source that can be employed for producing clean and sustainable hydrogen. This solar energy source has the potential to play a vital role in the transition from conventional to renewable energy sources.

A recent study[71] conducted the energetic, exergetic, and economic assessment of a solar energy-driven hydrogen production system. The designed system consisted of photovoltaic panels, proton exchange membrane (PEM) electrolyzer, PEM fuel cell, and hydrogen storage unit, and the proposed system was investigated using a potential software package TRNSYS. The net photovoltaic (PV) panel area was 300 m^2 integrated with the fuel cell of 5 kW capacity, and hydrogen is compressed at 55 bars pressure for storage. The significant objective of this research study was to validate that system covered emergency electricity demand without undergoing a shortage. The analysis was performed for the whole year, and exergetic and energetic efficiencies were found to be 4.25% and 4.06%.

The consumption of solar energy has been increasing significantly around the globe due to the reduced cost, flexible products and incentives for applications. Fig. 3.1 exhibits the global consumption of solar energy in the different regions of the world (in Mega Watts (MW)). Continuous and significant growth of solar energy sources can be depicted in all regions of the world including North America, South and Central America, Europe, Commonwealth Independent States, Middle East, Africa, and the Asia Pacific from 2008 to 2018. Solar energy consumption in North America increased from 805 to 57,118 MW in the years 2008–18, increased from 39 to 7206 MW in South and Central America, increased from 10,522 to 128,758 MW in Europe, increased from 0 to 600 MW in the Commonwealth Independent States, increased from 10 to 3181 MW in the Middle East, increased from 65 to 6093 MW in Africa and increased from 2955 to 284,873 MW in the Asia Pacific region.

The intermittent nature of solar energy urges the requirement of energy storage systems that are described in Chapter 1. Fig. 3.2 exhibits the routes of solar energy-based hydrogen production.

A recent study[72] was directed to the hydrogen production using renewable energy. The cost comparison of hydrogen with other fuels is the challenging point to achieve the comparison of gasoline with hydrogen for passenger vehicles. Water

Renewable Hydrogen Production. https://doi.org/10.1016/B978-0-323-85176-3.00009-3

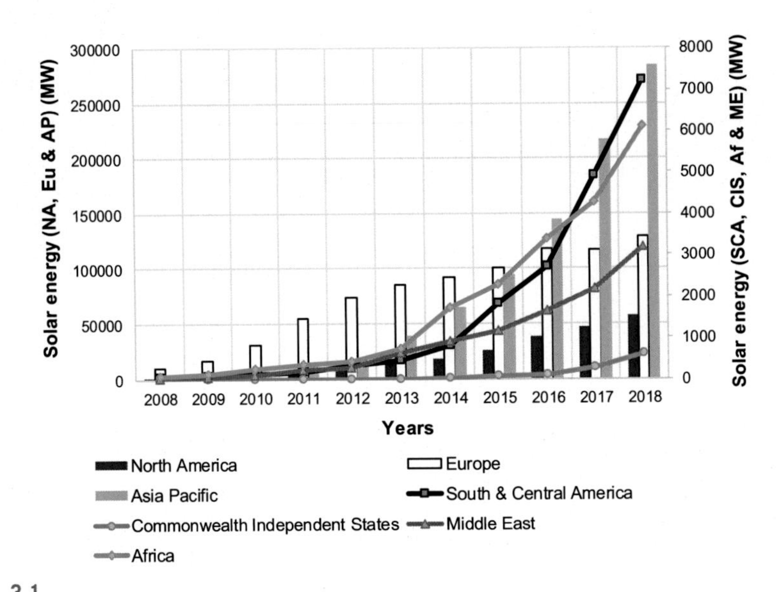

FIG. 3.1

Global solar energy consumption in the different regions of world (MW).

Data from [48].

electrolysis is a typical hydrogen production commercial technology that is converted to the renewable energy-based system if integrated with solar and wind energy sources. The biomass processes carry the potential of producing hydrogen using municipal sewage and forest residue employing the processes of gasification, fermentation, and pyrolysis.

A recent book chapter[73] was directed on the future of hydrogen infrastructure. Hydrogen can be employed as an efficient and clean source in various applications, namely fuel cells, transportation, portable applications, power, combustion, and heating. Hydrogen can be produced through environmentally benign resources using renewable sources such as wind, biomass, solar, geothermal and hydropower, and nuclear energy and conventional sources such as natural gas and coal. Furthermore, hydrogen helps to utilize the intermittent nature of renewable energy sources, namely solar and wind. Nevertheless, hydrogen undergoes significant technical, infrastructure, economic and societal challenges for large-scale implementation. This study evaluated current and projected hydrogen infrastructure status evolving production of hydrogen, delivery systems as well as infrastructure design problems. They also provided a comparative study of hydrogen with alternative fuels for transportation concerning greenhouse gas (GHG) emissions reduction and cost-effectiveness to build hydrogen infrastructure.

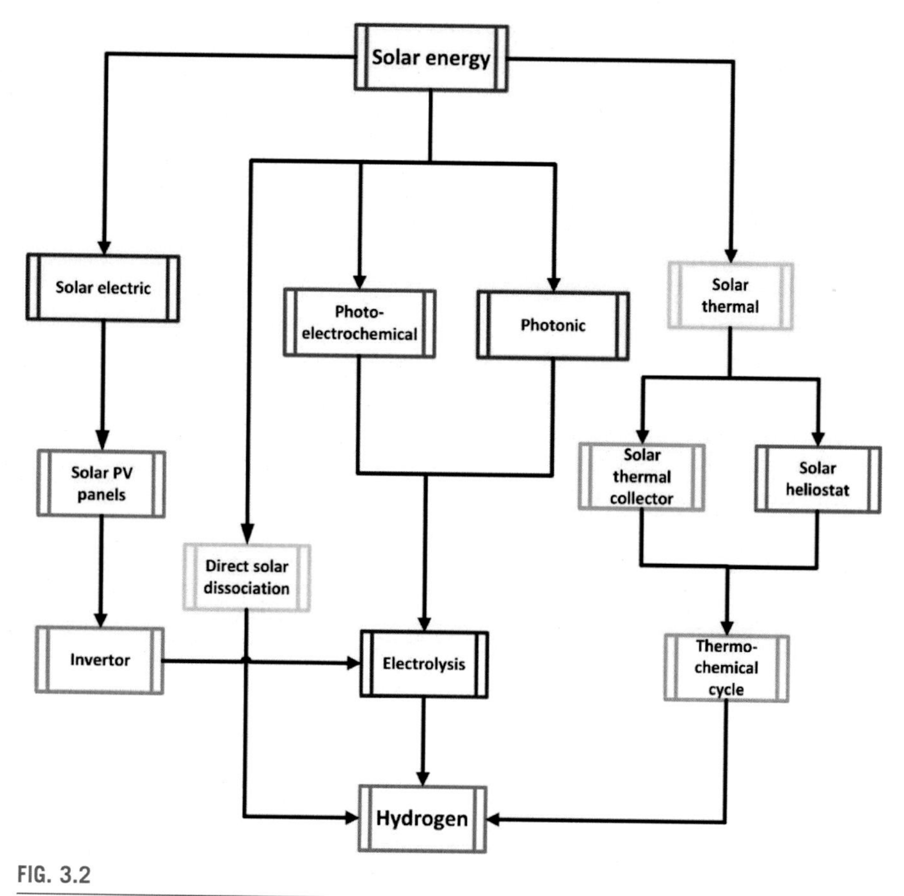

FIG. 3.2

Routes of solar energy-based hydrogen production.

Modified from [50].

The solar energy source is distributed among four significant routes that can be employed for hydrogen production, and these four significant routes are further divided into some categories. The significant elements for hydrogen production methods employing different routes of solar energy are listed as follows:

- Photoelectrochemical energy option:
  - Solar PV panel(s)
  - Battery bank
  - Electrolysis
  - Hydrogen
- Photonic energy option:
  - Electrolysis
  - Hydrogen

- Solar photovoltaic energy option:
  - Solar PV panels
  - Battery bank
  - Invertors
  - Electrolysis
  - Hydrogen
- Solar thermal option:
  - Solar thermal collector
  - Solar heliostat
  - Thermochemical cycle
  - Hydrogen

The concentrated solar thermal energy systems can directly be used to produce hydrogen using multiple routes namely; photoelectrochemical energy, photonic energy, photovoltaic energy, and solar thermal energy using the thermochemical cycle. The photoelectrochemical energy and photonic energy are capable of producing hydrogen directly.

The photovoltaic option generates electrical power that is employed to the electrolysis for hydrogen production. The thermal energy extracted through the solar thermal collector and solar heliostat field can be either used to produce power or employed to the thermochemical cycle for hydrogen production. Table A-2 in the appendix tabulates the global solar energy generation in different regions.

A dynamic approach in the analysis is important to explore the unpredictable nature of the solar energy source. Table 3.1 displays the average day of the month taken

Table 3.1 Average Days of Each Month Used for the Dynamic Simulation of Wind and Solar energy sources.

Month	For Average Day of Month		Date
	n	δ	
January	17	−20.92	17/01/2019
February	47	−12.95	16/02/2019
March	75	−2.42	16/03/2019
April	105	9.42	15/04/2019
May	135	18.79	15/05/2019
June	162	23.09	11/06/2019
July	198	21.18	17/07/2019
August	228	13.45	16/08/2019
September	258	2.22	15/09/2019
October	288	−9.60	15/10/2019
November	318	−18.91	14/11/2019
December	344	−23.05	10/12/2019

Source: [74].

into consideration for the dynamic analysis along with declination angle δ. Toronto is the geographical location chosen for a dynamic (time-dependent) simulation.

The dynamic analysis for solar energy source employing solar radiation intensities considered for every hour of average days of each month presented in Table 3.1. This section encloses the modeling equations that are employed to perform dynamic analysis. The direct normal irradiance helps to determine the incoming beam radiation that is significant to calculate the total solar power that can be extracted using solar PV source or solar thermal energy that can be harvested using solar energy. The equation employed to determine the total solar power output can be expressed as follows:

$$\dot{P}_{PV} = \eta_{PV} \dot{Q}_{PV_{SI}} \tag{3.1}$$

The correlation that is used to calculate the solar heat extracted through solar energy source can be represented as follows:

$$\dot{Q}_{PV_{SI}} = A_{PV} \dot{I}_{beam} \tag{3.2}$$

where $\dot{Q}_{PV_{SI}}$ indicates solar heat input, η_{PV} represents the efficiency of the PV cell, $\dot{I}_{beam}$ denotes incoming beam radiation, and A_{PV} symbolizes PV cell area. The direct normal irradiance helps to determine the incoming beam radiation and correlation can be represented as follows:

$$\dot{I}_{beam} = \dot{I}_{normal} \cos \theta_{zenith} \tag{3.3}$$

where θ_{zenith} characterizes the zenith angle. The correlation to quantify direct normal irradiance can be expressed as follows:

$$\dot{I}_{normal} = 0.9715 E_{ecc} \dot{I}_{solar\ const} \tau_{Rayleigh} \tau_{ozone} \tau_{gas} \tau_{water} \tau_{aerosol} \tag{3.4}$$

where τ signifies scattering transmittance of Rayleigh, ozone, gas, water, and aerosol, respectively, and E_{ecc} represents the eccentricity factor. The correlation to determine the eccentricity factor can be represented employing day angle.

$$E_{ecc} = 1.00011 + 0.034221 \cos(\mathrm{DA}) + 0.00128 \sin(\mathrm{DA}) + 0.000719 \cos(2\mathrm{DA}) + 0.000077 \sin(2\mathrm{DA}) \tag{3.5}$$

$$\mathrm{DA} = (n-1)\frac{360}{365} \tag{3.6}$$

where ST designates solar time that helps in determining the hour angle can be determined using the following equation:

$$\text{Solar time} = 4\left(L_{SM_{lt}} - L_{longitude}\right) + E_{ecc} + \text{standard time} \tag{3.7}$$

The correlation that is used to calculate the zenith angle cosine can be represented as follows:

$$\cos\theta_{zenith} = (\cos(\omega))(\cos(\delta))(\cos(\phi)) + (\sin(\phi))(\sin(\delta)) \quad (3.8)$$

where δ signifies declination angle, ω represents hour angle, ϕ denotes latitude, and the correlations that are employed to determine these factors can be represented employing the following equations:

$$\omega = 15(12 - \mathrm{ST}) \quad (3.9)$$

$$\delta = 23.45\sin\left(360\frac{n+284}{365}\right) \quad (3.10)$$

The dynamic analysis of the solar energy source helps in calculating the total energy extracted using the solar source. The step followed to calculate the direct normal irradiance are listed as follows:

- The declination angle δ, hour angle ω, and latitude ϕ are calculated that help to determine the zenith angle
- The standard time along with the latitude and longitude are calculated to determine the solar time
- The eccentricity factor is determined through day angle
- The direct normal irradiance is calculated using scattering transmittance of Rayleigh, ozone, gas, water, and aerosol and eccentricity factor
- The incoming beam radiation is calculated using the direct normal irradiance and zenith angle
- The solar heat input is calculated using the incoming beam radiation and solar cell area.
- The solar power is calculated using the solar heat input and efficiency of the solar cell.

After following the mentioned steps on average days of each month, one will be able to quantify the direct normal irradiance that helps to determine the solar heat that can be extracted using the solar thermal collectors and also the solar power that can be produced using the solar photovoltaic panels. Fig. 3.3a displays the direct normal irradiance from January to June. The maximum values of direct normal irradiance are displayed by the month of June from months January to June. Fig. 3.3b displays the direct normal irradiance from July to December. The maximum values of direct normal irradiance are displayed by the month of July from months July to December.

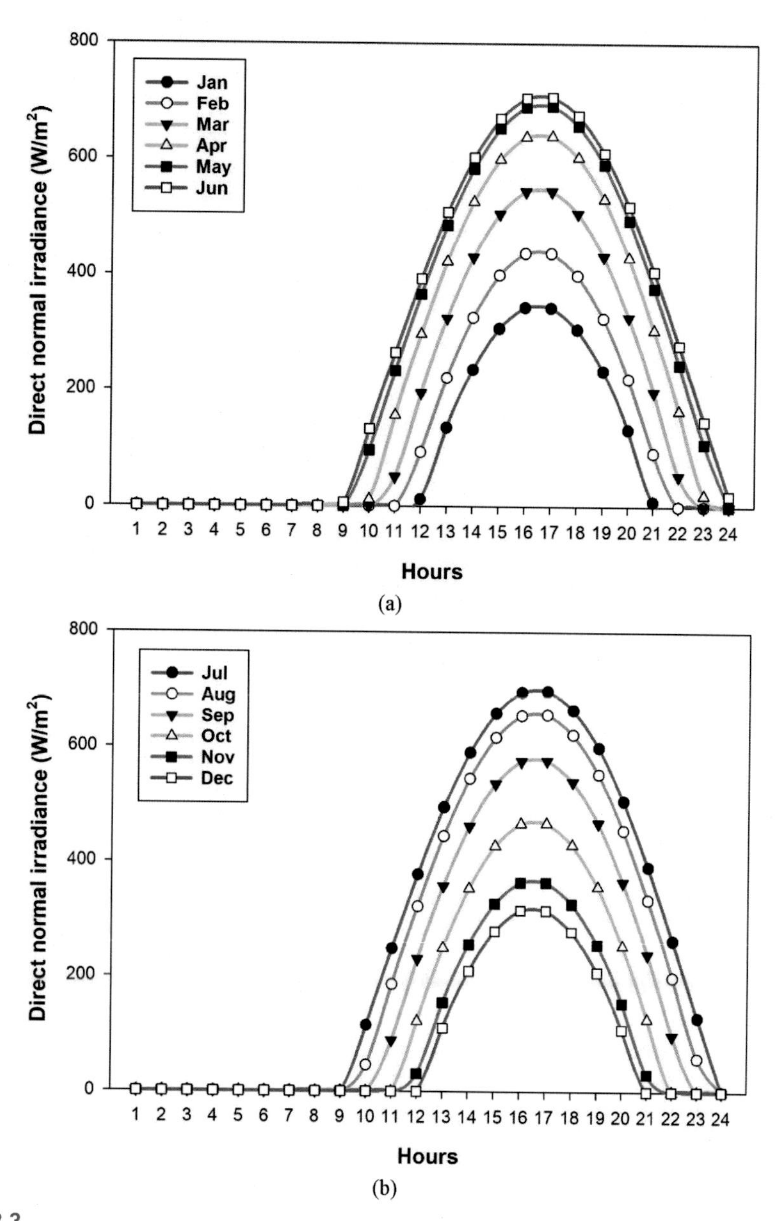

FIG. 3.3

Direct normal irradiance on average days of each month: (a) Jan–Jun; (b) Jul–Dec.

3.1 Photoelectrochemical Hydrogen Production

In the photoelectrochemical water-splitting (PECWS) process, sunlight is labored to split water into its components of oxygen and hydrogen employing specialized semiconductors named as photoelectrochemical materials that employ the light energy for direct water dissociation into its components. The PECWS process employs semiconductor material that converts solar energy straight into the chemical energy in the hydrogen form. The semiconductor materials that are employed in PECWS process are like those that are employed to generate photovoltaic solar electricity nevertheless for PECWS applications, the semiconductor is absorbed in water-based electrolytes and sunlight energizes the water-splitting process. Fig. 3.4 displays the conceptual schematic of photoelectrochemical hydrogen production.

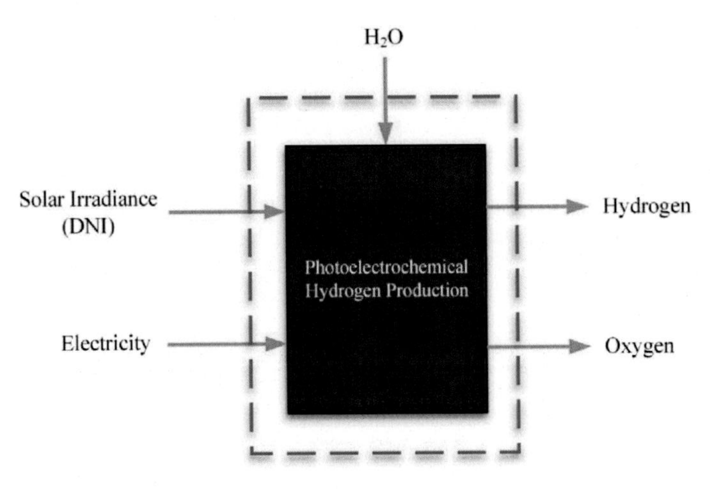

FIG. 3.4

Photoelectrochemical hydrogen production.

A recent handbook[75] has offered a detailed investigation of solar photoelectrochemical hydrogen production. This handbook considered electricity and hydrogen as an everlasting energy system and also discussed specific actual information and technical details. The text focused on the variety of applications, namely hydrogen fuel cells and catalytic hydrogen combustion. This book also included material and discussion on inversion curves, thermodynamic and physical tables, specific heat data, storage materials properties, temperature—entropy, and compressibility charts and many more. They did not only analyze the hydrogen energy production principles, storage, and utilization but also investigate different methods, namely thermolysis, electrolysis, photolysis, biomass-based hydrogen production, thermochemical cycles, and other hydrogen production techniques. This handbook was published to offer a source of hydrogen energy to the reference of scientists investigating the solar photoelectrochemical hydrogen production.

A recent study[45] in the literature provided a comparative investigation of thermochemical, photoelectrolytic, electrolytic, and photochemical technologies to produce hydrogen using solar-to-hydrogen technique. The solar energy-based hydrogen produced is a promising technique that has the potential to play a significant role in contributing to sustainable energy supply. This study discussed the exceptional properties of hydrogen production using solar energy as compared with other energy forms. They also explored the advanced research and improvement of several solar-to-hydrogen techniques of electrical, thermal, and photon energy. A comparative analysis was also provided to evolve methods of water splitting, energy efficiency, solar energy forms, and engineering systems, among others. It was revealed that the energy losses can be minimized in thermochemical cycles by employing large production scales. The solar energy sources, namely electrolysis, photochemical, and photoelectrochemical techniques can be further beneficial for hydrogen fueling stations applications as these processes require fewer processes.

The standard photovoltaic cells involve the negative charge (electrons) excitation in the semiconductor medium and negatively charged electrons are eventually extracted to generate power. The photoelectrolytic cell state is relatively dissimilar. In water-splitting PEC cell, the electron excitation in a semiconductor using light leaves a hole that draws electron through adjacent water molecule:

$$H_2O\ (l) \rightarrow 2H^+\ (aq) + \frac{1}{2}O_2 \tag{3.11}$$

This leaves positively charged carriers (H^+ ions) in the solution that bond with another positively charged ion, and two electrons are combined to form hydrogen gas using the following chemical reaction.

$$2H^+\ (aq) + 2e^- \rightarrow H_2(g) \tag{3.12}$$

Fig. 3.5 exhibits the process schematic of the hydrogen production system using the photoelectrochemical process. In the PECWS-based hydrogen production process, the existing photoelectrode in the photoelectrochemical cell captures the solar irradiance though electricity is supplied to the photoelectrochemical cell as well. The water is supplied to a photoelectrochemical cell that splits into negative (O^{-2}), and hydrogen gas is formed when two positives (H^+) ions react with each other.

The free energy change is linked with water molecule conversion into 1 mol of hydrogen, and a half mole of oxygen is 237.2 kJ/mol that matches the electrolysis cell voltage of 1.23 V per transferred electron under standard conditions. The general schematic of water-splitting photoelectrochemical cell is shown in Fig. 3.4 that includes cathodic and anodic electrodes. The anode is a photoactive semiconducting material-based electrode that is exposed by a beam of light through water splitting. The cathode is counter electrode that is also termed as photocathode even though it is not exposed through light. To run this reaction, photoactive material on photoanode surface absorbs radiant light to boost electrode potential from 1.23 V thus, water molecules are oxidized to produce protons (H+) and oxygen while protons are instantaneously reduced to produce hydrogen at the cathode. Consequently, the water-splitting sluggish kinetics, energy loss occurs during the electrons transfer at the photoanode interface. Therefore, the energy required for photoelectrolysis is often provided as 1.7–2.4 eV. The photoinduced process generates four electron–hole pairs on photoanode to form oxygen. If the photoanode is exposed to the light that carries energy more than a bandgap of photoactive material, at that moment, the

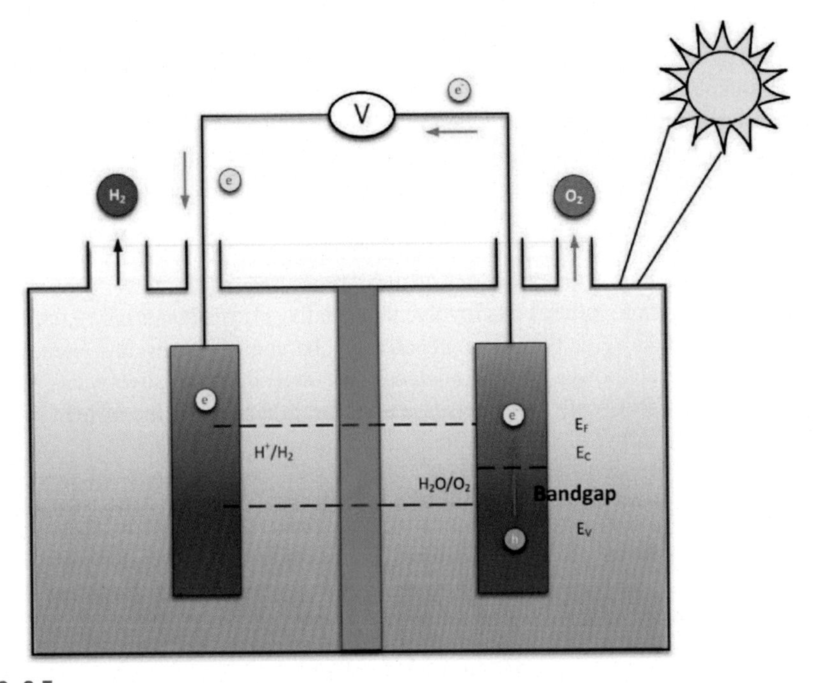

FIG. 3.5

General schematic of water-splitting photoelectrochemical cell.

Modified from [76].

valence band electrons become excited into the conduction band although the holes are kept in the valence band. The photogenerated electrons then pass over the external wire and range the cathodic surface to react with positively charged ions (H+), forming hydrogen and holes existing at photoanode oxidize $H_2O(O^{2-})$ to produce oxygen.

3.2 Photonic Hydrogen Production

Hydrogen production using renewable energy as primary source appears to be one of the most auspicious methods to protect future energy generation. Photocatalysis is a hydrogen production approach that involves sunlight and semiconducting material interaction. Consequently, semiconducting material desires to greatly absorb photons and make a metal layer (high work function) to work as Schottky barrier that creates an electrons reservoir to produce molecular hydrogen by the reduction of hydrogen cations and also forms metal oxide layer in some cases to produce molecular oxygen using oxygen anions oxidation.

A recent study[77] offered a comprehensive review study directed to the hydrogen production options. Different sources and systems for producing hydrogen and hydrogen storage options were relatively investigated in this study. Economic, social, environmental, and technical reliability and performance were employed to provide a comparison among the designated options. This study considered biomass, hydro, geothermal, solar, nuclear, and wind energy sources were considered in this study utilizing the different hydrogen production methods of biological, photonic, thermal, and electrical were investigated in this study. Furthermore, they also provided some case studies along with basic research that is required to improve the hydrogen energy systems performance and to deal with the significant hydrogen economy challenges. They revealed that solar source offered a maximum environmental performance of (8/10), the total average ranking was found to be (7.4/10), geothermal offered lowermost total average ranking of (4/10), and nuclear offered a lowermost environmental performance of (3/10) among designated sources of hydrogen production.

Photocatalytic water splitting (PCWS) is attracting substantial attention in recent literature as it employs the most abundant, renewable, clean, and natural source of energy. Hydrogen production using the photocatalytic process is termed as one of the potential solutions to address limited reserves, excessive utilization, and negative environmental influence carried by fossil fuels. The probable advantages of photocatalytic water splitting into oxygen and hydrogen can be organized as environmental and economic benefits as the process utilizes solar energy and forms hydrogen using a clean approach undergoing zero GHG emissions.

A photosensitizer that can absorb solar radiation should be dissolved in solution to employ solar irradiance directly for PCWS as water is visible spectrum transparent. The categorization of hydrogen production systems using photonic energy is displayed in Fig. 3.6.

In PCWS-based hydrogen production, the photons are required that carry comparatively higher energy than photocatalyst band gaps to produce electron—hole pairs for water splitting. The less energetic photons cannot generate these electron—hole pairs; thus, they are not used for photocatalytic production of hydrogen as their water-splitting requirements and bandgap of the most catalysts can employ photons. The key categories of photonic hydrogen production methods employing different routes are listed as follows:

- Heterogeneous photocatalysis
  - Photoelectrochemical cell
    - Photocathode
    - Photoanode
    - Combined
  - Die synthesized tandem cell
- Homogeneous photocatalysis
  - Supramolecular photocatalysis
    - Single electron
    - Multielectron
- Hybrid photocatalysis

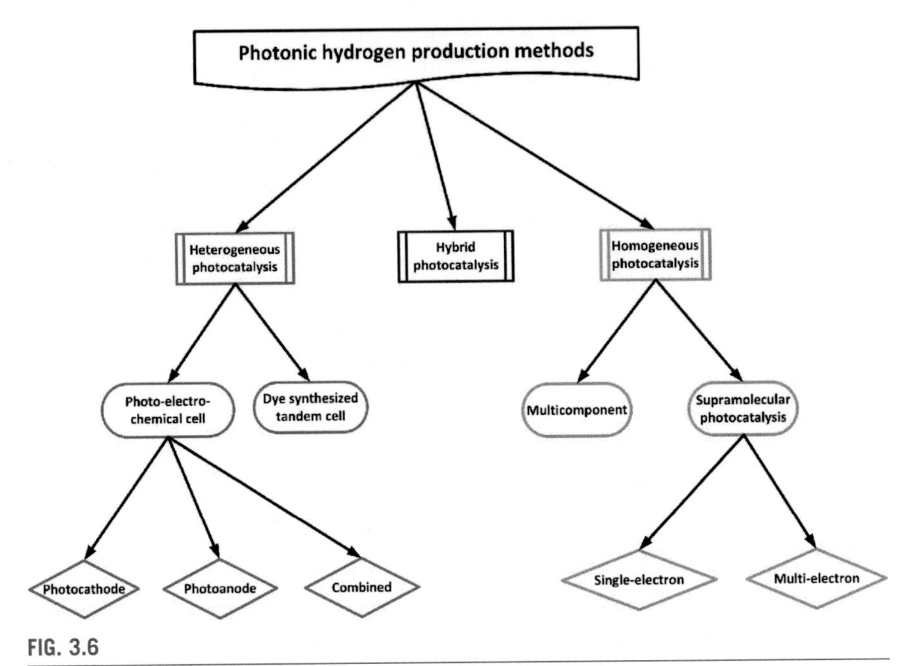

FIG. 3.6

Classification of photonic hydrogen production methods.

Modified from [78].

3.3 Solar Photovoltaic Energy

Solar photovoltaic energy is a renewable and clean energy source that employs the direct normal irradiance from a solar source to generate electricity. Solar photovoltaic energy source undergoes a photoelectric effect that uses a certain material to absorb photons that are light particles and release electrons that generate electric current. Table A-3 in the appendix displays the global solar PV energy consumption by different regions.

A recent study[79] was published on solar photovoltaic electrolytic cell-based hydrogen production. Among the numerous solar energy-driven hydrogen production methods, the integration of solar photovoltaic sources with the electrolytic cell is considered as a promising candidate for clean and renewable hydrogen production that is reviewed in this study. Principally, the significant focus was drawn on the water oxidation electrocatalysts and their fabrication techniques appropriate for monolithic PV electrolytic cell. A recent review study[30] presented a comprehensive review on the solar and hydrogen/fuel cell-based energy systems for immobile applications. This study offered a comprehensive review study on the solar energy-based hydrogen production methods accompanied with the advancements and their current status.

Solar photovoltaic cells absorb the sunlight and employ this light energy to produce an electrical current. There are multiple PV cells in a single solar panel. The net current produced by all the PV cells can be set depending upon the power requirement. Solar panels generate a direct current that is converted to the alternating current using an inverter. The PV cells absorb the sunlight and extract the DC electricity. The I–V characteristics of the photovoltaic source describe the maximum power that can be extracted from the PV array. The following are the significant benefits associated with the solar energy source:

- Renewable energy source
- Produces clean energy
- Reduces electricity bills
- Various applications
- Undergoes zero carbon emissions
- Maintenance cost is low
- Advanced and improved technology development

To analyze the photovoltaic energy-based hydrogen system, Yilanci et al.[30] conducted an experimental study that is adopted in this study. The experimental setup was installed at Clean Energy Center in 2007 at Pamukkale University located in Turkey. The designed system was driven by 5 kWe PV panels. Fig. 3.7 exhibits the schematic of the solar photovoltaic energy-based water electrolysis system.

For performance assessment, half photovoltaic modules were mounted tilt while the other half were installed on solar trackers. The photovoltaic modules fixed tilt at 45° south were located on the building rooftop. Individual tracker comprises 10 modules undergoing 1.25 kWe of nominal power. Table 3.2 exhibits the energetic and exergetic efficiencies of photovoltaic and solar thermal hydrogen production systems.

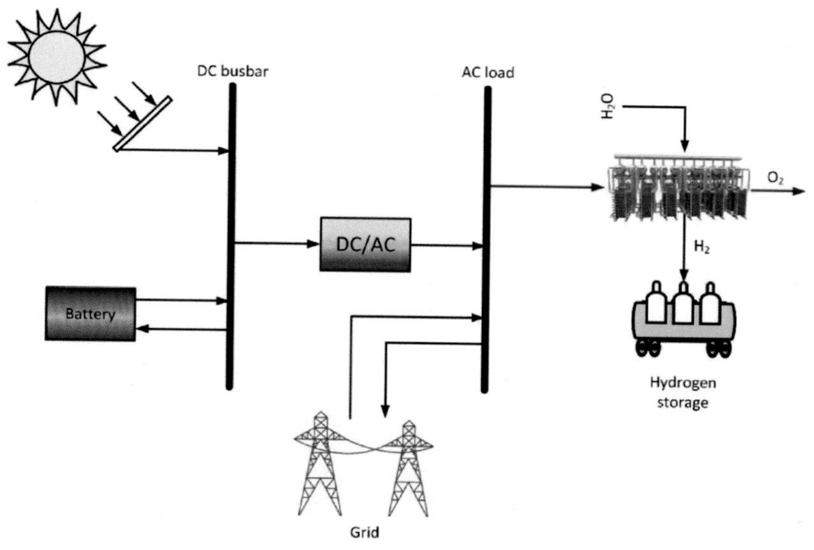

FIG. 3.7

Solar PV energy-based water electrolysis system.

Modified from [30].

Table 3.2 Energy and Exergy Efficiencies of Photovoltaic and Solar Thermal Hydrogen Production Systems [30].

Photovoltaic hydrogen production				
Efficiencies	***Photovoltaic cells***	***Charge regulators***	***Inverter***	***Electrolyzer***
Energy efficiency	11.2–12.4	85–90	85–90	56
Exergy efficiency	9.8–11.5	85–90	85–90	52
Solar thermal hydrogen production				
Efficiencies	***Solar collector***	***Heat engine***	***Generator***	***Electrolyzer***
Energy efficiency	58.33	63.8	50–62	56
Exergy efficiency	39.18	30–50	50–62	52

3.3.1 Case Study 1

The photovoltaic source generates electrical power that is employed to the electrolysis process for hydrogen production. A dynamic analysis is significant to explore the unpredictable nature of the solar energy source. Table 3.1 displays the average day of the month taken into consideration for the dynamic analysis along with declination angle δ. Toronto is the geographical location chosen for dynamic simulation. This case study does not only represent the dynamic analysis of the solar PV system but also exhibits the experimental results conducted in the recent study.[80]

Fig. 3.8 displays the solar PV energy-based hydrogen production system using water electrolysis. The sunlight is absorbed by the photovoltaic cells to produce an electrical current. There are multiple PV cells in a single solar panel. In this case study, 120 solar PV panels are employed and an efficiency of 16% is assumed. The AC/DC converter is employed to convert the DC power that is extracted from the solar PV source into AC power. The power harvested from the solar PV source is employed to the PEM electrolyzer that splits water into its components. The hydrogen produced using a PEM electrolyzer is stored in the hydrogen storage tank. The solar PV power that is extracted using the solar PV source is determined under different weather forecast conditions for the described geographical location. For the mentioned geographical locations, 43.85°N latitude and 79.38°W longitude is employed for dynamic analysis. Table 3.3 displays the design parameters of the dynamic simulation of solar PV sources along with latitude and longitude of geographical location.

The dynamic analysis is conducted on the solar PV source to produce the electrical power and use it for hydrogen production. To calculate the solar PV power, the

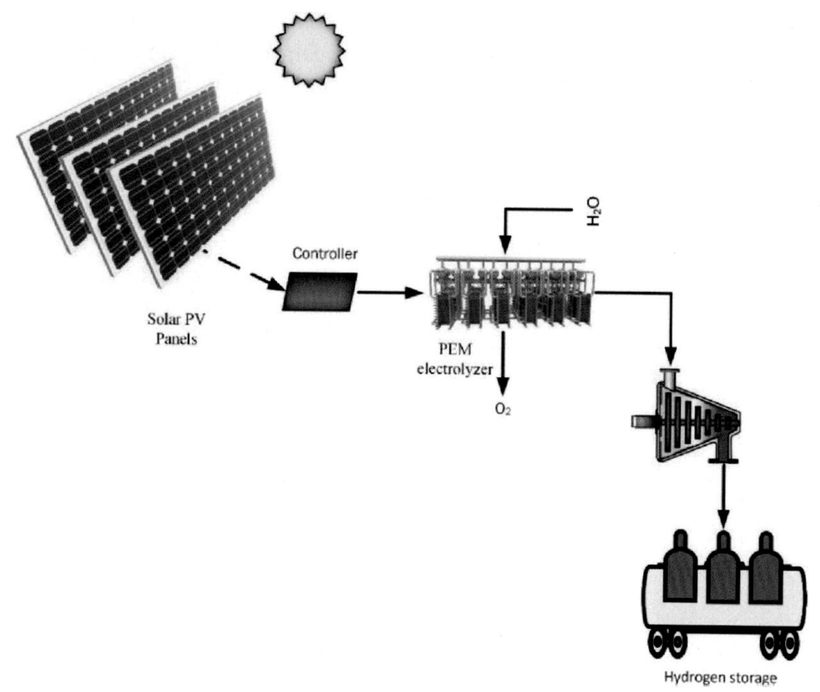

FIG. 3.8

Solar PV energy-based hydrogen production system using water electrolysis.

Table 3.3 Design Parameters of the Dynamic Simulation of Solar PV Source.

	Parameter	Value
Location	Geographical latitudes	Toronto Latitude = 43.85° N Longitude = 79.38° W
Solar PV	PV cell type	Poly-Si_CSX-310
	Number of cells	120
	Efficiency	16%

direct normal irradiance is shown in Fig. 3.3a and b on average days of each month is employed. It is significant to investigate the designed system under different operating environments and sunlight conditions.

3.3.2 Case Study 2

Fig. 3.8 displays the solar PV energy-based hydrogen production system using water electrolysis. This case study employs the SunWize Model: SC3−6V type solar PV cells for the experimental investigations with 20 cells of 22 cm × 7 cm dimensions. This case study evaluates the voltage and power against the electrical current. The solar radiation intensity applied to solar PV source is 870 W/m^2. The proton exchange membrane electrolyzer is employed to produce hydrogen using the electrical power extracted using a solar PV source with Nafion PFSA membrane. The PEM electrolyzer consisted of four electrolysis cells and cell diameters were 13.8. The designed solar PV-powered hydrogen production system undergoes the hydrogen production capacity of 16.7 cm^3/s. Characteristic operating conditions of the experimental solar PV-based water electrolysis system are arranged in Table 3.4. The solar PV source is investigated for current, power, and energetic and exergetic efficiencies. The PEM electrolyzer efficiencies and solar to hydrogen efficiencies are also calculated in this case study.

The solar PV power that is extracted using the solar PV source is determined under different weather forecasts for the described geographical location. Following the direct normal irradiance, the solar PV power is calculated using the dynamic analysis. Fig. 3.9a exhibits the solar PV power that can be extracted using solar PV panels on the average day of each month from January to June. The maximum solar PV power values are displayed by June from months January to June as displayed in Fig. 3.9a. Fig. 3.9b represents the solar PV power that can be extracted using solar PV panels on the average day of each month from July to December. The maximum solar PV power values are displayed by July from months July to December as displayed in Fig. 3.9b.

Solar PV cells and PEM electrolyzer are the most significant components of the designed system shown in Fig. 3.8, and it is important to explore the performance of these substantial components. Fig. 3.10 depicts the effect of single PV cell voltage on the current and power output. Also, Fig. 3.11 exhibits the performance of the single solar PV cell in terms of energetic and exergetic efficiencies. This case study is

Table 3.4 Characteristic Operating Conditions of the Experimental Solar PV-Based Water Electrolysis System [80].

Parameter	Value
Proton exchange membrane electrolyzer	
Cell diameter	13.8 cm
No. of electrolysis cells	4
Membrane type	Nafion PFSA
Hydrogen production capacity	16.7 cm^3/s
Type of PV cells	SunWize Model: SC3–6V
Cell dimensions	22 cm × 7 cm
Solar radiation intensity applied	870 W/m^2
Number of PV cells	20

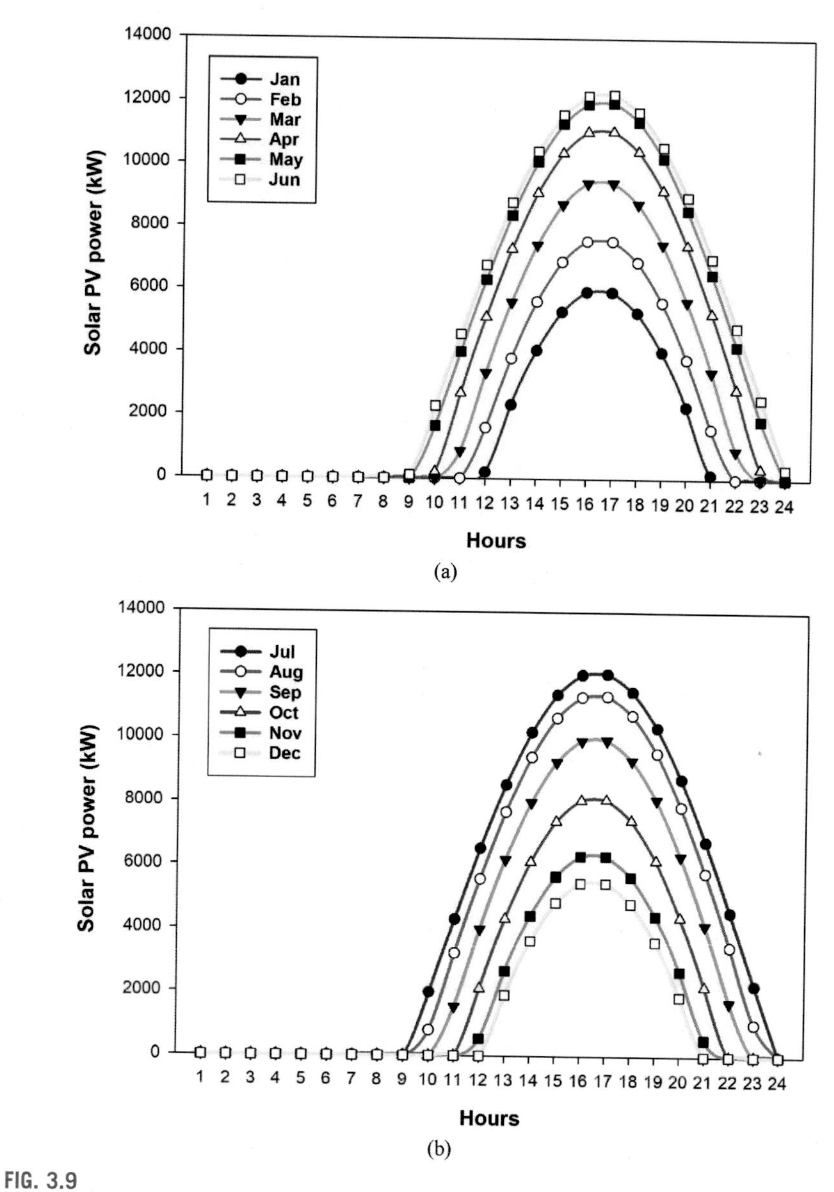

FIG. 3.9

Solar PV power on average days of each month: (a) Jan–Jun; (b) Jul–Dec.

modified from a study presented in the literature by Siddiqui et al.[80] For a single cell, the maximum power point was found as 2.37 W that exists at 8.8 V voltage 0.27 A current. Table 3.4 exhibits the characteristic operating conditions of the experimental solar PV-based water electrolysis system.

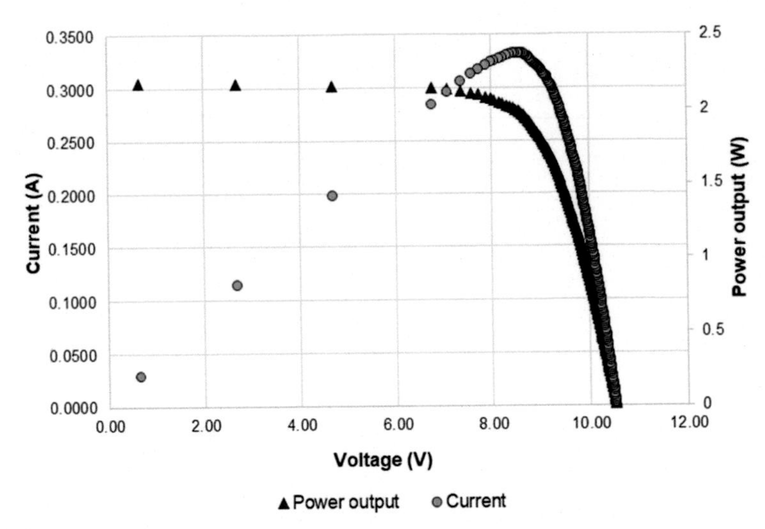

FIG. 3.10

Effect of single PV cell voltage on the current and power output.

Modified from [80].

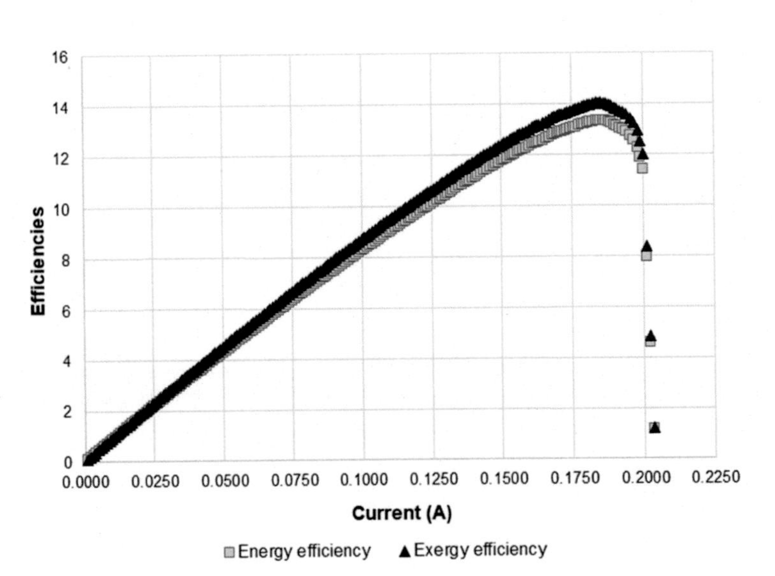

FIG. 3.11

Effect of varying single PV cell currents on the energetic and exergetic efficiencies.

Modified from [80].

Moreover, the PV cell short-circuit current was found to be approximately 0.31 A, and open-circuit voltage was found to be 10.5 V. It is significant for a solar PV cell to be operated at maximum power point that achieves maximum useful output along with maximum efficiencies. The number of PV cells that are connected in series was determined based on the current and voltage requirement to PEM electrolyzer, and PV cells were operated at the maximum power-point. The PV cells connection configuration can be improved dependent upon current and voltage requirements of PEM electrolyzer. Additionally, a power electronic system is recommended to be installed to control the output current from PV cells and voltage. Such an integrated solar PV-based water electrolysis system can be established where PV cells output voltage and current at different operational conditions are suitable for electrolyzer input requirements.

Fig. 3.11 exhibits the effect of varying single PV cell currents on the energetic and exergetic efficiencies. The maximum energy efficiency of 13.3% and maximum exergy efficiency of 14.0% were found at the maximum power-point. The efficiencies of PV cells can be enhanced by new semiconductor materials development that will offer comparatively higher electrical outputs. Furthermore, the solar radiation intensities also affect the peak power, and it is significant to explore the system performance at different solar intensities that are investigated in case study 1.

The energetic and exergetic efficiency of each subsystem and the overall system are assessed to determine system performance (Fig. 3.12). The correlations for energetic and exergetic efficiency of solar PV cells can be written as follows:

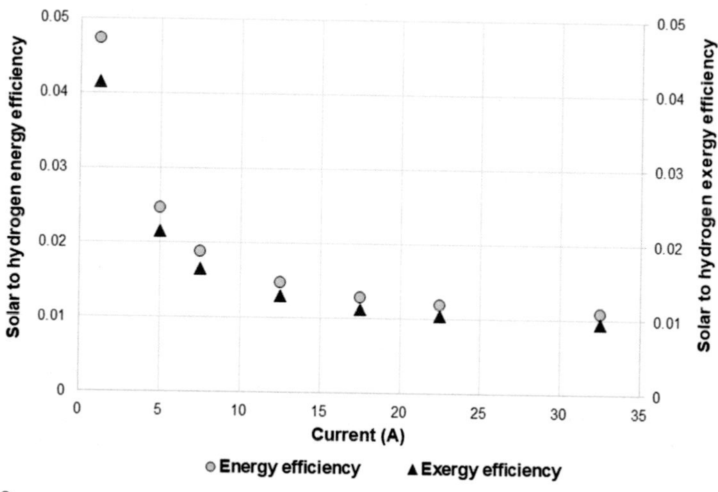

FIG. 3.12

Energetic and exergetic efficiency results of the solar to hydrogen process at varying current inputs.

Modified from [80].

$$\eta_{PV} = \frac{\dot{P}_{PV}}{\dot{q}_{in,sol} A_{PV}} \tag{3.13}$$

$$\eta_{ex_{PV}} = \frac{\dot{P}_{PV}}{\dot{q}_{in,sol}\left(1 - \frac{T_0}{T_S}\right) A_{PV}} \tag{3.14}$$

where $\dot{q}_{in,sol}$ represent solar radiation intensity, A_{PV} denotes area, T_0 signifies ambient temperature, T_S indicates sun temperature, and $\dot{P}_{PV}$ represents PV power output that is determined using current (I_{PV}) and voltage (V_{PV}) of corresponding PV cells:

$$\dot{P}_{PV} = V_{PV} I_{PV} \tag{3.15}$$

The correlations for energetic and exergetic efficiency of proton exchange membrane electrolyzer can be expressed as follows:

$$\eta_{PEM} = \frac{\dot{N}_{H_2} \overline{LHV}_{H_2}}{\dot{P}_{in}} \tag{3.16}$$

$$\eta_{ex_{PEM}} = \frac{\dot{N}_{H_2} \overline{ex}_{H_2}}{\dot{P}_{in}} \tag{3.17}$$

where $\dot{N}_{H_2}$ signifies molar flow rate of hydrogen, $\overline{LHV}_{H_2}$ denotes molar lower heating value (LHV), $\dot{P}_{in}$ signifies the input power, and $\overline{ex}_{H_2}$ symbolizes specific molar exergy. The energetic and exergetic efficiencies of solar to hydrogen production can be expressed as follows:

$$\eta_{SH} = \frac{\dot{N}_{H_2} \overline{LHV}_{H_2}}{\dot{q}_{in,sol} A_{PV}} \tag{3.18}$$

$$\eta_{ex_{SH}} = \frac{\dot{N}_{H_2} \overline{ex}_{H_2}}{\dot{q}_{in,sol}\left(1 - \frac{T_0}{T_S}\right) A_{PV}} \tag{3.19}$$

3.4 Solar Thermal Energy

Solar thermal power systems employ concentrated solar energy. In such systems, a working heat-transfer fluid is employed to utilize the heat that is used to produce steam. The produced steam is fed to the turbine that converts it into mechanical work and mechanical energy is converted to the electrical energy using a generator. The thermal energy extracted from the solar source can be employed in numerous applications such as air conditioning, space heating, drying, hot water, industrial process heat, electrical power, and hydrogen.

3.5 Solar Thermal Collector

The solar thermal energy can be employed for hydrogen production using different techniques. The most significant methods of using solar thermal energy for hydrogen production extract the solar thermal energy using solar thermal collector and solar heliostat field. The thermal energy extracted using these sources can be employed to the steam power cycles for power production, and it can also be employed to the thermochemical Cu–Cl cycle for hydrogen production. This section includes one example from each type of producing power using solar thermal collectors and producing hydrogen using the thermochemical Cu–Cl cycle. Fig. 3.13 exhibits the system schematic of a solar concentrating collector-based hydrogen production system. The significant benefits of solar heating are as follows:

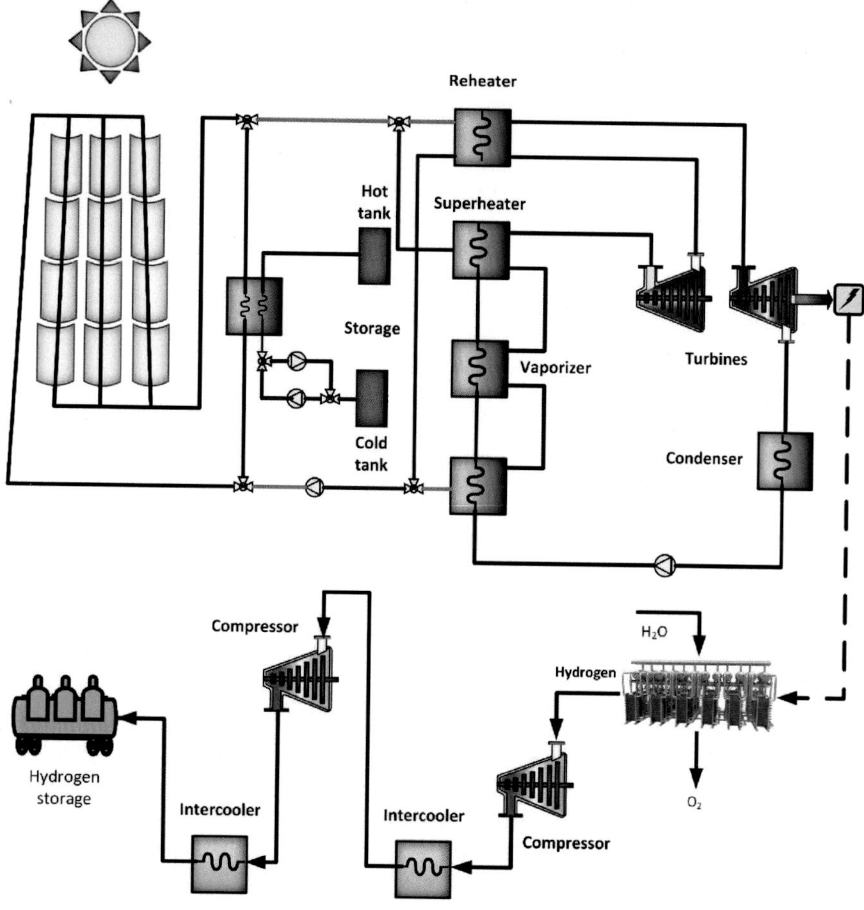

FIG. 3.13

System schematic of a solar concentrating collector-based hydrogen production system.

- Endless free energy
- Solar energy-based systems can be easily integrated into currently existing systems
- Cost savings
- Zero operational CO_2 emissions
- Reduced fossil fuels consumption

The main characteristic of the flat-plate collector is that the black surface absorbs the solar light. The coating of the absorber surface is designed in a way that it can absorb maximum radiation and reflects a very low energy amount. The energy that is absorbed using the solar collectors is transferred to the working fluid that works as a heat carrier and circulates in tubes under the absorber surface. The significant advantages of solar collectors are as follows:

- Low purchase price
- Low repair and maintenance cost
- Needs less roof area
- High energy efficiency
- Can be integrated with high-temperature heating systems

The two significant types of solar collectors are as follows:

- Flat-plate collectors
- Evacuated tube collectors

The solar light that is around 70–100 times concentrated offers the operating temperatures ranged from 350 to 550°C. This technique can be employed to run the power plants with a capacity of 200 MW or more. Under normal operating conditions, the design of the solar collector field can provide additional energy than the turbine limit. This increases the solar operating hours beyond daylight times and stored excess energy can be employed to feed the turbine with required energy during insufficient solar radiation periods.

Heat storage technique comprises two large tanks (cold tank and hot tank), each individual containing molten salt that is employed as a storage medium with required heat capacity. A heat exchanger is used to transfer the heat to/from the collector fluid. The pump is used to flow molten salt from cold tank to hot tank through heat exchanger during discharging and charging periods. Fig. 3.13 exhibits the system schematic of a solar concentrating collector-based hydrogen production system. The heat exchanger transfers the heat to the water and converts it into steam that is employed to the turbine to generate power. The electrical power generated by the designed system is employed to the PEM electrolyzer for hydrogen production. The produced hydrogen passes through a multistage compression and intercoolers that help in storing hydrogen in the storage unit at high pressure.

3.6 Photocatalysis

Photocatalytic splitting of water is an artificial process of photosynthesis that undergoes photocatalysis reaction in the photoelectrochemical cell and dissociates H_2O into its constituents of H_2 and O_2 employing solar light. Theoretically, the photocatalytic splitting of water requires catalyst, water, and solar energy (photons). Hydrogen has gained a global research focus as a fuel due to the rapidly growing global warming issues. Photocatalytic water-splitting method is being explored and investigated to produce clean-burning hydrogen fuel. Photocatalytic water-splitting technique grasps particular potential and scientific attraction as it employs a naturally available renewable source and water. As denoted by the name, photocatalytic splitting of water employs sunlight and catalyst to produce hydrogen using water.

Fig. 3.14 displays the schematic of the photocatalytic water-splitting process generating hydrogen. The photocatalysis process includes the semiconductor photocatalyst light absorption, generation, recombination and separation of excited charge carrier (holes and electrons), migration of holes and electrons, charge carriers

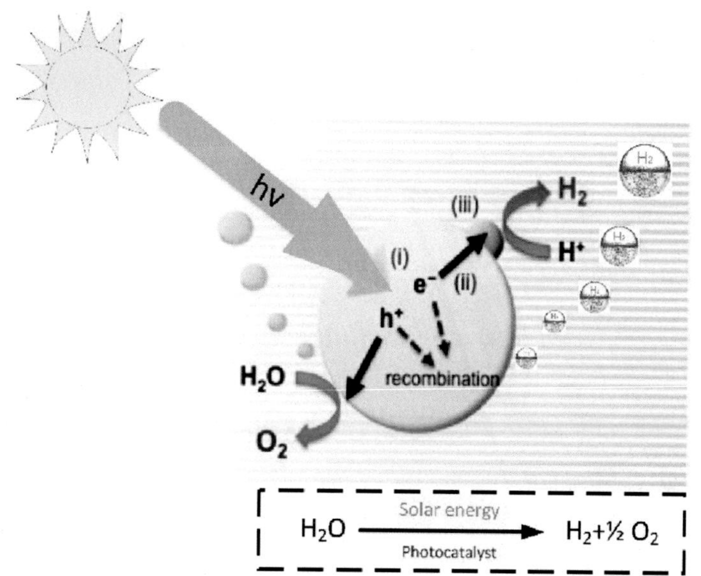

FIG. 3.14

Schematic design of photocatalytic water-splitting process.

Modified from [82].

transfer to water, and charge carrier trapping. All of the mentioned processes affect the hydrogen generation through semiconductor photocatalyst system. The amount of the excited electrons at the photo/water–catalyst interface helps in determining the total hydrogen generation amount. Once the hole/electron pairs are formed, charge separation and recombination processes are two imperative processes that mainly affect the photocatalytic water-splitting process efficiency.

3.7 Thermolysis

The conventional water thermolysis is a single-step process that employs solar energy-driven heat source of more than 2227°C to accomplish a reasonable water dissociation degree. The reasonable degree of water dissociation at 1 bar is usually 9% and around 25% at 0.05 bar. Furthermore, the high temperature also prevents the explosions by keeping the generated oxygen and hydrogen separate from each other. A schematic of the solar thermal energy-driven conventional thermolysis process for hydrogen production is shown in Fig. 3.15.

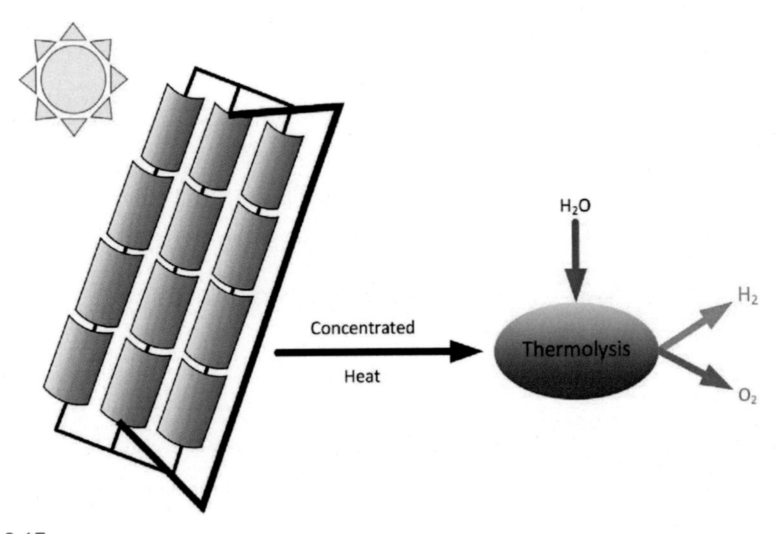

FIG. 3.15

Schematic of the solar thermal energy-driven conventional thermolysis process for hydrogen production.

3.8 Solar Heliostat

Solar thermal energy can also be extracted using a solar heliostat field. In such systems, a working heat-transfer fluid is employed to utilize the heat that is used to produce steam. Heat storage technique comprises two large tanks (cold tank and hot tank), each individual containing molten salt that is employed as a storage medium with required heat capacity. A heat exchanger is used to transfer the heat to/from the collector fluid. The pump is used to flow molten salt from cold tank to hot tank through heat exchanger during discharging and charging periods. The following equation is used to calculate the solar heat input depending upon the direct normal irradiance on average days of each month as shown in Fig. 3.3a and b.

$$\dot{Q}_{\text{solar}} = \eta_{\text{he}} \dot{I}_{\text{b}} A_{\text{he}} N_{\text{he}} \tag{3.20}$$

The heat losses caused by the emissivity, conduction, reflection, and convection can be expressed by the following equations:

$$\dot{Q}_{\text{em}} = \frac{\varepsilon_{\text{average}} \sigma \left(T_{\text{CR,Sur}}^4 - T_{\text{o}}^4 \right) A_{\text{SHF}}}{C} \tag{3.21}$$

$$\dot{Q}_{\text{conv}} = \frac{((h_{\text{FCI}}(T_{\text{sur}} - T_{\text{o}})) + (h_{\text{NCI}}(T_{\text{sur}} - T_{\text{o}})))A_{\text{SHF}}}{C \times F_{\text{r}}} \tag{3.22}$$

$$\dot{Q}_{\text{reflec}} = \frac{\dot{Q}_{\text{CR}} F_{\text{r}} \rho}{A_{\text{SHF}}} \tag{3.23}$$

$$\dot{Q}_{\text{cond}} = \frac{(T_{\text{sur}} - T_{\text{o}})A_{\text{SHF}}}{\left(\frac{\vartheta_{\text{I}}}{\lambda_{\text{I}}} + \frac{1}{h_{\text{air}}} \right) C \times F_{\text{r}}} \tag{3.24}$$

where $\dot{Q}_{\text{solar}}$ signifies solar heat input, h represents heat transfer coefficient, $\varepsilon_{\text{average}}$ denotes average emissivity, A_{he} implies heliostat area, FCI and NCI represents forced and natural convection insulation, λ shows thermal conductivity, N_{he} exhibits heliostat number, η_{he} signifies heliostat efficiency, constant σ represents 5.67×10^{-8} W/m^2·K^4 value, $\dot{I}_{\text{b}}$ indicates irradiance, and F_{r} denotes view factor. Fig. 3.16 exhibits the schematic of solar heliostat-based hydrogen production system using the thermochemical Cu–Cl cycle.

3.8.1 Case Study 3

A case study is presented to explore the integration of the solar heliostat field with the thermochemical Cu—Cl cycle for clean hydrogen production as shown in Fig. 3.16. The heat exchanger (B1) transfers the heat from the molten salt to the water and converts into steam. The steam extracted from the solar thermal energy is employed to the thermochemical Cu—Cl cycle for clean hydrogen production. The designed case study is simulated using Aspen Plus V11, and Fig. 3.17 exhibits the Aspen Plus simulations of solar heliostat-based hydrogen production system using the thermochemical Cu—Cl cycle.

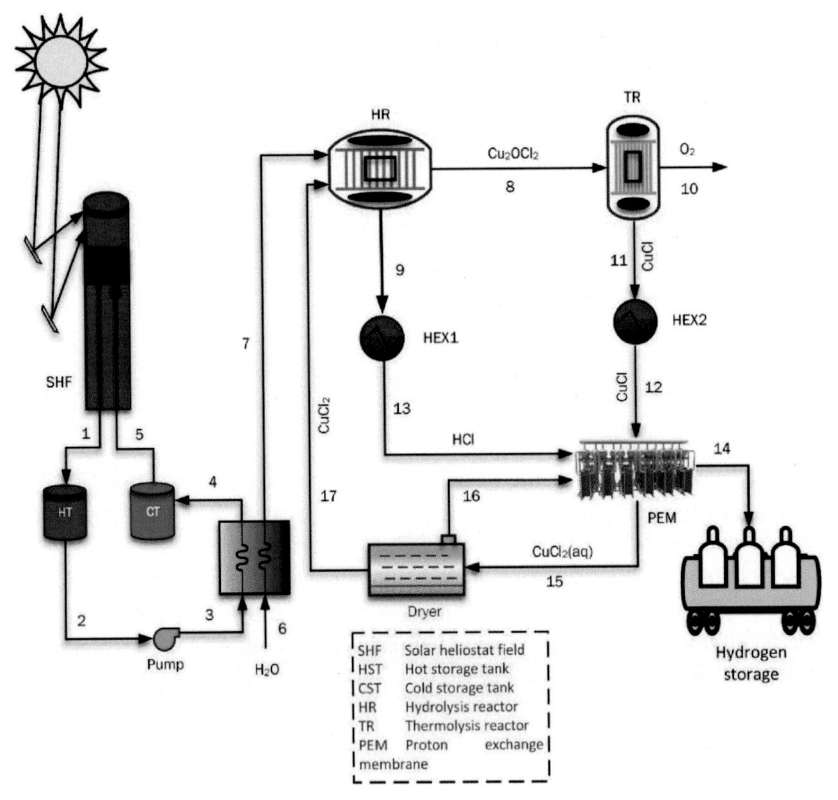

FIG. 3.16

Solar heliostat based hydrogen production system using thermochemical Cu—Cl cycle.

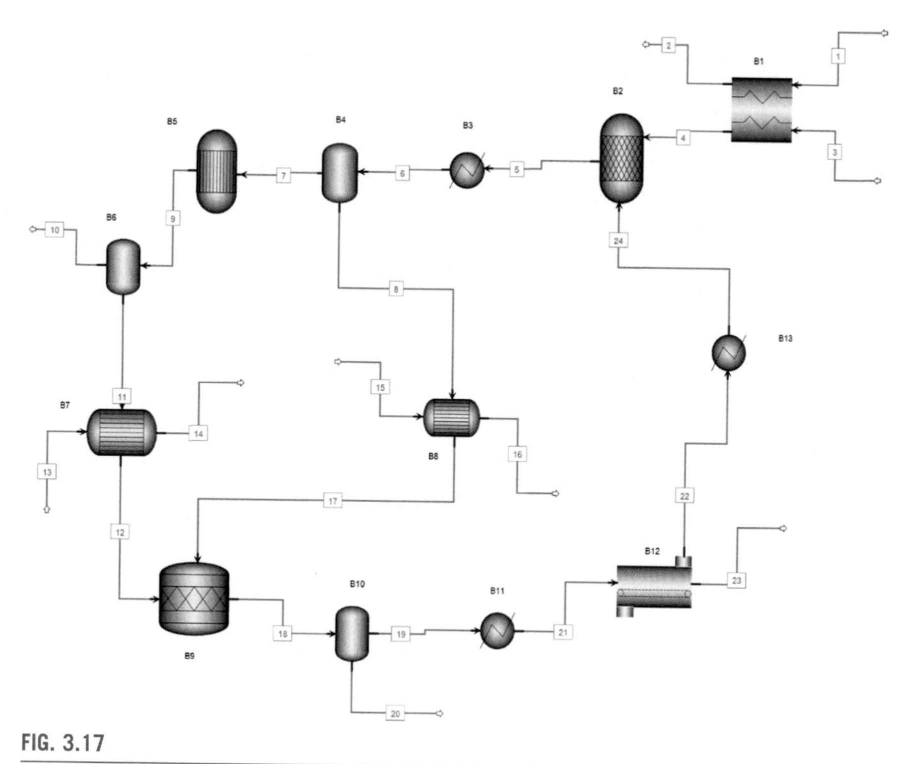

FIG. 3.17

Aspen Plus simulations of solar heliostat based hydrogen production system using thermochemical Cu–Cl cycle.

Solar heliostat field

The solar heliostat field is incorporated with the designed system as shown in Fig. 3.16. The molten salt is employed as a working fluid to absorb the heat from the solar heliostat field, and the thermal management unit is used to facilitate the supply of this heat to the steam using a heat exchanger. The water is fed to the heat exchanger that is converted to the steam using the solar thermal energy, and steam is employed to the thermochemical Cu–Cl cycle for thermochemical hydrogen production. The thermochemical cycle only employs water to split into its components, while all other constituents are recycled throughout the cycle. The significant operating parameters are displayed in Table 3.5.

The incoming beam and direct normal irradiance are calculated using the dynamic analysis, which is shown in Fig. 3.3. These incoming and direct normal radiation intensities are employed to determine the solar input heat for average days of each month during every single hour. The solar heliostats employed to generate the solar thermal energy comprises an area of 100 m^2 and the number of mirrors is 10. The efficiency of solar heliostat is assumed to be 90%. Once the beam radiation is determined, the incidence of solar radiation on solar heliostat field can be

Table 3.5 Operating Parameters of the Designed Case Study.

Design Parameters	Value
Solar heliostat	
Working fluid	Molten salt
Solar radiation intensity $\dot{I}_b$	800 W/m^2
Number of heliostats N_{he}	10
Dimensions of heliostat mirror area	100 m^2
Heliostat efficiency	90%

determined using incoming beam radiation that is represented as $\dot{I}_{beam}$ and area of the solar heliostat field expressed as A_{SHF} using the following expression:

$$\dot{Q}_{PV_{SI}} = A_{SHF}\dot{I}_{beam} \quad (3.25)$$

The correlation used to calculate the solar heat that is absorbed by the solar heliostat field can be described as follows:

$$\dot{Q}_{solar} = \eta_{he}\dot{I}_b A_{he} \quad (3.26)$$

Here, $\dot{Q}_{solar}$ is the solar heat, A_{he} denotes heliostat area, η_{he} is heliostat efficiency, N_{he} is the number of heliostats, and $\dot{I}_b$ is irradiance. Fig. 3.18a displays the solar heat input from January to June, and Fig. 3.18b shows the solar heat input from July to December.

The thermochemical cycle consists of four significant steps of electrolysis, drying, hydrolysis, and thermolysis. In hydrolysis reaction (B2), cupric chloride ($CuCl_2$) is hydrolyzed to form HCl gas and copper-oxychloride (Cu_2OCl_2), which is decomposed into oxygen and cuprous chloride (CuCl) using thermolysis reactor (B5). Oxygen is separated and CuCl reacts with HCl gas is electrolysis step (B9) and forms hydrogen gas with aqueous $CuCl_2$. Hydrogen is separated and water is removed from the aqueous $CuCl_2$ is drying step (B12) and circulated to the hydrolysis reactor. Following are the chemical equations representing each significant step:

$$2CuCl(aq) + 2HCl(aq) \xrightarrow{25^\circ C} H_2(g) + 2CuCl_2(aq) \quad (3.27)$$

$$CuCl_2(aq) \xrightarrow{80^\circ C} CuCl_2(s) \quad (3.28)$$

$$2CuCl_2(s) + H_2O(g) \xrightarrow{400^\circ C} Cu_2OCl_2(s) + 2HCl(g) \quad (3.29)$$

$$Cu_2OCl_2(s) \xrightarrow{500^\circ C} 0.5\ O_2(g) + 2CuCl(l) \quad (3.30)$$

Several parametric studies are conducted to analyze the performance of the solar heliostat field for thermochemical hydrogen production using the Cu–Cl cycle.

Molten salt is employed in the storage tanks for thermal energy storage. The heat is transferred from the molten salt to the water and converts into steam through heat

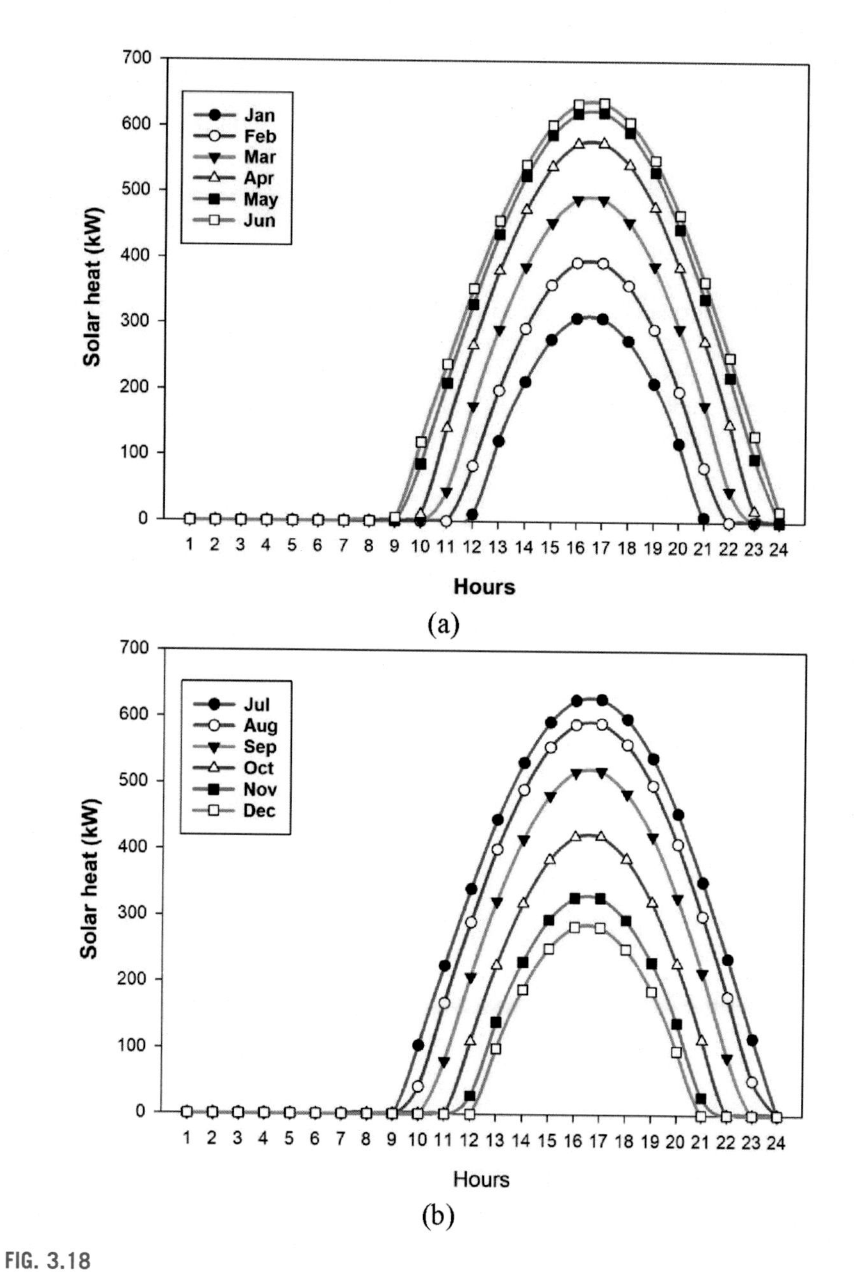

FIG. 3.18

Solar heat input using incoming beam radiation: (a) Jan–Jun; (b) Jul–Dec.

exchanger B1. It is significant to explore the performance of the solar energy-based thermochemical cycle using system significant input parameters such as molten salt input and output temperature and steam and cupric chloride input flow rates. Fig. 3.19 displays that the flow rate varies with the molten salt input and output temperature. It can be depicted that mass flow rate rises with the upsurge in molten salt output temperature and mass flow rate drop-down with the upsurge in molten salt input temperature. It is substantial that the input and output temperatures of the molten salt affect the mass flow rate significantly.

The cupric chloride is hydrolyzed in the hydrolysis reaction and water is the only constituent that is consumed during this thermochemical cycle while all other constituents are recycled throughout the cycle. Thus, it is significant to explore the variation in the hydrogen production rate with input parameters. Fig. 3.20 exhibits the effect of input steam and cupric chloride flow rates on thermochemical hydrogen production. It can be depicted that the hydrogen production flow rate increases with the rise in input cupric chloride and steam flow rates. The input steam flow rate is ranged from 0.1 to 14 mol/s while the input cupric chloride flow rate ranged from 0.2 to 28 mol/s and the hydrogen production flow rate increases from 0 to 4.2 mol/s.

The hydrogen chloride gas leaves the hydrolysis reactor at a high temperature of 400°C and cuprous chloride exits the thermolysis reactor at a high temperature of 500°C, and both of these constituents meet at the electrolysis unit that operates at ambient temperature. Thus, plenty of room is available to recover the heat from these two streams. Fig. 3.21 displays the effect of the steam input flow rate on

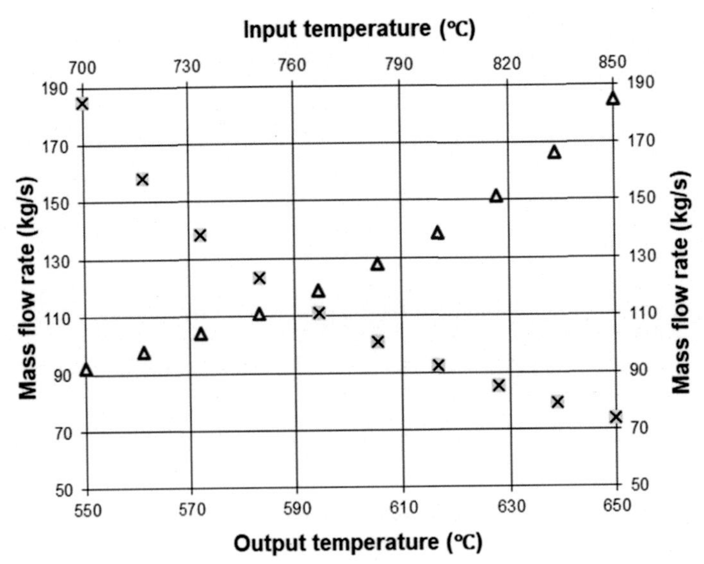

FIG. 3.19

Flowrate variation with molten salt input and output temperature.

Modified from [83].

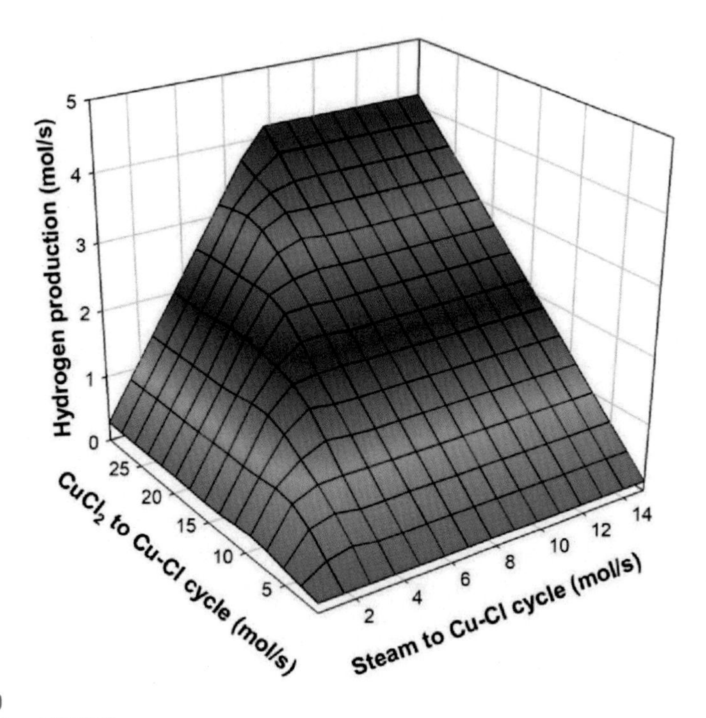

FIG. 3.20

Effect of input H_2O and $CuCl_2$ flowrates on thermochemical hydrogen production.

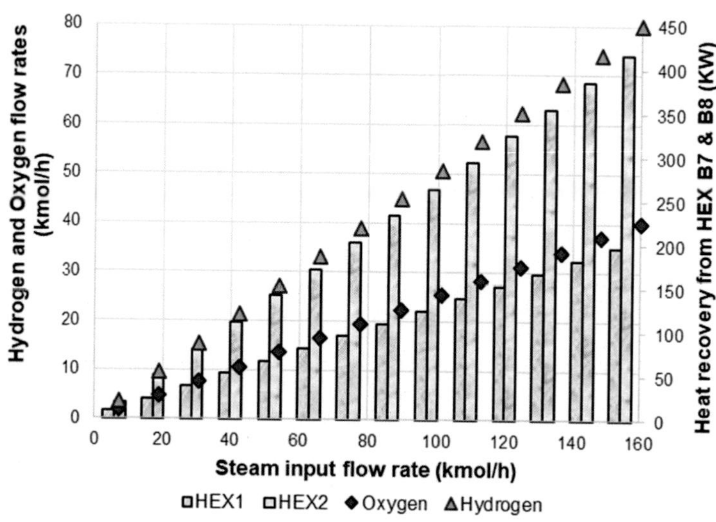

FIG. 3.21

Effect of steam input flow rate on hydrogen and oxygen flow rates and heat recovery through heat exchangers HEX1 and HEX2 represented as B7 and B8.

hydrogen and oxygen flow rates and heat recovery through heat exchangers HEX1 and HEX2 represented as B7 and B8. The input steam flow rate ranged from 0 to 160 kmol/h. It can be depicted that both hydrogen and oxygen flow rates and heat recovery through heat exchangers B7 and B8 increase with a rise in the input steam flow rate.

3.9 Closing Remarks

Solar energy is an enormous, infinite, and clean renewable energy resource. This primary energy source can be employed directly for heating, electricity generation, and hydrogen production. Natural gas reforming is the conventional method that is used for hydrogen production although this process costs harmful emissions. This process does not only produce greenhouse gas emissions but also costs the depletion of fossil fuels and carbon emission taxes based on carbon emission footprints. The solar resource is massive, and the sunlight amount that strikes the surface of the earth in 1.5 h is sufficient to knob the global energy consumption for 1 year. Due to the intermittent nature of the solar energy source, it is mandatory to integrate the storage system to meet the demand during night times.

Solar energy can be employed using three different routes for hydrogen production, namely solar energy-based water electrolysis, solar thermal energy-based thermochemical hydrogen production, and direct solar splitting of water. All of these methods undergo clean and sustainable hydrogen production. There are numerous benefits of employed solar energy sources such as it is a renewable energy source that provides endless free energy, solar energy-based systems can be easily integrated into currently existing systems, reduces fossil fuel consumption, needs low maintenance costs, undergoes zero CO_2 emissions, and offers diverse applications.

The solar energy resource is categorized among four significant routes that can be employed for hydrogen production. The photoelectrochemical energy and photonic energy can directly be employed for hydrogen production. The electricity generated using solar photovoltaic energy is employed to the electrolyzer for water splitting. Solar thermal energy can be extracted using solar thermal collector and solar heliostat field and can be employed to the power cycle to produce electricity, and power can be employed for hydrogen production or can be used to produce hydrogen directly using the thermochemical cycle. Solar energy-based hydrogen can be employed to the hydrogen fuel cells that undergo zero carbon emissions. The electricity generation using solar panels, solar thermal collectors, or solar heliostat field that is employed to produce hydrogen undergoes zero greenhouse gas emissions.

CHAPTER

Wind Energy-Based Hydrogen Production

4

The wind power or wind energy terminologies both designate the process of employing wind for generating mechanical work and hence into electrical energy which is known as power. The generated power can be employed for a large variety of explicit tasks, such as water pumping, or a generator can be integrated to convert mechanical into electrical work. Wind energy-based electricity generation appears to be among the fastest-growing global electrical generation methods. The kinetic energy associated with air is converted to mechanical energy using the mounted wind turbine, and the generator converts mechanical power into electrical power. Wind energy is among the promising global energy resources to produce clean and environmentally benign hydrogen. The electrical power extracted using wind energy is employed in the electrolyzer that breaks water into its constituents of hydrogen and oxygen.

Wind energy-driven hydrogen production in Western Canada has recently been investigated in a recent study.[84] Conventionally, hydrogen production is carried out using natural gas reforming (NGR), which generates significant greenhouse gas (GHG) emissions. Wind energy-driven hydrogen production systems can eliminate the GHG footprint and also assist in upgrading the bitumen industry. They aimed to develop a comprehensive technoeconomic model to analyze and assess the wind energy-driven hydrogen production through water electrolysis. They extended an already existing 1.8 MW wind farm with the options of electricity and hydrogen production. The electrolyzer with capacities of 240 and 360 kW were found to offer optimal results for variable and constant electrolyzer flow rates. The mentioned electrolyzer sizes offered a minimum price of hydrogen production as \$10.15 and \$7.55/kg H_2. The price of hydrogen production became equal to NGR with the addition of \$0.13/kWh Feed-in-Tariff with a 24% rate of return.

A recent study[85] estimated worldwide hydrogen production using wind energy. It is prospective that renewable energy sources like solar and wind offer great opportunities for large-scale production of hydrogen in the future. Wind energy is projected by employing a 2 MW turbine over the earth's surface globally. Wind power production is harvested using the mean of monthly wind speeds. The low-pressure electrolysis process was carried out for hydrogen production while transmission through high-pressure gas pipelines. Wind energy-driven hydrogen production system was considered offering not only hydrogen but energy as well for the consumption of electricity at the local production site. The practical hydrogen production potential was projected to be 116 EJ.

Renewable Hydrogen Production. https://doi.org/10.1016/B978-0-323-85176-3.00015-9

Renewable energy for sustainable urban mobility and hydrogen production has been investigated in a recent study.[86] Recently, renewable energy-driven power plants have been growing continuously around the globe, and hydrogen has turned out to be an appropriate and right fit for a medium-long-term storage solution as an energy carrier and a fuel for fuel-cell and hybrid electric vehicles as it offers carbon-free solutions. They aimed at carrying out both economic and environmental analyses for a renewable energy-based start-up plant employing the simulation results they offered in a previous study. The location of the South of Italy was selected for renewable energy-driven hydrogen production plants. Particularly, an 850 kW-capacity wind turbine was employed in the hydrogen production plant.

Another recent study[87] has offered a stand-alone renewable energy-driven hydrogen production system. A solar and wind hybrid energy system integrated with a suitable hydrogen storage system is an auspicious solution to meet the intermittency limitations. It also included short and long-term devices to store energy along with fuel cells. The objective of their study was to provide the demonstration of an autonomous renewable energy-driven system with long-term hydrogen storage.

A model of wind energy-driven hydrogen storage system for immense curtailment of wind energy was proposed in a recent study.[46] The rise in wind energy penetration causes the curtailment of wind energy severe in some wind turbines. Thus, a huge energy storage technique is mandatory to overcome these challenges. In their study, a new integrated model of curtailed wind energy-based hydrogen production and storage was published considering real-time data of 10 min average throughout the year. Their study considered that two scenarios of wind energy-driven hydrogen production using water electrolysis were modeled, and the impact on the payback period by water electrolysis and the price of hydrogen were established along with model validity.

This chapter primarily focuses on wind energy based hydrogen production methods and their applications. It further covers some fundamental aspects, working principles, types of turbines and operating conditions and details, etc. There are also illustrative examples and case studies presented to better illustrate the importance of the subject in hydrogen production.

4.1 Working Principle and Advantages of Wind Energy

The rotor sits at the tower top whose significant components are wind turbine blades. The wind turbine blades are connected with a horizontal shaft that is further attached to a generator positioned in the nacelle. The basic phenomenon that occurs in wind turbines is that they employ the kinetic energy of the air to generate electrical power. The kinetic energy of the air rotates the wind turbine in terms of turning energy, and the rotating wind turbine generates mechanical power. The generator is positioned in the wind turbine that converts mechanical power into electrical power. The electricity generated by the wind turbine is an alternating current (AC) that is converted to the direct current (DC) using AC/DC converter.

For the wind turbines, the Betz limit specifies maximum theoretical efficiency that was established by Albert Betz (a German physicist) in 1919. Betz revealed that 59.3% is the maximum efficiency that can be achieved ideally by a wind turbine to employ wind power for producing electricity.

The blowing wind rotates the blades, and this rotation produces electricity. As the wind turbine blades start rotating, the gearbox positioned in the nacelle is activated that operates the generator. The generated electricity is then provided to the consumers through the grid. The produced energy amount is dependent primarily on three significant factors as follows:

- Wind speed
- Area swept by the blades
- Air density

Wind turbines employ the kinetic energy of atmospheric air and convert it into a rotary motion that is used to generate mechanical power using wind turbine blades. The produced mechanical power is converted to electrical power employing a generator. The produced electric power can be connected with the distribution grid for distribution to the consumers.

Unlike fossil fuels including natural gas and coal-powered power plants that are commercially employed to generate power, wind energy does not contaminate or pollute the air. They do not undergo any process that can cause GHG emissions or acid rain. Wind energy offers numerous advantages as follows:

- Becomes cost-effective
- Helps industrial growth
- Wind sector creates jobs
- Enables competitiveness
- Can be installed on existing farms
- Serves as clean fuel source
- Provides a domestic source of energy
- Is sustainable

Small-sized wind turbines can be employed to offer a cost-effective route to produce renewable electricity to meet household requirements. Nevertheless, several residential properties may not be appropriate to install wind turbines due to limited reasons. For example, wind turbines need to be installed at a windy location to produce enough electricity for the household and to justify a worthwhile investment.

The wind turbines are frequently installed on residential, commercial, community, agricultural, and industrial sites and can be ranged in capacities from a 5-kW turbine that is usually located at home to the turbines undergoing multimegawatt that are usually installed at manufacturing facility. Wind turbines that are installed nearby houses generally undergo 1–10 kW range and can also go beyond this range. During the wind intervals, the installed wind turbine generates more electrical power that can be metered and accredited to the customer as it is supplied back to the distribution grid. Following are some facts about wind energy:

- Wind power includes turning energy using the kinetic energy of air that is employed to rotate the turbine and converted from wind to other useful energy forms.
- Wind energy can be extracted employing different ways.
- Windmills have been used since 200 B.C. when it was originally in Persia.
- Wind power is renewable and clean.

Wind power offers numerous advantages that justify why wind energy is among the fastest-growing global energy sources. Research works are being directed to address the challenges associated with wind energy. Even though wind power plants undergo a moderately diminutive environmental impact as compared with conventional power plants, some challenges are associated with the turbine blades' noise produced as well as landscape visual impacts. A large number of significant advantages associated with wind energy can be described as follows:

- Wind power does not cause environmental pollution as it is a green energy source that employs a turbine for power production.
- The wind power potential is massive and predicted to be around 20 times the whole population requires.
- Wind energy is classified as renewable, and it is not possible to run out of wind power as it is originated from the sun.
- Wind turbines are space-efficient as well. The wind turbine with maximum capacity can supply power to 600 US houses.
- Horizontal-axis turbines offer high efficiencies, and they convert 40%–50% of wind energy into electricity.
- Wind energy only holds around 2.5% of global electricity generation although rising continuously at a promising 25% per annum rate.
- The prices of wind energy have been decreased by nearly 80% since 1980 that are predictable to drop further down.
- The operational costs of wind power source are low.
- The domestic potential of wind energy is upright, as wind turbines installed at residential locations neglects power outages and offer energy savings.

The consumption of wind energy is increasing significantly around the globe. Fig. 4.1 exhibits the global wind energy consumption in a different region of the world (MW). A continuous and significant growth of wind energy resources can be depicted in all regions of the world including North America, South and Central America, Europe, Commonwealth Independent States, Middle East, Africa, and the Asia Pacific from 2008 to 2018. Wind energy consumption in the North America increased from 27,088 to 111,987 MW in the years 2008–18, increased from 613 to 20,388 MW in South and Central America, increased from 64,227 to 190,118 MW in Europe, increased from 13 to 196 MW in the Commonwealth Independent States, increased from 72 to 612 MW in the Middle East, increased from 552 to 5464 MW in Africa and increased from 22,798 to 235,584 MW in the Asia Pacific.

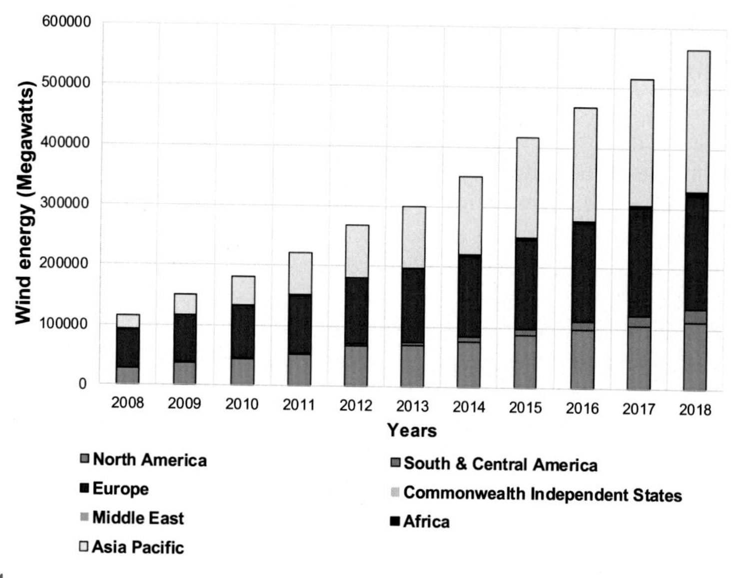

FIG. 4.1

Global wind energy consumption in a different region of the world (in Mega Watts (MW)).

Data from [48].

4.2 Types of Wind Turbines

Categorically, wind turbines can mainly be categorized into two of the following classes based on the installation and geographical location: onshore wind turbines and offshore wind turbines. In terms of design, the wind turbines can be categorized into two following significant types that are employed generally for power production:

- Horizontal-axis wind turbines
- Vertical-axis wind turbines

Among the following two wind turbine types, namely horizontal- and vertical-axis turbines, the horizontal-axis wind turbine characteristically consists of a vertical propeller having three blades that face the wind while the vertical-axis wind turbine includes a blade set that spins through the vertical axis.

4.2.1 Horizontal-Axis Wind Turbines

Horizontal-axis turbines comprise a key rotor shaft as well as an electrical generator at the tower top that should be directed toward the wind. Small-sized turbines employ wind vanes for pointing while large-sized turbines usually employ wind sensors.

Horizontal-axis turbines consist of a set of blades similar to airplane propellers, and the set of blades usually includes three blades. The biggest horizontal-axis wind turbines can be tall as equivalent to the 20-story building, and blades can be 100 ft (30 m) long or more. The higher the turbine with large blades, the more will be the electricity. Approximately all of the wind turbines presently are horizontal-axis wind turbines. Horizontal-axis turbines are the most common turbines that offer numerous characteristics as follows:

- Variable blade pitch that offers the blades of the turbine with an optimized attacking angle. Permitting the attacking angle to be accessible and adjusted remotely offers superior control. Thus, the turbine harvests the maximum wind energy quantity for all times of the day and season.
- Taller the tower base, the more will be the exposure toward stronger wind in wind shear sites. In sites with wind shear, the rise in 10-m height can raise the wind speed by 20% and output power by 34%.
- Horizontal-axis wind turbines offer high efficiency, and the reason is that blades continuously rotate perpendicular to the wind that enables that power receiving through the entire rotation. In comparison, vertical-axis wind turbines undergo numerous reciprocating actions, needing surfaces of an airfoil to backtrack against the wind that leads to integrally lower efficiency.
- Taller blades and poles are further problematic for transportation as well as installation. The approximate cost to transport and install costs of around 20% of total equipment costs. Strong and strengthened construction of the tower is obligatory to back heavy blades, generators, and gearboxes.

The horizontal-axis turbines offer four following significant advantages:

- High output power
  - Horizontal-axis turbines are normally constructed to offer high capacity ranging from 2 to 8 MW dependent upon usage. The output wind turbine power is subject to the size of turbine power, blades, and wind speed. A regular onshore wind turbine offering a capacity ranged from 2.5 to 3 MW can generate the power around 6 million kWh per annum that can meet the electricity requirement of 1500 households.
- Highly efficient
  - The energy loss is mandatory to occur in any energy conversion form. The improvement in energetic conversion efficiency is considered among the biggest challenges and emphases of wind turbine advancement in the wind power industry. At present, horizontal-axis turbines offer high efficiencies, and they convert 40%–50% of wind energy into electricity.
- High reliability
  - The research and advancements in horizontal-axis turbines are mature enough and dominant models of wind turbines for decades. The reliability of existing market products is accompanied by the usage and applications of horizontal-axis turbines that are thoroughly investigated.
- High operating wind speed
  - Because of rotor height, horizontal-axis turbines become able to harvest electricity using greater wind speed that shows that horizontal-axis turbines probably run at high speed of the wind that assists to achieve optimum performance. Subsequently, such heights offer comparatively stable air flow that supports these turbines to offer high consistency in output power.

4.2.2 Vertical-Axis Wind Turbines

Vertical-axis wind turbines consume blades that are connected to the bottom and top of the vertical rotor. The Darrieus wind turbines are the most common category among vertical-axis turbines that is termed after a French engineer named Georges Darrieus who published a patent back in 1931 that looked like a two-bladed giant egg blade. Some vertical-axis turbine types can be 50 ft in width and 100 ft tall. The insufficient number of vertical-axis turbines are employed nowadays due to the lower performance as compared with horizontal-axis turbines.

Even though vertical-axis wind turbines generate low energy as compared with horizontal-axis turbines, they can still generate power and might be a suitable possibility contingent on the application. Vertical-axis turbines are more appropriate with limited space and accompany fewer risks and challenges to maintain. The air that passes through the blades of a wind turbine is converted to the rotational momentum that operates the generator that results in a less-efficient vertical-axis turbine rotor as compared with the horizontal-axis turbine rotor. There are numerous advantages and disadvantages of vertical-axis wind turbines. A large number of advantages of vertical-axis wind turbines are as follows:

- Wind turbines employ fewer parts as compared with turbines that position the blades horizontally, which designates fewer break down of components. Similarly, the supportive strength of the tower is not required to be much as the generator and gearbox are close to the ground.
- The turbine is not either required to face the right direction of the wind. In the vertical-axis turbine system, the flow of air through any speed or direction can rotate blades. Consequently, the vertical-axis turbines are employed to produce power under gusty winds.
- Maintenance workers are not required to climb high to range the tower parts that justifies the safety of workers. Vertical-axis turbines involve some significant components closer to the ground. Maintenance of significant components, such as gearboxes, generators, and mostly electrical and mechanical parts of the structure are not required to be scaled as mentioned components are not mounted at the top. The climbing gear and lifting equipment are also not required.
- The design of vertical-axis turbines is possible to be scaled-down to even cover an urban rooftop. There might not be space for other renewable technologies in cities but vertical-axis turbines offer a feasible substitute to traditional hydrocarbon sources.

Some additional benefits of vertical-axis turbines include the following:

- Less expensive manufacturing costs as compared with horizontal-axis turbines
- Easy installation installed as compared with horizontal-axis turbines
- Can easily be transported to the desired sites undergoing less transportation cost
- Low-speed blades reduce the risk to birds
- Operates under extreme weather through flexible winds
- Suitable at the locations where higher constructions are prohibited
- Creates less noise as compared with horizontal-axis turbines

The vertical-axis turbines are more suitable to be mounted in denser arrays. The vertical-axis turbine model approximately 10 times shorter as compared with horizontal models can be grouped into arrays that create turbulence that supports increasing flow around blades. Consequently, the increased wind speed causes the generated power to rise. A low gravity center justifies the suitability of such models in offshore installations.

The vertical design permits the installation of wind turbines close to each other, and this is one of the significant benefits of vertical-axis turbines. The group of vertical-axis turbines is not obligatory to be set apart, so they do not require a large ground area. The immediacy of vertical-axis turbines to each other can generate turbulence and wind speed reductions, which also affect the power extracted from neighboring units.

All the blades of vertical-axis turbines do not generate torque at the same time that restricts the vertical-axis turbine system efficiency to produce energy and accompanied blades are pushed along. The blades also face more drag on rotating even though vertical-axis turbines can operate in gusty winds which is not the case at all times. Low dynamic stability issues and preliminary torque limit the turbine functionality under diverse conditions.

Subsequently, the vertical-axis wind turbines are installed lower to the ground. Thus, such turbines cannot extract the power from the high wind speeds frequently observed at high levels. The vertical-axis turbines are further hard to be installed if the designers favor erecting structure. Nevertheless, it is further common and practical to mount a vertical-axis turbine system on base level like the ground or building top.

The vibrations and increased noise due to the wind turbine installation can be a problem from time to time. The flow of air at the ground or lower levels can upsurge the turbulence as well as increased vibration that can probably exhaust bearings. Now and then, this issue can lead to the more required maintenance that upsurges the cost linked to it. Among prior model designs, blades were inclined to cracking and bending that can cause the turbine to fail.

Even though vertical-axis turbines generate less energy as compared with horizontal-axis turbines, they still generate power and can offer better possibilities contingent on the application. Vertical-axis turbines are suitable at places with limited space that can afford limited challenges and maintenance risks. This vertical-axis turbine design is popular in specific urban areas following small-scale installations. There is still potential and room for improvement for scientific innovations to enhance the efficiency of vertical-axis turbines to generate power and intensification of benefits they can propose in different applications. Table A-4 in the appendix displays the global wind energy consumptions and installed wind capacities from 1980 to 2018, and a gradual increase can be depicted using the data arranged in the table.

The vertical-axis wind turbines also undergo several disadvantages, and some of the significant drawbacks are as follows:

- Lower rotation efficiency
  - Vertical-axis turbines offer lower rotation efficiency. The rotor design does not support all blades on the vertical-axis rotor to collect input wind. The blades facing the wind only operate by the wind while following along. The vertical-axis rotors also face more aerodynamic resistance on blades or drag during rotation. This phenomenon is exclusively exposed by Savonius type wind turbines as they offer broader surfaces of blades.
- Less available wind speed
  - As vertical-axis turbines are characteristically installed on lower levels, they are not able to extract power using the high speeds of wind that is originated at high levels. Therefore, lower wind energy can be harvested using vertical-axis turbines. Such issues are generally resolved by installing the turbines on building rooftops. The rotor designs have been improved to be mounted on pole top to address this issue. The advanced rotor design installs the pole and rotor at the height of 10 m after the power electronics and generator to be positioned at the height of 4 m.
- Wearing down of component
  - As vertical-axis turbines are normally installed in populated environments on the ground level, they face higher vibrational and turbulence issues. During the operation, the design does not only need the blades to endure more force but also bearing between pole and rotor are expected to withstand high pressures. In prior design models, it was probable for blades to crack or bend, which can cause more maintenance followed by increased cost. To reinforce the vertical-axis turbines, wear-down calculations are considered for manufacture and design. The specifically designed LuvSide vertical-axis turbines are robust enough to withstand high wind speeds corresponding to a robust tropical storm.
- Lower efficiencies
  - Vertical-axis turbines are recognized to offer lower efficiencies in comparison with horizontal-axis turbines, which are followed by the design nature and operational characteristics of such turbines. Normally, the horizontal-axis wind turbines are 40%–50% efficient, which means the turbine converts 40%–50% of kinetic energy into actual power. On the contrary, a special type of Savonius vertical-axis turbine offers an efficiency ranging from 10% to 17% while Darrieus type vertical-axis turbines offer efficiencies ranging from 30% to 40%. Nonetheless, a Savonius-type vertical-axis wind turbine can generate sufficient power to meet the normal two-person household consumption if a suitable environment is provided.
- Mechanism of self-starting
  - As horizontal-axis and Savonius-type wind turbines offer self-starting features upon wind being received, Darrieus-type wind turbines frequently depend on starting mechanism, and this starting mechanism is required due to the wing design in Darrieus-type wind turbines that do not at all times lead the wind to generate sufficient torque to support rotation. Currently, this is the most significant challenge that is being faced to deal with these types of turbines.

4.3 Onshore and Offshore Wind Turbines

At the moment, wind energy is characterized into two different forms depending on the installation location as follows: (1) onshore and (2) offshore.

The onshore wind farms are largely installed on the land location including urban and rural areas while offshore wind farms are installed on water bodies locations, such as ocean, sea, rivers, lake, dam and waterfalls, etc. Onshore wind turbines employ wind energy to generate electrical power employing large blades that rotate and drive the generator positioned with the gearbox at the tower top behind blades. Onshore wind turbine-based power is one of the fastest rising renewable energy utility-scale technologies.

Offshore wind turbines offer a higher tendency of being more efficient as compared with onshore for the reason that direction and wind speed are more reliable and consistent. Credibly, a lesser number of turbines are required to deliver the same electricity amount as onshore turbines. Offshore wind is ample and plentiful even in the locations nearly the main coastal load centers. The blowing wind flows over the turbine blades that are usually designed in airfoil shape spins the turbine blades. The airfoil-shaped wind turbine blades are coupled with a drive shaft that operates the electric generator to generate electrical power.

Some significant features of onshore wind turbines are summarized as follows:

- Onshore wind turbine farms are naturally installed at locations with low habitat or conservation value.
- The required infrastructure for electricity transmission through onshore turbines stands substantially less as compared with the offshore turbine followed by the less drop in voltage between consumer and wind turbine.
- The installation of the onshore wind turbines is quicker as compared with the offshore turbine. The significant parameters of transportation and installation ease along with other additional factors affect the total capital cost. Thus, onshore wind turbines are cheaper to be installed than offshore wind turbines.
- Furthermore, the scientifically proven technology of onshore wind turbines faces less wear and tear due to moisture offered by the onshore installation area consumes very minor erosion. Thus, onshore technology offers lower maintenance costs in comparison with offshore wind farms.

Some minor drawbacks of onshore wind turbines are listed as follows:

- Public may criticize wind turbines to be an eyesore.
- Noise may become another minor issue associated with wind turbines.
- Onshore speeds of wind are not foreseeable as compared with offshore wind speeds. Likewise, the direction of onshore wind varies more frequently. Subsequently, turbines are optimally designed to operate at maximum power-generating wind speeds, this inconsistency in wind speeds can result in efficiency limitation.

Some significant features of offshore wind turbines are listed as follows:

- Offshore wind turbine farms are built and installed in water bodies that offer high wind speeds.
- The wind speeds followed by the offshore locations incline to be faster that offer the room to generate more energy.
- Offshore wind turbine farms have no physical limitations and restrictions like buildings or hills that could cause blockage in wind flow
- As wind turbines are optimally designed to operate at maximum power-generating wind speeds, the direction and speed of wind offered by the offshore locations are consistent thus, offer a more efficient and reliable energy source
- The offshore wind turbine farms can be constructed as taller as desired to harvest more electrical power
- As offshore wind turbine farms are installed at sea, they incline to a have minor than usual noise influence on humans.
- Offshore wind turbine farm does not even occupy, restrict, confine, or interfere with land usage.
- The offshore wind might even assist marine ecosystems. In reality, research works show that offshore wind turbine farms offer sea life protection by narrowing the accessibility to certain waters and growing artificial habitats

However, owing to thc complex logistics and larger structures of tower installation, offshore wind turbine farms require much more capital as compared with onshore wind turbine farms. Some minor drawbacks of offshore wind turbines are as follows:

- Generally, offshore wind turbine farms require a 20% higher cost for the towers and maintenance while approximately 2.5 times the cost than the same sized onshore farm is required for the foundation of an offshore wind turbine farm.
- The costs regarding offshore installations, construction, and grid connectivity are likewise substantially higher as compared with onshore.
- After constructing the facility, maintenance and operations also cost much more for offshore facilities. It attributes the fact that even with efforts and measures done to reduce corrosion impacts, seawater origins further maintenance that is not essential for onshore design. Any offshore turbine maintenance is a significant responsibility that includes a helicopter fully loaded with extremely trained technicians and important maintenance entails the high rental expense of a jack-up rig.

The generation of wind energy is increasing significantly around the globe. Fig. 4.2 displays the wind energy generation (in GWh) in different significant regions around the world. A continuous and significant upsurge in wind energy generation can be depicted in all the significant regions of the world including Canada, China, India, the United Kingdom, and the United States from 2000 to 2018. Wind energy generation in Canada increased from 0.264 to 32.17 GWh in the years 2000–18, increased from 0.589 to 366 GWh in China, increased from 1.582 to 60.311 GWh in India, increased from 0.946 to 57.116 GWh in the United Kingdom, and increased from 5.65 to 277.729 GWh in the United States.

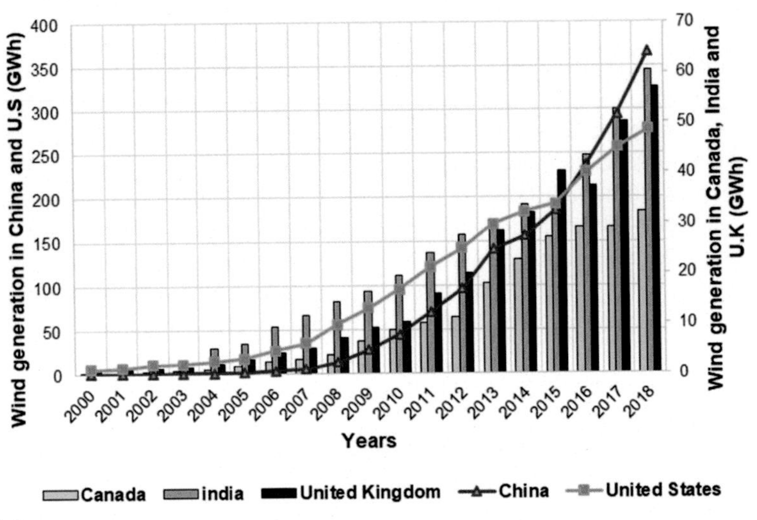

FIG. 4.2

Wind energy generation in different regions.

Adapted from [48].

4.4 Wind Turbine Configuration

Designing a wind turbine is the procedure to define the form, specifications, and methods that are followed by a wind turbine to generate power using wind energy. The installation of a wind turbine comprises the required system and structure that are desired to capture wind energy, to direct the turbine in the direction of the wind, to convert mechanical work into electrical and control panel that is used to start, control, and stop the system.

Moreover, to aerodynamic design blades, complete system design of wind turbine also needs to be addressed involving the controls, hub, generator, foundation, and supporting structure designs. Additional design inquiries rise with the integration of a wind turbine system with electrical power grids. Fig. 4.3 shows the wind turbine configuration with significant components. The significant components of a wind turbine are as follows:

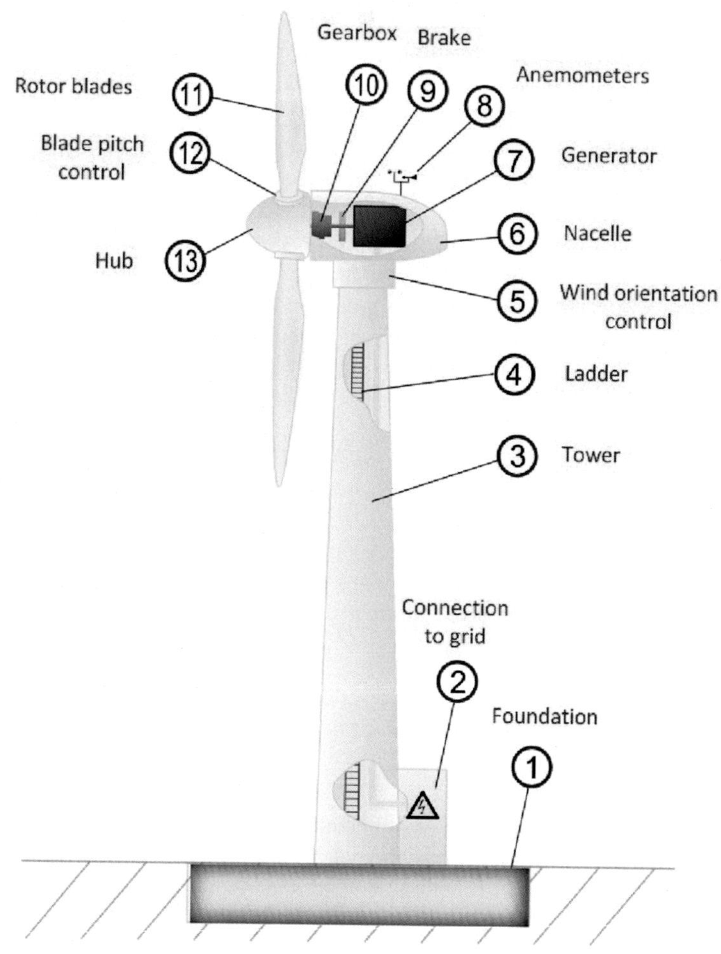

FIG. 4.3

Wind turbine configuration with significant components.

Modified from [89].

- Tower
- Access ladder
- Nacelle
- Generator
- Wind orientation Yaw control
- Electric or Mechanical Brake
- Anemometer
- Rotor blade
- Gearbox
- Rotor hub
- Blade pitch control
- Foundation
- Connection to grid

The blades of wind turbines are coupled with a horizontal shaft that is further connected with the generator positioned in the nacelle. The basic phenomenon that occurs in wind turbines is that they employ the kinetic energy of air to generate electrical power. The kinetic energy of the air produces turning energy, and a rotating wind turbine generates mechanical power. The generator is positioned in the wind turbine that converts mechanical power into electrical. The electricity generated by the wind turbine is AC that is converted to DC using AC/DC converter. Wind energy is one of the promising global resources for producing clean hydrogen. The electrical power extracted using wind energy is employed in the electrolyzer to produce hydrogen. The significant components of the wind turbine are briefly explained to provide the conceptualization of how electrical power is extracted using wind energy.

Anemometer

Anemometer measures wind speed at each interval and communicate with the controller to provide wind speed data.

Blades

Blades are lifted and rotated when blowing wind flows over them and causes the spinning motion in the rotor. Wind turbines consist of two or three blades.

Brake

Brake is employed to stop the rotor electrically, mechanically, or hydraulically in case of emergencies.

Controller

Controller starts the mechanism at designed optimum wind speeds normally ranged from 8 to 16 mph and also shuts down the mechanism at wind speed around 55 mph. Turbines are not designed to function at wind speeds higher than 55.mph as they might damage due to high winds.

Gearbox

Gearbox interconnects the low-speed and high-speed shafts and plays a vital role in increasing rotational speed from around 30–60 rpm to around 1000–1800 rpm, which is rotational speed essential for the generator to generate electricity. Gearbox is one of the heavy and costly wind turbine parts, and researchers are working on direct-drive generator types that can eliminate the gearbox required for low rotational speeds.

Generator

Generator is employed to produce 60-cycle AC power that is later converted to DC power depending upon the application.

High-Speed Shaft

High-speed shaft is employed to drive a generator.

Low-Speed Shaft

Low-speed shaft turns low-speed shaft at a rotational speed of around 30–60 rpm.

Nacelle

Nacelle is installed at the tower top and carries gearbox along with low- and high-speed shafts, controller, generator, and brake. Some designs of the nacelles are sufficiently large to land a helicopter.

Pitch

Pitch turns the blades with wind flow to control rotor speed as well as preserve rotor from turning in wind speeds that are too low or high to generate power.

Rotor

Rotor is the combination of blades and hub which is recognized as a key mechanical part.

Tower

Towers are usually constructed using tubular steel, steel lattice, or concrete which is required to support the turbine structure. As the speeds of wind rise with height, towers taller in height permit turbines to extract more energy to produce higher electricity.

Wind Vane

Thc wind vane is employed to detect the direction of the wind and to communicate with yaw drive to adjust and position the turbine respective to the wind.

Yaw Motor

Yaw motor powers yaw drive.

Yaw Drive

Yaw drive is employed to position the upwind turbines to orient them wind facing during the intervals of change in wind direction. Yaw drive is not required in downwind turbines as the wind blows the rotor far from it.

4.5 Wind Energy-Based Hydrogen Production

The wind is a plentiful and an abundant resource for electricity generation but accompanies the intermittent nature. It is mandatory to analyze the wind energy source using dynamic analysis for average days of each month. The electricity extracted using the wind energy resource can be employed directly in the water electrolysis process for hydrogen production, and produced hydrogen can be employed for numerous applications, namely to fuel hybrid and electric vehicles, combustion, heating, as fuel and stored hydrogen can be employed to the fuel cells to meet the low wind speed intervals.

The wind power terminology designates the process of employing wind for generating mechanical work and hence electrical power. The generated power can be employed for a large number of explicit tasks, such as water pumping or a generator can be integrated to convert mechanical into electrical work and electrical work can be employed in the water electrolysis process for producing hydrogen. Wind energy-based electricity generation is among the fastest-growing global electrical generation methods. Wind energy is among the promising global energy resources to produce clean and environmentally benign hydrogen. The electrical power extracted using wind energy is employed in the electrolyzer that breaks water into its constituents of hydrogen and oxygen.

Wind-to-hydrogen stands as an auspicious and a promising route to produce clean and environmentally benign hydrogen and reduces GHG emissions. Consequently, the fossil fuel reduction potential by wind-to-hydrogen can be projected considering geographical, meteorological, and technical constraints.

Fig. 4.4 exhibits the steps followed to employ wind power for hydrogen production. Wind energy reaches the wind turbine in the first step where the wind turbine employs the kinetic energy of air and yield rotational energy. Once, the turbine starts rotating, the kinetic energy of air is converted to mechanical energy. A generator is positioned inside the turbine hub with the gearbox that converts mechanical energy into electrical. The electric power generated by the wind turbine is in the form of AC that is converted to the DC using AC/DC converter. The DC power from the converter is employed in the water electrolysis process that splits water into its constituents and produces clean hydrogen.

In general, 41.4 kWh electrical power per kg of hydrogen along with 8 L water is required to produce 1 kg hydrogen which represents 149 MJ/kg hydrogen. This represents an efficient process as hydrogen offers 141.8 MJ/kg of energy. Considering an electrolysis process undergoing 100% efficiency involves 39 kWh electrical power per kg of hydrogen while considering the process irreversibilities and losses, currently, existing devices entail around 48 kWh electrical power per kg of hydrogen. Consequently, considering the cost of electricity to be 0.05 US$/kWh, the electrolysis process power cost will be 2.4 US$/kg of hydrogen.

The hydrogen produced by the wind energy resource can be stored and employed in a wide range of applications. Different methods of hydrogen storage are discussed in detail in Chapter 1. Hydrogen fuel cells generate electrical power by combining oxygen and hydrogen atoms. In the electrochemical cell, hydrogen atoms react with oxygen atoms similar to the battery to generate electricity, heat, and water. Different fuel-cell types depending upon the applications are presented and discussed in Chapter 1.

Wind power plays a significant role to reduce GHG emissions and can also assist to encounter electricity demand. Fig. 4.5 displays a wind energy-driven hydrogen production system. The electricity generated by the wind power passes through

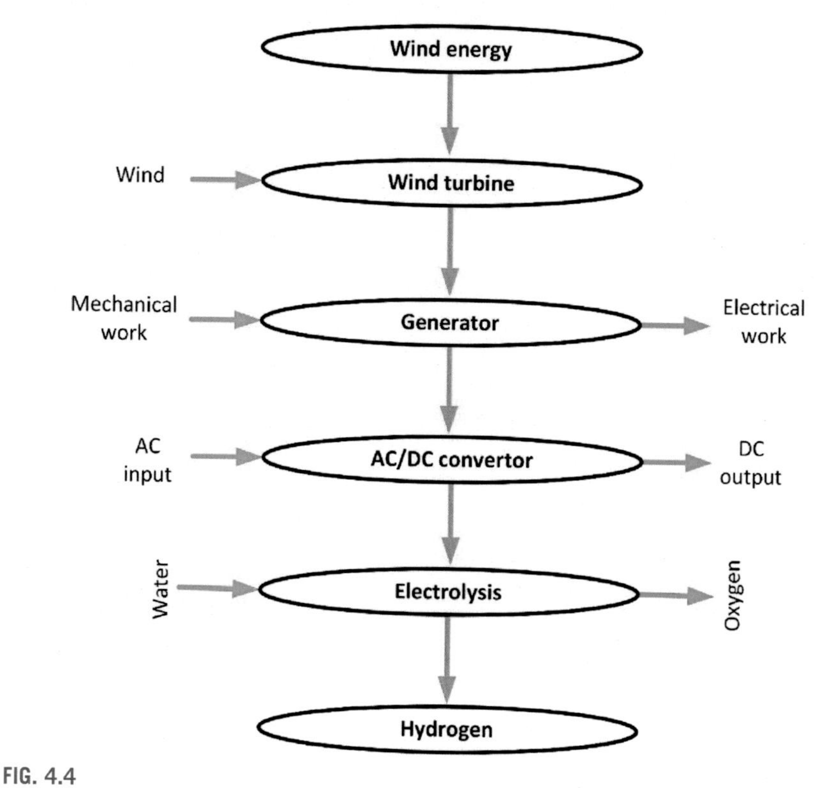

FIG. 4.4

The steps followed to employ wind power for hydrogen production.

the controller. The generated power can be controlled using the controller during the intervals of high and low wind speeds. The electric power generated by the wind turbine is in the form of an AC, which is converted to the DC using an AC/DC converter. The DC power from the converter is employed in the water electrolysis process that splits water into its constituents and produces clean hydrogen. A compressor is followed by water electrolysis that compresses the produced hydrogen as desired pressure for the storage.

4.5.1 Wind Turbine Thermodynamic Analysis

The thermodynamic analysis of wind turbines can be depicted from a study[90] published in the literature that is employed to describe the thermodynamic model equations of the wind turbine. The expression of wind chill is adopted using the

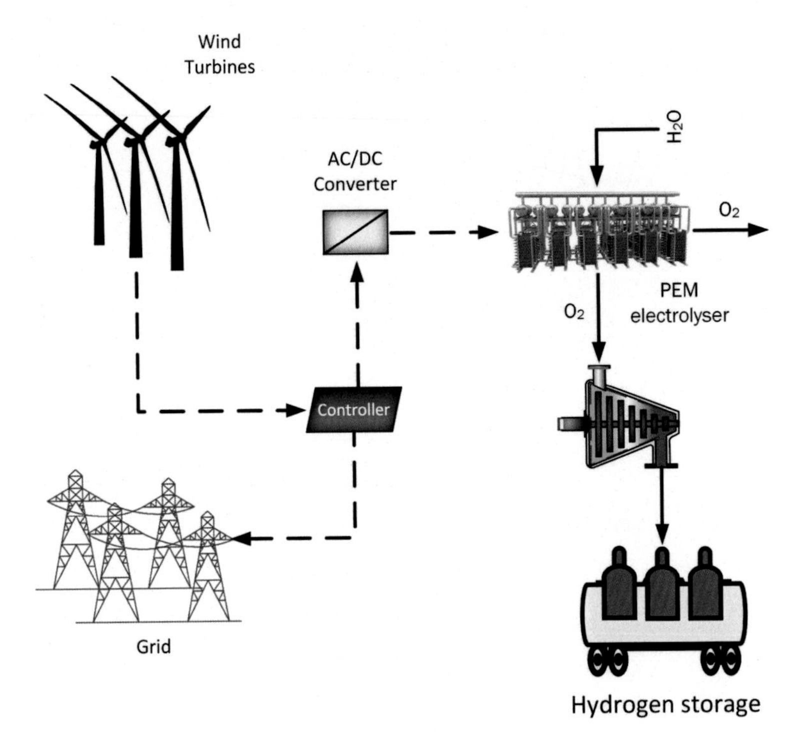

FIG. 4.5

Wind energy-driven hydrogen production system.

Osczevski[91] approach. The JAG/TI correlation employs the advancements in science, computer, and technological modeling to offer a precise, useful, and understandable formula to calculate the risks and hazards accompanied by freezing temperatures and winter winds. Moreover, scientific trials and results were verified and improved the expression accuracy that can be expressed as follows:

$$T_{\text{wind}} = 35.74 + 0.6215T_{\text{air}} - 35.75\left(V^{0.16}\right) + 0.4274T_{\text{air}}\left(V^{0.16}\right) \tag{4.1}$$

where V signifies wind speed measured in mph and T_{wind} denotes wind chill temperature measured in °F.

Energy analysis

Wind energy E represents the air kinetic energy at wind speed V. The mass is somehow hard to quantify; thus, it is expressed using the volume approach using the density $\rho = \frac{m}{V}$. The horizontal length and cross-sectional area lead to the volume using the expression $V = AL$. Physically, the horizontal length is $L = Vt$; thus, the wind energy expression can be written as

$$\dot{E} = \frac{1}{2}\rho A_T V^3 \tag{4.2}$$

Here, Betz applied the windmill momentum theory that has been established by Froude[92] intended for ship propellers. The wind retardation flowing through the windmill exits in two stages: after and before the winding passage over the windmill rotor in that work. Employing the air mass flowing over rotor per unit time, the rate of change in momentum is expressed as $\dot{m}(V_1 - V_2)$ that is equivalent to the resulting thrust. The absorbed power can be stated as

$$\dot{P} = \dot{m}\,(V_1 - V_2)\overline{V} \tag{4.3}$$

Here, V_1 represents upwind speed and V_2 denotes downwind speed at an extensive distance from the rotor. The rate of change in wind kinetic energy can be written as

$$\dot{E}_{K.E} = \frac{1}{2}\dot{m}\left(V_1^2 - V_2^2\right) \tag{4.4}$$

The expressions represented in the last two equations representing the wind retardation before and behind the rotor that are expressed as $V_1 - \overline{V}$, and $\overline{V} - V_2$ should be equal assuming the axial wind velocity direction and uniform velocity over the area. Thus, the power achieved by the rotor can be expressed as follows:

$$\dot{P} = \rho A\overline{V}\,(V_1 - V_2)\overline{V} \tag{4.5}$$

Moreover,

$$\dot{P} = \rho A\overline{V}\,(V_1 - V_2)V = \rho A\left(\frac{V_1 \quad V_2}{2}\right)^2 (V_1 - V_2) \tag{4.6}$$

and

$$\dot{P} = \rho\frac{AV_1^3}{4}\left[(1+\alpha)\left(1-\alpha^2\right)\right] \tag{4.7}$$

where $\alpha = \dfrac{V_2}{V_1}$

After making the necessary mathematical treatment, one may obtain that $\alpha = \frac{1}{3}$ exhibits the maximum power that represents the final wind velocity as one-third of upwind velocity.

Exergy analysis

Exergy analysis is recognized as an approach that employs the second law of thermodynamics along with the principles of conservation of mass and energy to analyze, design, and improve energy systems. Exergy can be defined as the maximum work amount that can be produced through a system or energy or flow of matter as it reaches equilibrium concerning the reference environment. Exergy is not subjected to the law of conservation unlike energy excluding reversible or ideal processes, relatively, exergy is destroyed or consumed due to the process irreversibilities. The consumption of exergy in a process is proportionate to the entropy generation associated with process irreversibilities. Exergy quantifies the quality of energy that is not conserved in any real process but rather lost or destroyed.

The reference environment conditions are necessary to be specified for exergy analysis that is done by providing the specific reference environment chemical composition and temperature and pressure. The exergy analysis results are qualified for a specific reference environment that is modeled based on the actual local reference environment in most applications. A system that is at equilibrium under the reference environment undergoes zero exergies. This draws the implications between environment and exergy that has an environmental impact.

The general equations of energetic and exergetic balances can be written as

$$\sum_{in}(h+K.E+P.E)_{in}\dot{m}_{in}-\sum_{out}(h+K.E+P.E)_{out}\dot{m}_{out}+\sum_{uc}Q_{uc}-W=0 \tag{4.8}$$

$$\sum_{in}ex_{in}\dot{m}_{in}-\sum_{out}ex_{out}\dot{m}_{out}+\sum_{r}Ex^{Q}-Ex^{W}-Ex_{d}=0 \tag{4.9}$$

where $\dot{m}_{in}$ denotes input mass flow rate and $\dot{m}_{out}$ signifies output mass flow rate, Q_{un} represent heat transfer amount into the system across the region under consideration on system boundary, W denotes work transferred from the system, Ex^{Q} represents exergy transfer linked with Q_{un}, I is system exergy consumption, Ex^{W} represents exergy associated with W and $P.E$, $K.E$, h, and ex signify the specific values of potential energy, kinetic energy, enthalpy, and exergy, respectively. It is to be noted that exergy consumption for an irreversible process is greater than zero and for a reversible process, it is equal to zero.

For a closed system:

$$\dot{m}_{in}=\dot{m}_{out}=0 \tag{4.10}$$

Eqs. (4.8) and (4.9) can be simplified as

$$\sum_{uc}Q_{uc}-W=0 \tag{4.11}$$

$$\sum_{r}Ex^{Q}-Ex^{W}-I=0 \tag{4.12}$$

To express the exergy of a flowing matter stream, consider a matter flow at μ_j chemical composition of j species, T temperature, P pressure, m mass, x_j mass fraction, h specific enthalpy, and s specific entropy. The equilibrium environment state is considered as with intensive properties at P_0, T_0, and μ_{j00}. The large enough environment that it undergoes negligible effect by any system interactions is considered. Employing the above-mentioned considerations, the flow matter-specific exergy can be stated as

$$ex_{ph}=[K.E+P.E+(h-h_0)-T_0(s-s_0)] \tag{4.13}$$

$$ex_{ch}=\left[K.E+P.E+\left\{\sum_{j}(\mu_{j0}-\mu_{j00})x_j\right\}\right] \tag{4.14}$$

The expressions of physical and chemical exergies can be stated as

$$ex = ex_{ph} + ex_{ch} \tag{4.15}$$

$$ex = [K.E + P.E + (h - h_0) - T_0(s - s_0)] + \left[\sum_j (\mu_{j0} - \mu_{j00})x_j\right] \tag{4.16}$$

Note that if $K.E = P.E = 0$, the physical exergy expression becomes $(h - h_0) - T_0(s - s_0)$ that denotes maximum available work as the system is brought to the environmental condition and chemical exergy expression becomes $\left\{\sum_j (\mu_{j0} - \mu_{j00})x_j\right\}$ that denotes maximum available work extracted as the system is brought from environmental condition to dead state.

The wind energy exergy can be simply assessed using the following expression of wind energy as it carries no chemical and heat components.

$$Ex^W = W \tag{4.17}$$

To quantify the exergy consumption during a process due to irreversibilities, the following expression is used:

$$I = T_0 S_{gen} \tag{4.18}$$

The wind is a plentiful resource for electricity generation but accompanies the intermittent nature that names the dynamic analysis of wind power mandatory to explore the wind power under different intervals of time. Table 4.1 displays the design parameters, such as geographical location and latitudes, wind turbine area, heat capacity, number of turbines, and wind turbine efficiency considered for the dynamic simulation of wind energy source. The same average days of each month are considered for the dynamic analysis as shown in Table 3.1.

The electric power generated by the wind turbines can be quantified using air density, wind speed, specific heat capacity, cross-sectional area, number of turbines, and wind turbine efficiency employing the following expression:

$$\dot{E}_{out,WT} = \frac{1}{2}\rho A_T \eta_{gen} V^3 C_p N_T \tag{4.19}$$

Table 4.1 Design Parameters of the Dynamic Simulation of Wind Energy Source.

	Parameter	Value
Location	Geographical latitudes	Toronto Latitude = 43.85°N Longitude = 79.38°W
Wind farm[93]	Specific heat capacity	0.49
	Area	1200 m^2
	Efficiency of wind turbine	45%
	Number of turbines	10

Here, E represents the air kinetic energy at wind speed V. The mass can be expressed in terms of area A, density ρ, and velocity V using the following expression:

$$\dot{E} = \frac{1}{2}\rho A_T V^3 \tag{4.20}$$

It is significant to conduct the dynamic analysis of the wind energy resource-based power production system to investigate the performance under different wind speeds. Fig. 4.6a and b display the wind speed data according to the geographical location of Toronto for average days of each month. The latitudes of the considered geographical location are displayed in Table 4.2. Fig. 4.6a exhibits the wind speed that is measured using the controller. The generated power is employed in the generator that converts mechanical into electrical work and electrical work is employed in the water electrolysis process for producing hydrogen. Fig. 4.6a displays the maximum wind speed values in the month of April from months January to June in 11th–13th hours as 13.1 m/s. Fig. 4.6b shows the maximum wind speed values in the month of December from months July to December in the 16th hour as 10 m/s.

The dynamic analysis of wind energy resource-based is also investigated to explore the performance of wind turbine farms to generate power. Fig. 4.7a and b displays the wind power that is extracted using the wind speed for average days of each month. To calculate the wind power, Eq. (4.19) is employed. Fig. 4.7a exhibits the wind power that is measured using the air significant parameters of density, specific heat capacity, wind speed, number of turbines, cross-sectional area, and wind turbine efficiency. Fig. 4.7a displays the maximum power that is extracted using the wind speed values in the month of April from months January to June in 11th–13th hours as 3521.5 kW. Fig. 4.7b shows the maximum wind power that is extracted using the wind speed values in the month of December from months July to December in the 16th hour as 1566.4 kW.

As noted above, the key driver in wind energy applications is the kinetic energy which is a result of wind speed (velocity). The dynamic analysis of the kinetic energy in terms of wind energy input is also investigated to explore the performance of wind turbine farms to generate power. Fig. 4.8a and b displays the wind energy input that is extracted using the wind speed for average days of each month. Eq. (4.20) is employed to calculate the wind energy input. Fig. 4.8a exhibits the wind energy input that is measured using the air significant parameters of cross-sectional area A, density ρ, and velocity V. Wind speed data according to the geographical location of Toronto for average days of each month is employed to determine the kinetic energy in terms of wind energy input. Fig. 4.8a displays the maximum wind energy input that is determined using the wind speed values in the month of April from months January to June in 11th–13th hours as 11,324.6 kW. Fig. 4.8b shows the maximum wind energy input that is quantified using the wind speed values in the month of December from months July to December in the 16th hour as 5400 kW.

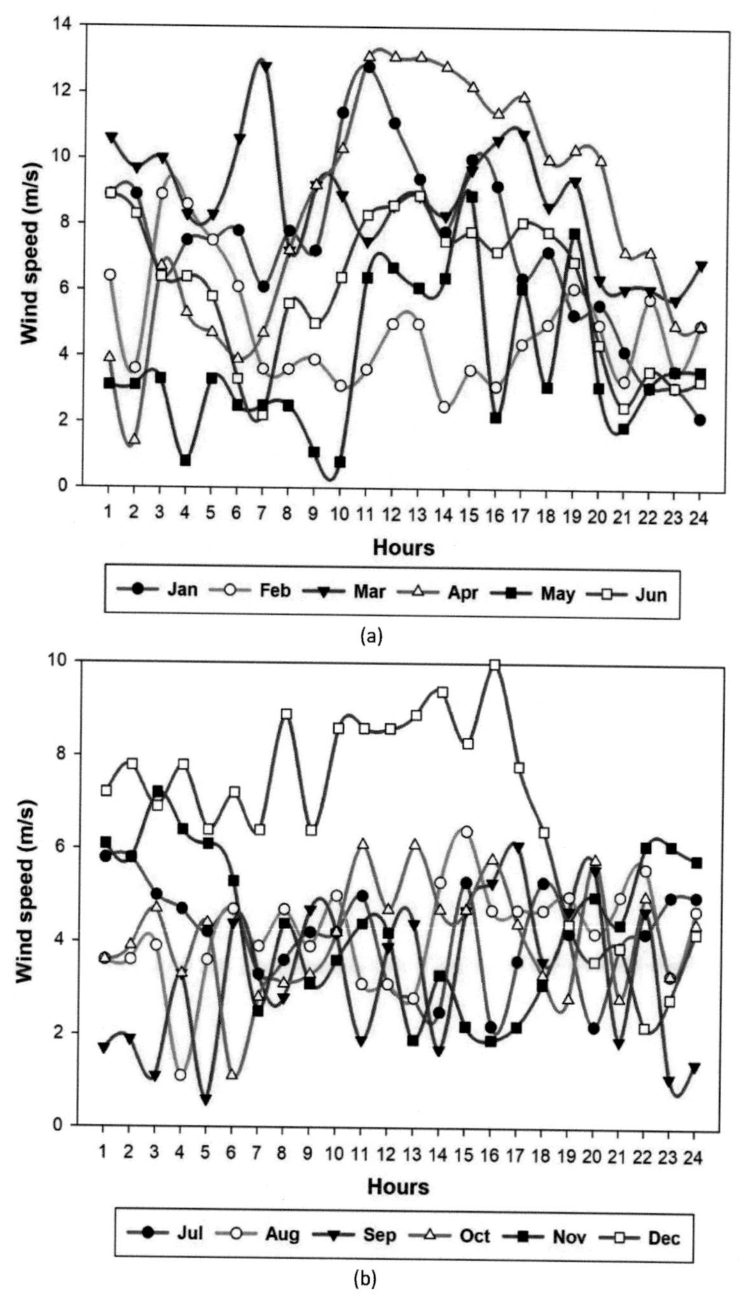

FIG. 4.6

Wind speed data for average days of each month: (a) January–June; (b) July–December.

Table 4.2 Design Parameters of the Wind Energy-Driven Hydrogen Production System.

	Parameter	Value
Wind farm[93]	Specific heat capacity	0.49
	Area	1200 m^2
	Efficiency of wind turbine	45%
	Number of turbines	10
PEM electrolyzer[95]	Membrane thickness	100 μm
	Faraday number	96,486 C/mol
	Operational temperature	80°C
	Cathodic preexponential $\left(J_c^{ref}\right)$	46 × 10 A/m^2
	Anodic preexponential $\left(J_a^{ref}\right)$	17 × 10^4 A/m^2
	Cathodic membrane interface (ϕ_o)	10
	Anodic membrane interface (ϕ_o)	14
	Cathodic activation energy E_{act_c}	18,000 J/mol
	Anodic activation energy E_{act_a}	76,000 J/mol
PEM fuel cell	Current density	11,500 Am^{-2}
	Cell operating pressure	100 kPa
	Cell operating temperature	80°C
Hydrogen compression and storage	Pressure ratio of compressors	5
	Hydrogen storage pressure	12,500 kPa
	Dispensed pressure of hydrogen P_i	2500 kPa
	Dispensed temperature of hydrogen T_i	60°C
	Filling time	120 s

4.5.2 Case Study 4

It is significant to investigate the wind energy-driven hydrogen production system. After exploring the dynamic analysis of the wind energy resource, it is mandatory to explore the wind-to-hydrogen system. This case study is designed to establish the wind energy-driven clean hydrogen and power production system to supply power to the community and store hydrogen to employ during the low wind speed intervals. The experimental results of the proton exchange membrane (PEM) electrolyzer are also displayed to investigate the water electrolysis performance. This case study is Fig. 4.9 displays the wind energy-driven hydrogen and power production system including high-pressure hydrogen storage. The kinetic energy of the wind turbine is converted to mechanical energy employing the wind blades and the generator installed in the wind turbine hub converts mechanical into electrical energy. The electrical energy extracted using the wind energy resource is supplied to an AC/DC converter to convert alternating current into direct current. A part of

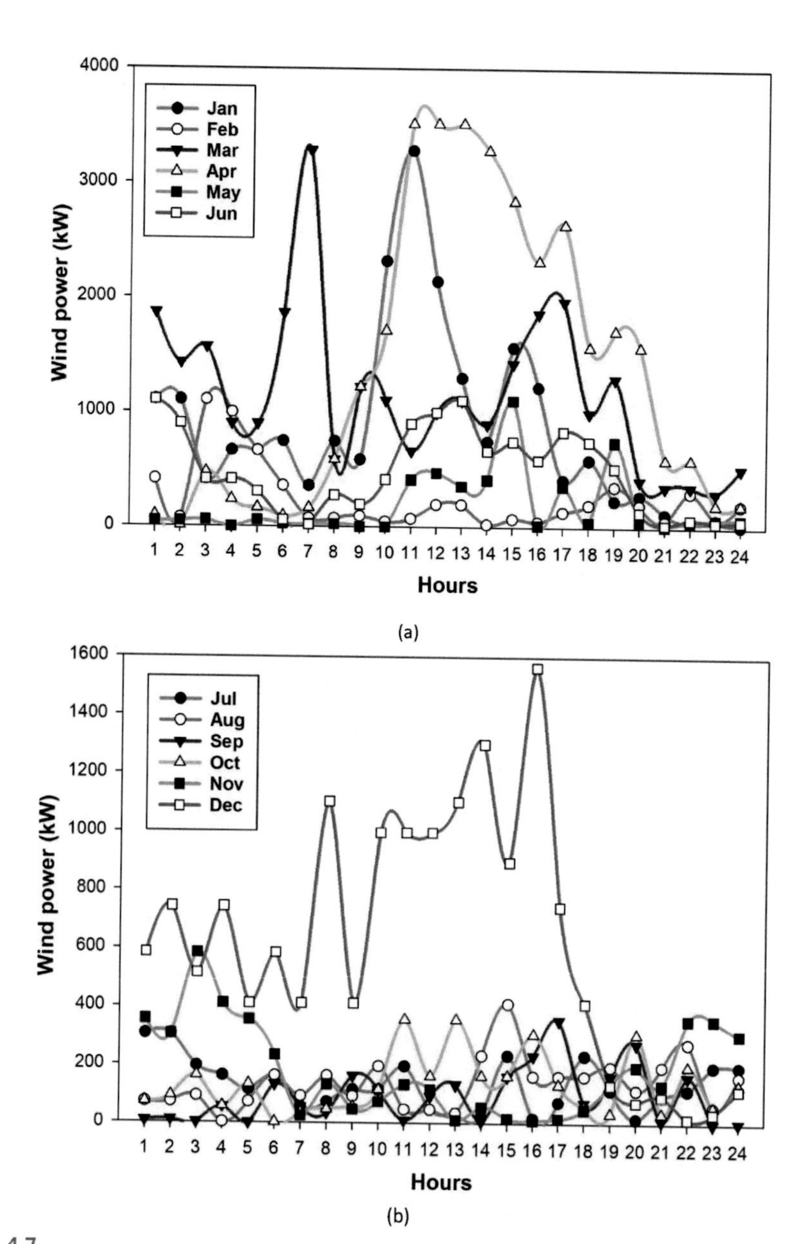

FIG. 4.7

Wind power data for average days of each month: (a) January–June; (b) July–December.

the power output is supplied to the community, and the remaining is employed in the PEM electrolyzer that splits water into hydrogen and oxygen. The hydrogen output from the PEM electrolyzer splits into two different streams. One stream leads to the multistage hydrogen compression unit while the other stream feeds hydrogen to the

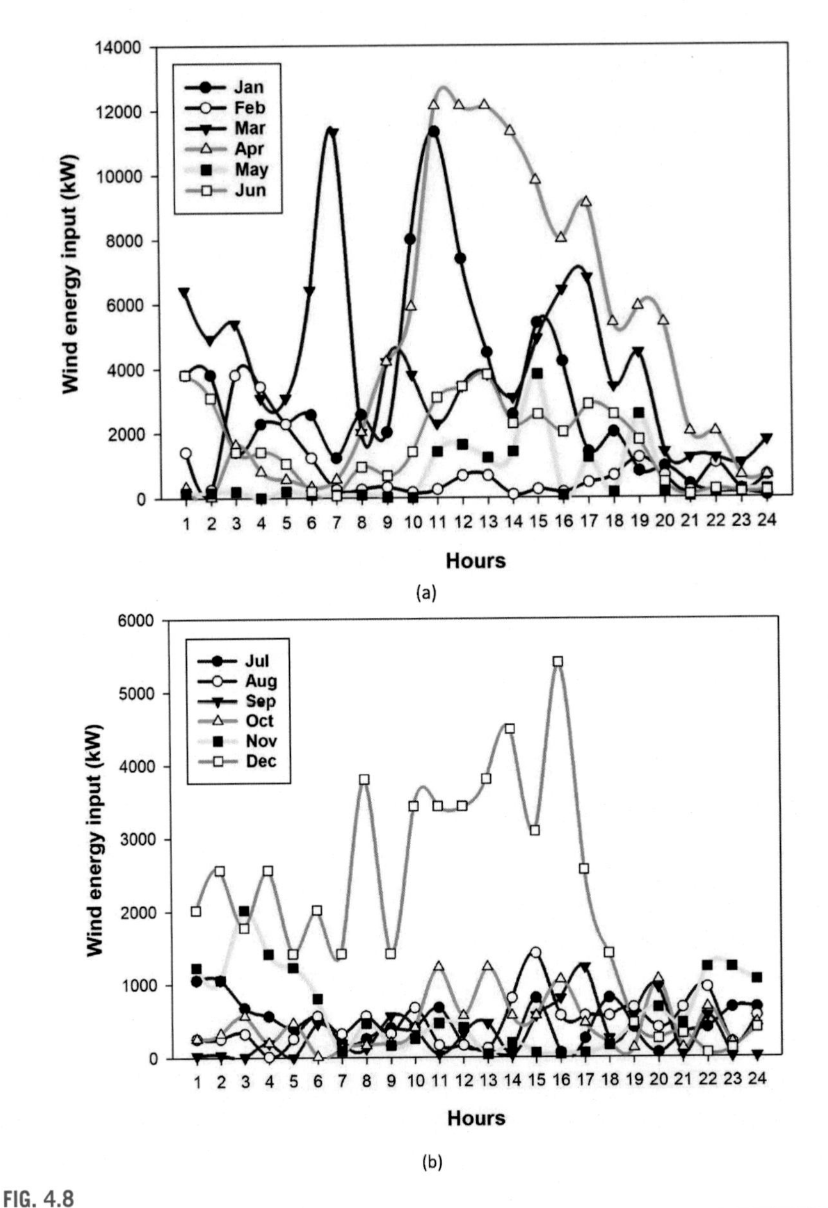

FIG. 4.8

Wind speed data for average days of each month; (a) January–June; (b) July–December.

PEM fuel cell. The multistage compression unit compresses the hydrogen at high pressure of 12,500 kPa. The wind turbine farm supplies electrical power to a community. During the low wind speed intervals, stored hydrogen is employed in the PEM fuel cell that generates electricity to meet the community requirements while

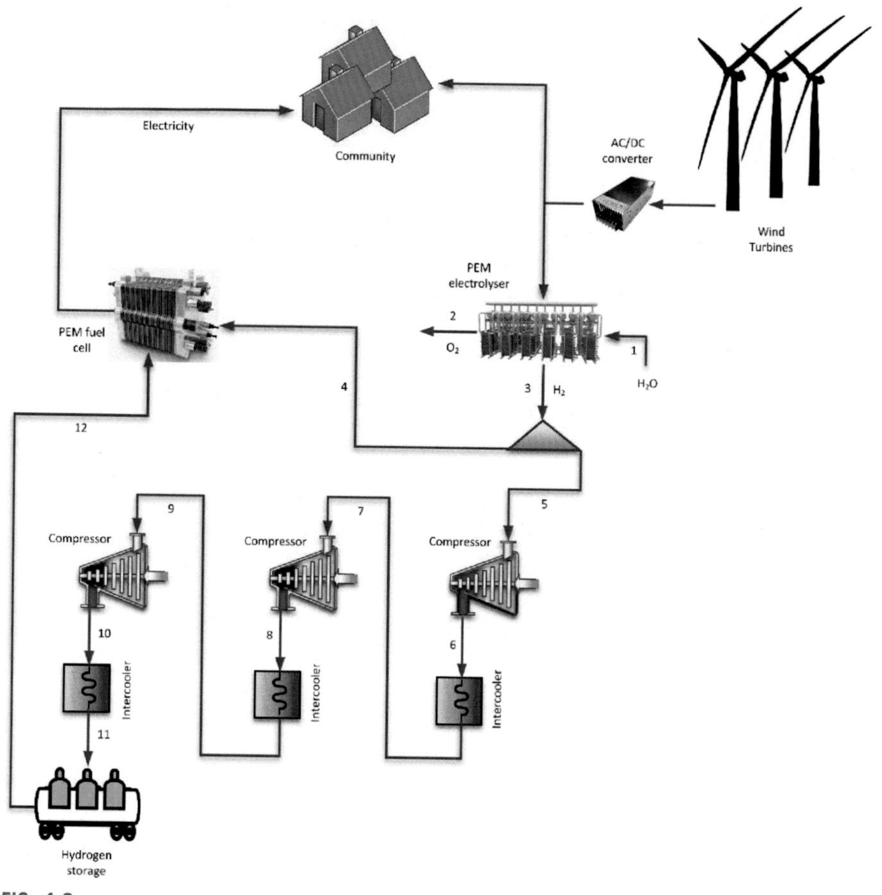

FIG. 4.9

Wind energy-driven hydrogen and power production system.

Modified from [94].

during the high wind speed intervals, additional electric power is supplied to the PEM electrolyzer to produce additional hydrogen that is stored using multistage compression unit. Table 4.2 displays the design parameters of the wind energy-driven hydrogen production system.

Wind turbine farm analysis

As described in the earlier section, the wind energy system employs the wind speed in terms of kinetic energy which rotates the turbine blades and generates mechanical energy and the generator converts mechanical into electrical energy. The kinetic power involved can be quantified using the air mass flow rate and wind velocity using the following expression:

$$\dot{P}_{\text{wind}} = \frac{1}{2}\dot{m}V^2 \tag{4.21}$$

The mass is somehow hard to quantify; thus, it is expressed using volume approach using the density $\rho = \frac{m}{V}$. The mass flow rate can be quantified using cross-sectional area A, air density ρ, and wind velocity V, and the previous equation can be further expressed as follows:

$$\dot{P}_{i,\text{WT}} = \frac{1}{2}\rho A_T V^3 \tag{4.22}$$

Nevertheless, the wind power that is measured using the air significant parameters of density, specific heat capacity, wind speed, number of turbines, cross-sectional area, and wind turbine efficiency can be expressed as follows:

$$\dot{E}_{\text{out,WT}} = \frac{1}{2}\rho A_T \eta_{\text{gen}} V^3 C_p N_T \tag{4.23}$$

The following expression is employed to determine the rate of exergy destruction of the wind turbine farm:

$$\dot{E}x_{\text{dest,wt}} = \left(\frac{1}{C_{p,\text{wt}}} - 1\right) P_{i,\text{wt}} \tag{4.24}$$

The rate of exergy destruction rate of the wind turbine can be assessed using the exergy rate balance and neglecting the heat losses:

$$\dot{E}x_{\text{in}} = \dot{E}_{\text{out,WT}} + \dot{E}x_{\text{dest}} \tag{4.25}$$

Similarly, the expression for the exergy efficiency is as follows:

$$\eta_{\text{wt}} = \frac{\dot{E}_{\text{out,WT}}}{\dot{P}_{i,\text{WT}}} \tag{4.26}$$

PEM electrolyzer and fuel cell

The PEM electrolyzer and PEM fuel cell work under the same process principles, but they both are reverse processes. Conductivity shows the capability of a material to conduct electric current. To relate the resistance of a conductor with conductivity, the following correlation is used:

$$R_c = \frac{l}{\sigma A} \tag{4.27}$$

where R_c signifies the conductor resistance, σ denotes the conductivity, A indicates area, and l represents the conductor length. Ohmic losses calculations uses the specific membrane electrical conductivity equations that are determined using the following correlation presented by Ni et al.[66]:

$$\sigma(c,T) = (0.5139c - 0.326)\exp\left(1268\left(\frac{1}{303} - \frac{1}{T}\right)\right) \tag{4.28}$$

Here, T signifies membrane temperature, and c denotes water molar concentration inside the membrane. The assumption of the linear change in molar water content across the membrane is reasonable.

$$c(x) = c_c + \left(\frac{c_a - c_c}{\delta}\right)x \tag{4.29}$$

The electrical conductivity is well defined as $\sigma = \frac{dx}{dR}$ where R represents electrical resistance, and the differential equation can be represented as $dR = \sigma^{-1}dx$:

$$R_{\Omega,\mathrm{PEM}} = \int_0^{\delta} \sigma^{-1}(c(x), T)\, dx \tag{4.30}$$

The correlation for the concentration overpotential can be expressed as follows:

$$\Delta E_{\mathrm{conc}} = J^2\left(\beta\left(\frac{J}{J_{\mathrm{lim}}}\right)^2\right) \tag{4.31}$$

Factor β can be defined as follows:

$$\beta = (7.16 \times 10^{-4}T - 0.622)P + (-1.45 \times 10^{-3}\, T + 1.68) \text{ if } P < 2 \text{ atm} \tag{4.32}$$

$$\beta = (8.66 \times 10^{-5}T - 0.068)P + (-1.6 \times 10^{-4}\, T + 0.54) \text{ if } P \geq 2 \text{ atm} \tag{4.33}$$

where $P = \frac{P_i}{0.1173} + P_{sat}$ is local cathodic or anodic pressure, P_i denotes cathodic or anodic partial pressure, and P_{sat} signifies water saturation pressure and index i is denoted as c for cathode and a for the anode.

A recent article by Ni et al.[66] helps determine the correlation between cathodic and anodic activation potentials as follows:

$$\Delta E_{\mathrm{act},i} = \frac{\mathrm{RT}}{F}\ln\left(\frac{J}{J_{0,i}} + \sqrt{\left(0.5\frac{J}{J_{0,i}}\right)^2 + 1}\right) \tag{4.34}$$

where $J_{0,i}$ is cathodic and anodic exchange current density.

High exchange current density indicates high electrode reactivity that outcomes lower overpotential. The Arrhenius equation can be used to express electrolysis exchange current density:

$$J_{0,i} = J_{\mathrm{ref},i}\ \exp\left(-\frac{\Delta E_{\mathrm{act},i}}{\mathrm{RT}}\right) \tag{4.35}$$

Performance assessment

In the designed case study, wind turbine farm generates electrical power: a part of which is supplied to the community and additional is fed to the electrolyzer. The system efficiency can be defined using three different scenarios. During the intervals of high wind speed, the excess power is employed in the PEM electrolyzer and the produced hydrogen is stored at high pressure. During the average wind speed, the wind turbine farm generates enough power to meet the community load while during the low wind speed, the stored hydrogen is employed in the PEM fuel cell to supply the community load. Thus, three different equations are shown to represent the system performance under three different scenarios. The net output power of the designed case study can be depicted as follows:

$$\dot{P}_{\text{net}} = \dot{P}_{\text{wt}} - \dot{W}_{\text{comp1}} - \dot{W}_{\text{comp2}} - \dot{W}_{\text{comp3}} \tag{4.36}$$

The system efficiency during high wind speed intervals can be expressed as follows:

$$\eta_{\text{ov}} = \frac{\dot{P}_{\text{net}} + \dot{m}_{H_2}\text{LHV}_{H_2}}{\dot{P}_{i,\text{WT}}} \tag{4.37}$$

The efficiency during the average wind speed:

$$\eta_{\text{ov}} = \frac{\dot{P}_{\text{net}}}{\dot{P}_{i,\text{WT}}} \tag{4.38}$$

The efficiency expressions during the low wind speed intervals when stored hydrogen is employed to meet the community load can be expressed as follows:

$$\eta_{\text{ov}} = \frac{\dot{W}_{\text{stack}} + \dot{Q}_{\text{PEMFC}}}{\dot{m}_{H_2}\text{LHV}_{H_2}} \tag{4.39}$$

$$\psi_{\text{ov}} - \frac{W_{\text{stack}} + \dot{Q}_{\text{PEMFC}}\left(1 - \frac{T_0}{T}\right)}{\dot{m}_{H_2}\text{LHV}_{H_2}} \tag{4.40}$$

Sensitivity analyses

Several sensitivity and parametric analyses are conducted on the designed case study to investigate the system functionality and performance under different operating constraints. This section does not only display the sensitivity analyses from the simulation results but also the experimental results of a lab-scale PEM electrolyzer to assess the performance and efficiencies of the design case study.

The performance assessment of the designed system is defined using three different scenarios. During the low wind speed, the stored hydrogen is employed to the PEM fuel cell to supply the community load while the wind turbine farm generates enough power to meet the community load during average wind speeds and the intervals of high wind speed, the excess power is employed to the PEM electrolyzer and the produced hydrogen is stored at high pressure. Thus, it is substantial to explore the effect of the variation in wind speed on the significant system parameters, namely wind power, hydrogen production rate, exergy destruction rate, and work rates of the compressor with the experimental investigations of PEM electrolyzer. Fig. 4.10 displays the effect of wind energy on wind turbine farm power and hydrogen production rates. Both, molar and mass production rates of hydrogen are shown in the figure. The wind speed is ranged from 1 to 10 m/s. The gradual increase in wind turbine farm power can be depicted with the rise in wind speed. During the intervals of high wind speed, the additional power after meeting the community load

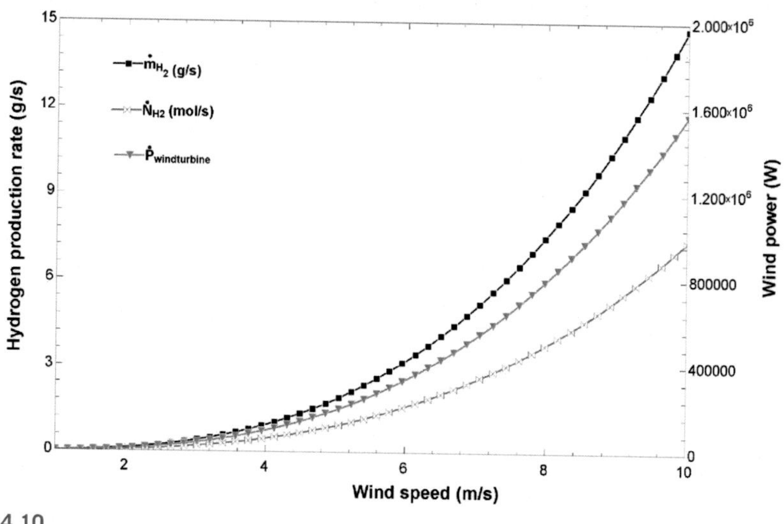

FIG. 4.10

Wind energy effect on wind power and hydrogen production rates.

is employed to the PEM electrolyzer to produce power. Thus, the increase in wind speed gives rise to the electrical power that is supplied to the PEM electrolyzer that produces higher mass and molar flow rates of hydrogen as shown in Fig. 4.10.

The complete kinetic energy amount cannot be converted to mechanical power as the system undergoes some losses, and wind turbine efficiency is assumed to be 45% in the designed case study. Thus, it is noteworthy to investigate the exergy destruction rates of all significant system components that also provide the gaps of possible improvement in the designed system. Fig. 4.11 displays the effect of wind energy on

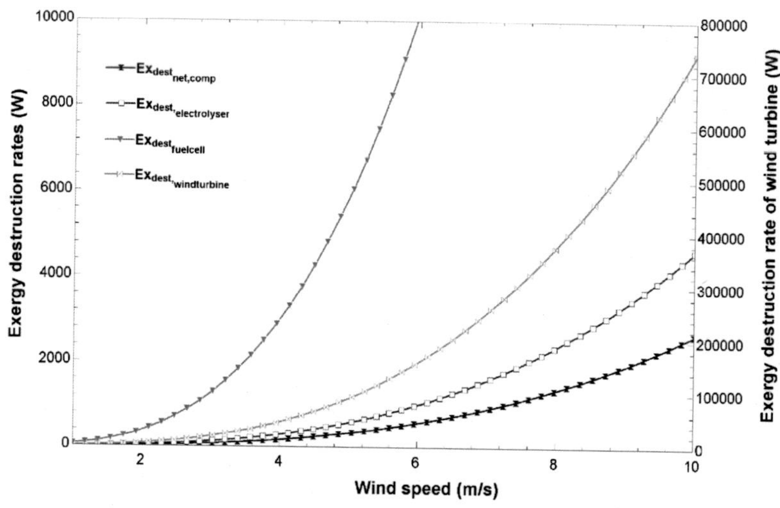

FIG. 4.11

Wind energy effect on the exergetic destruction rates of different system components.

the exergy destruction rates of electrolyzer, fuel cell, wind turbine, and compressors. As the rise in wind speed causes the system to deal with high electrical power that results in high hydrogen production and multistage compression unit also deals with high capacities, the increased exergy destruction rates are expected to deal with increased capacities. The increase in wind speed gives rise to the exergy destruction rates of all significant components of the designed case study as shown in Fig. 4.11.

In the designed system, the wind turbine farm generates electrical power a part of which is supplied to the community and additional is fed to the electrolyzer. The produced additional hydrogen is fed to the multistage compression system that compresses the hydrogen and stores it at 12,500 kPa. The increased wind speed carries a direct effect on the compressor work and exergy destruction rates as increased wind speed offers high additional power that produces more hydrogen, and the compressor deals with high hydrogen capacities. Thus, it is substantial to investigate the exergy destruction rate of compressor, and exergy destruction rate provides the gaps of possible improvement in the designed system. Fig. 4.12 displays the effect of wind energy on the work and exergy destruction rates of the compressor. As the rise in wind speed causes the system to deal with high electrical power that results in high hydrogen production, the hydrogen compressor compresses high capacities. Thus, an increase in wind speed gives rise to both work and exergy destruction rates of the compressor that is depicted in Fig. 4.12.

The PEM electrolyzer is a significant component of the designed system that employs the electrical power generated by the wind turbine farm to produce hydrogen. Thus, the water electrolysis system is investigated using both simulation and

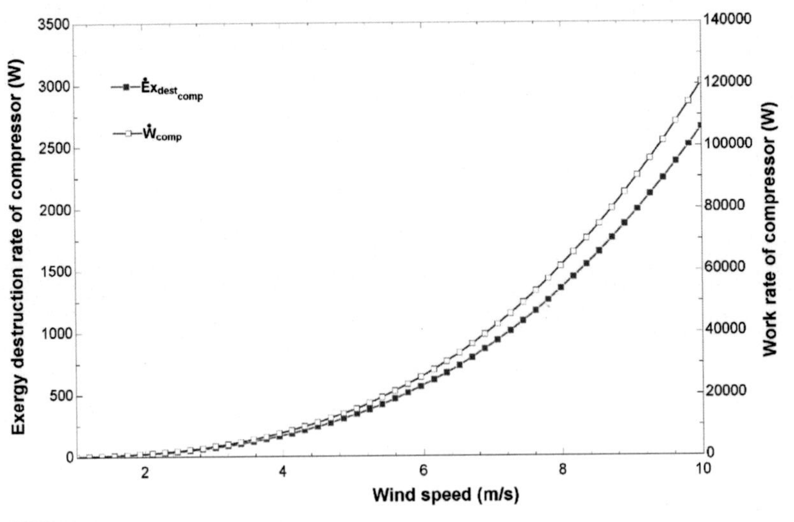

FIG. 4.12

Wind energy effect on the work and exergy destruction rates of compressor.

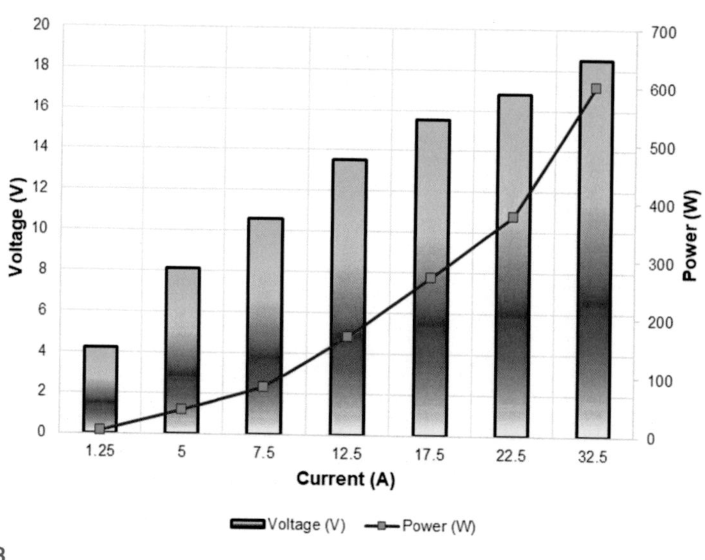

FIG. 4.13

PEM electrolysis process polarization curve.

experimental methods. Fig. 4.13 exhibits the polarization curve of PEM electrolyzer integrated into the designed case study displaying the power and voltage values at fluctuating currency values. The rising voltage slope with increasing currents is detected to drop through rising current input. In the beginning, low current values represent the activation polarization region where voltage rises due to the half-cell electrochemical reaction hindrances. Moreover, the increased input current causes higher Ohmic losses that boost the voltage. The increase in the input current from 1.25 to 32.5 A causes the voltage to rise from 4.5 to 18.5 V and power input increases from 4.25 to 601.25 with the rise in the input current to the PEM water electrolysis.

The PEM water electrolysis is a significant process in the designed system that employs the electrical power generated by the wind turbine farm to split water into its constituents. Thus, both simulation and experimental approaches are employed to investigate the water electrolysis process. Fig. 4.14 exhibits the energetic as well as exergetic efficiency of the PEM water electrolysis process at fluctuating input current. A reducing trend is detected in the energetic as well as exergetic efficiency with the rising input current. Moreover, the efficiency drop with the current rise is found comparatively higher at low values of current than large values of current. For example, the increase in input current from 1.25 to 5 A causes the energetic efficiency to drop from 30.1% to 15.6% and exergetic efficiency to reduce from 29.6% to 15.3%. Though, the increase in input current from 5 to 12.5 A causes the energetic efficiency to drop from 15.6% to 9.4% and exergetic efficiency to reduce from 15.3% to 9.2%. This drop in efficiencies can be accredited to high-activation polarization losses at comparatively lower currents and can be reduced by employing high electrochemical catalyst loading.

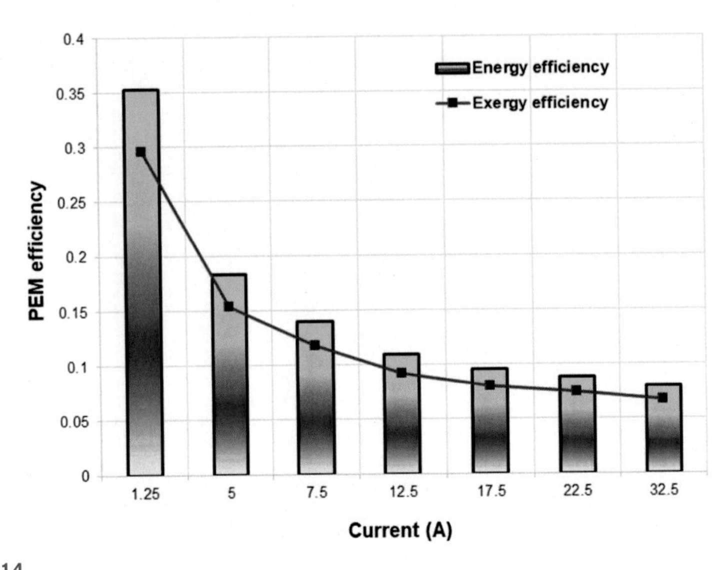

FIG. 4.14

Effect of current input to the PEM electrolyzer on energy and exergy efficiencies.

4.6 Closing Remarks

Wind energy is among the promising global energy resources to produce clean and environmentally benign hydrogen, and this promising route reduces GHG emissions as well. Wind energy-driven hydrogen production technology can play a significant role in providing the global transition to 100% renewable energy. The horizontal-axis wind turbine characteristically consists of a vertical propeller having three blades that face the wind while the vertical-axis wind turbine involves a blade set that spins through the vertical axis. Both of these wind turbine types can be employed using either onshore or offshore facilities to generate power. The wind power terminology designates the process of employing wind to generate mechanical work and hence electrical power. Wind energy-based electricity generation appears to be among the fastest-growing global electrical generation methods. The kinetic energy associated with air is converted to mechanical energy using the wind turbine blades, and the generator converts mechanical work into electrical power. The wind is a plentiful resource of power generation but accompanies the intermittent nature. It is obligatory to analyze the wind energy resource employing the dynamic analysis for average days of each month. The electricity generated by the wind energy resource can be employed directly to the water electrolyzer that breaks water into its constituents of hydrogen and oxygen, and the produced hydrogen can be employed for numerous applications, namely for combustion, to fuel hybrid and electric vehicles, heating, as fuel, and stored hydrogen can be employed to the fuel cells to generate power on demand.

Wind-to-hydrogen stands as a promising route to produce clean and environmentally benign hydrogen and undergoing zero carbon emissions. Subsequently, the reduction potential of fossil fuels by the wind-to-hydrogen can be predicted considering geographical, meteorological, and technical constraints. It is significant to conduct the dynamic analysis of the wind energy resource-based power production system to investigate the performance under different wind speeds. A case study is designed in this chapter that investigates the dynamic analysis of the wind energy source for average days of each month. The designed case employs a wind turbine farm to produce power that is employed in the water electrolysis process to produce clean hydrogen. The stored hydrogen can be employed in the fuel cells to generate power that is utilized for numerous applications.

CHAPTER

Geothermal Energy-Based Hydrogen Production

5

Geothermal energy is known as a source of driving heat within the earth's subsurface. The geothermal energy is carried through water or steam to the earth's surface. Contingent on the applications, geothermal energy can be employed for cooling and heating and also can be employed to harness clean electricity. Geothermal energy is extracted in the form of sustainable and clean heat from the earth. Geothermal energy resources range from the shallow ground on the way to hot rock and hot water that are found to be few miles underneath earth surface, and molten rocks called magma offering enormously high temperatures are found further down. Numerous experts have indicated that geothermal energy offers cleaner, consistent, environmentally benign, efficient, sustainable, and cost-effective solutions for clean hydrogen production as compared with fossil fuel burning that can reduce fossil fuel dependency. Geothermal energy source offers the cheapest solution of clean and sustainable hydrogen production followed by the wind energy source and gradually becoming cheaper, optimum, and cost-effective as compared with fossil fuel-fired energy.

Geothermal energy is recognized as a renewable energy resource as it uses earth core as the source that offers approximately unlimited heat generation. Geothermal areas are also dependent upon the hot water reservoir, and volume carried out can be reinjected back to the earth that makes geothermal energy a sustainable resource. The prime group of global geothermal power plants is situated at a geothermal California field named as Geysers. Geothermal power plants can also employ the heat that is extracted from the deep inside of the earth to produce electrical power using steam. Geothermal heat pumps are tapped into heat adjacent to earth surface to provide heat or water heating for buildings.

Fig. 5.1 displays the summary of the different geothermal direct-use categories of geothermal heat pumps, greenhouse heating, space heating, agricultural drying, aquaculture pond heating, industrial uses, cooling/snow melting, bathing and swimming, and others worldwide in terms of geothermal capacity and geothermal utilization from 1995 to 2015. In Fig. 5.1a, it can be depicted that geothermal capacity that is divided into different direct-use categories increased in most of the sectors. The direct usage of geothermal heat pumps increased from 1854 to 49,898 MW, space heating increased from 2579 to 7556 MW, greenhouse heating increased from 1085 to 1830 MW, aquaculture pond heating decreased from 1097 to 695 MW, agricultural drying increased from 67 to 161 MW, industrial uses increased from 544 to 610 MW, bathing and swimming increased from 1085 to 9140 MW, cooling/snow

Renewable Hydrogen Production. https://doi.org/10.1016/B978-0-323-85176-3.00003-2

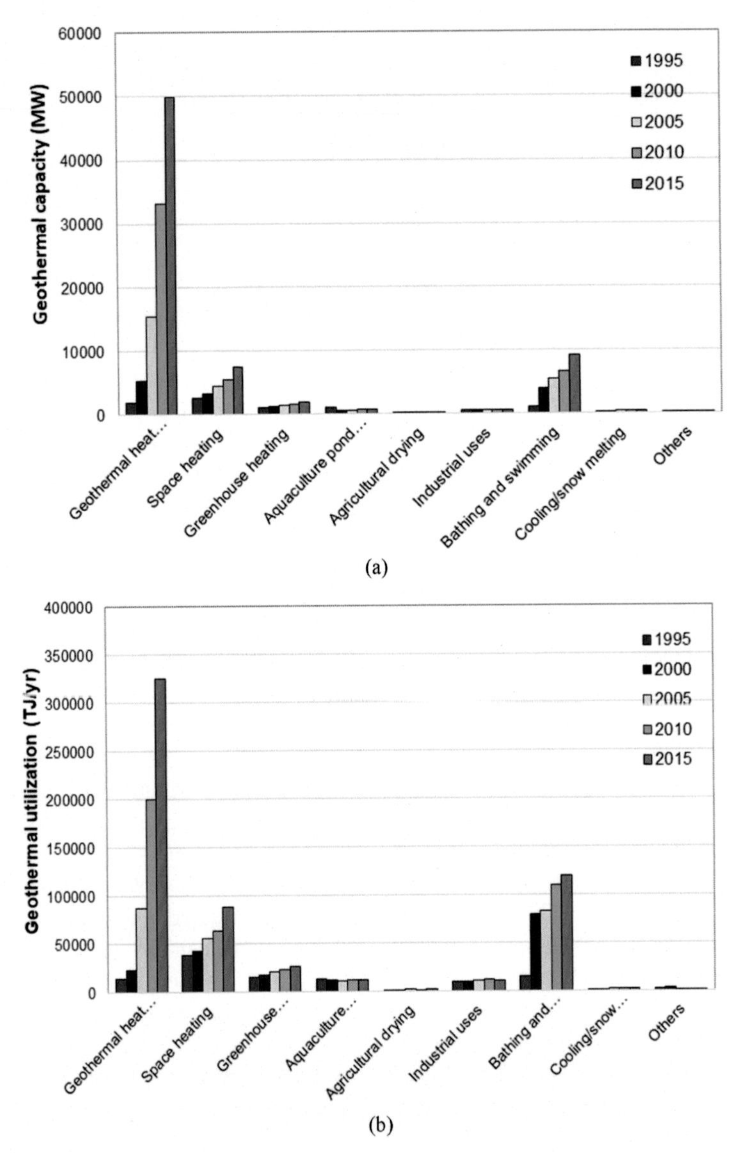

FIG. 5.1

Summary of the different geothermal direct-use categories worldwide (a) geothermal capacity (b) geothermal utilization.

Adapted from [96].

melting increased from 115 to 360 MW, and other direct usages reduced from 238 to 79 MW from the year 1995 to 2015. Fig. 5.1b displays the rise in geothermal utilization that is divided into different direct-use categories increased in most of the sectors.

The direct usage of geothermal heat pumps increased from 14,617 to 325,028 TJ/yr, space heating increased from 38,230 to 88,222 TJ/yr, greenhouse heating increased from 15,742 to 26,662 TJ/yr, aquaculture pond heating decreased from 13,493 to 11,958 TJ/yr, agricultural drying increased from 1124 to 2030 TJ/yr, industrial uses increased from 10,120 to 10,453 TJ/yr, bathing and swimming increased from 15,742 to 119,381 TJ/yr, cooling/snow melting increased from 1124 to 2600 TJ/yr, and other direct usages reduced from 2249 to 1452 TJ/yr from the years 1995 to 2015.

Geothermal energy by name represents the thermal energy produced and stored in the earth. The high pressure and temperature in the interior part of the earth result in melting some rocks and solid mantle which starts behaving plastically after the mantle portions convecting upward as it becomes lighter than adjacent rocks. Many confirm that geothermal energy offers cleaner, environmentally benign, efficient, sustainable, and cost-effective solutions, as compared with fossil fuels. One may further note that it can reduce the fossil fuel dependency. Geothermal energy-based power plants are also more reliable as compared with nuclear and coal plants as they can be operated consistently, 365 days each year. The geothermal resource offers the cheapest energy source of clean energy followed by the wind energy and gradually becoming cheaper, optimum, and cost-effective than fossil fuel-fired energy. Geothermal energy offers some significant advantages described briefly as follows:

- Geothermal energy is considered a renewable energy resource as it uses earth core as the source that offers approximately unlimited heat generation.
- Geothermal energy is an environment friendly renewable energy source that does not cause noteworthy pollution.
- Geothermal resources and reservoirs are naturally restocked and consequently, this renewable energy source cannot be consumed.
- The enormous potential and approximations reveal 2 TW of worldwide potential.
- An excellent and outstanding choice to encounter base-load energy demand as other renewables like solar and wind are intermittent by nature.
- An excellent source for cooling and heating and cooling and small households can also benefit from this resource.
- Geothermal energy harvesting does not include any fuels. Thus, it offers fewer cost fluctuations and consistent electricity prices.
- It carries minor land footprints and can be constructed partially underground.
- Recent technological developments and advancements have offered more exploitable resources and lowered costs as well as more feasible solutions.

5.1 Geothermal Energy Advantages and Disadvantages

It is important to discuss the advantages and disadvantages of geothermal energy and its utilization for various applications. Below is a brief discussion of such aspects.

5.1.1 Advantages

The detailed advantages of geothermal energy that can be harvested in terms of geothermal power plant or heat pumps are needed to be discussed in detail. The following are the significant advantages of geothermal energy.

Environment friendly

Geothermal energy is an environment friendly energy resource, and polluting aspects to harness geothermal energy are very minor as compared with conventional sources, such as coal, natural gas, and other fossil fuels, and geothermal power plants carry minimal carbon footprint. More development and advancement in geothermal resources are considered obliging in fighting against global warming. A regular geothermal power plant emits 122 kg CO_2 equivalent for each megawatt-hour (MWh) of generated electricity that accounts for one-eighth of CO_2 emissions accompanied by a regular coal power plant.

Renewable nature

Geothermal energy resources and reservoirs are naturally restocked and replenished. Thus, geothermal energy is among renewable energy sources. Geothermal energy resource sustains and withstands the consumption rate distinct to the conventional energy sources that make it sustainable. According to scientific research works, the energy that is harnessed through geothermal reservoirs will last for billions of years.

Massive potential

The global energy consumption that is around 15 terawatts (TW) is not even near to the geothermal energy amount that can be harnessed from the earth. Nevertheless, most geothermal reservoirs are utilized to employ a tiny portion of the total available potential. Realistic estimations of geothermal power plants potential ranged between 0.035 and 2 TW. Geothermal power plants across the globe are currently delivering about 10,715 MW of electricity and the installed capacity of geothermal heating is approximately 28,000 MW.

Sustainable development

Geothermal energy offers a reliable, consistent, and sustainable energy source. The geothermal power plant output power can be predicted with outstanding accuracy as the weather does not play a huge part in electricity generation unlike wind and solar energy sources. Thus, geothermal power plants are an excellent and outstanding choice to encounter base-load energy demand. Geothermal power plants offer a high capacity factor, and practical output power is close to installed capacity. The worldwide average output power was established as 73% back in 2005, which has now been demonstrated to be 96%.

Suitability for cooling and heating

The steam temperature of higher than 150°C or greater is mandatory to turn the turbines effectively and produce an efficient amount of electricity employing geothermal energy. An additional approach is available that is employed to benefit from the comparatively small temperature difference of ground source and surface. Earth is largely resistant to the seasonal temperature variation as compared with air. Therefore, the ground underneath a couple of meters to the surface can perform as a heat source or sink through a geothermal heat pump similar to the electric heat pump. An incredible growth can be observed in the household utilizing geothermal cooling and heating as well as geothermal power plants and geothermal heat pumps.

Reliability

The geothermal power plant offers a consistent and reliable energy source, and output power can be determined with outstanding accuracy as the weather does not play a huge part in electricity generation unlike wind and solar energy sources. This exhibits that the output power can be predicted with high accuracy for geothermal plants.

No fuel requirement

As geothermal energy is a naturally driven resource, no fuel is required, unlike the fossil fuels that require mining or other extraction techniques.

Quick evolution

Although the worldwide average output power has now been demonstrated to be 96%, the exploration and investigation of geothermal energy is a great deal to establish the advanced technologies that can make the process more efficient. Table A-5 in the appendix displays the geothermal capacity in different world regions.

5.1.2 Disadvantages

It is also substantial to explore the disadvantages accompanied by geothermal energy that can be harvested in terms of geothermal power plants or heat pumps. The following are the main disadvantages of geothermal energy.

Environmental issues

An abundant amount of greenhouse gases (GHGs) exists underneath the earth's surface and some are mitigated to the earth's surface and atmosphere. Such GHG emissions are predicted to be more nearby geothermal power plants. Geothermal power plants are also accompanied by silica and sulfur-dioxide emissions but irrespective of this drawback, the pollution that is linked with geothermal power is not even near to the fossil fuels and coal-powered plants. A common geothermal power plant emits 122 kg CO_2 equivalent for each MWh of generated electricity that accounts for one-eighth of CO_2 emissions accompanied by a regular coal power plant.

Surface instability (earthquakes)

Geothermal power plant construction can affect land stability. Earthquakes can be prompted owing to the hydraulic fracturing and cracking that is a fundamental step of evolving improved geothermal power plants.

Expensive

Commercial projects of geothermal power plants are quite expensive to construct. The exploration, survey, and drilling of reservoirs are accompanied by a steep price tag almost half the costs. In contrast, geothermal energy systems are probable to save money in the long run and consequently looked upon as long-term investments.

Location specific

The geothermal reservoirs highly depend on the location. If geothermal heat pumps are constructed to extract long distanced hot water but no electricity, they can undergo significant energy losses that need to be considered.

Sustainability issues

Rainwater soaks over the earth's surface and is obsessed with geothermal reservoirs over the years. Research studies indicated that geothermal reservoirs can undergo depletion if fluid removal is faster than replacement. Thus, it is substantial to reinject the fluid into a geothermal reservoir subsequent to the thermal energy extraction. Geothermal energy offers a reliable, consistent, and sustainable energy source that makes it an attractive source but costs associated with geothermal power plants and heat pumps need to be reduced further.

Gradual growth in the geothermal power plants is seen over the last few decades, which reveals a continuous rise in the installed capacities and electricity generation. Fig. 5.2 exhibits the installed geothermal energy capacity and electricity generation trends. A gradual increase in electricity generation can be depicted from 68,454 to 85,978 GWh from 2010 to 2017 and meanwhile, installed capacity increased from 9992 to 12,931 MW.

Fig. 5.3 exhibits the classification of geothermal energy utilization for hydrogen production. The geothermal energy source can be employed in two significant forms of direct source and thermal energy and thermal energy can be employed to different routes, such as thermochemical cycle that can operate hybrid cycles of produce direct hydrogen, electrical energy production using hybrid cycles or water electrolysis process and direct water electrolysis. The following is the classification of geothermal energy source:

- Geothermal energy source
  - Thermal energy (heat)
    - Thermochemical cycles
      - Hybrid cycles
      - Direct hydrogen
    - Electrical energy
      - Hybrid cycles
      - Electrolysis
    - Electrolysis
  - Direct source

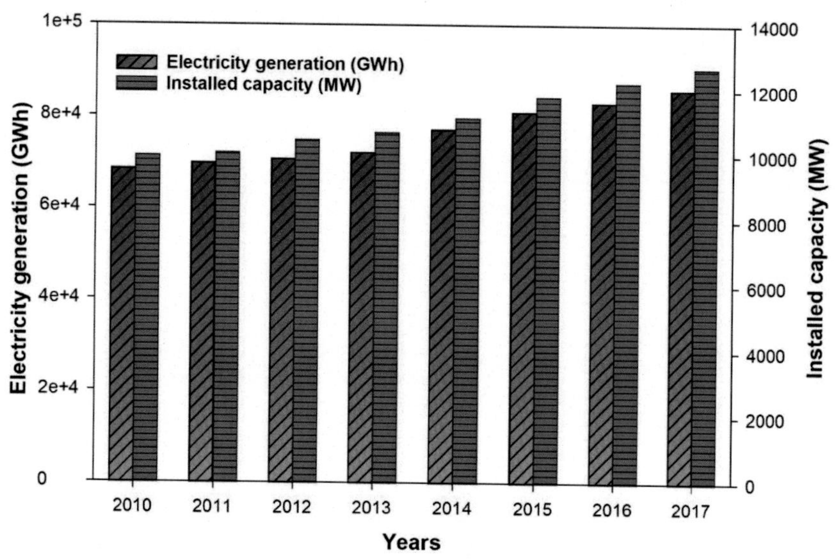

FIG. 5.2

Installed geothermal energy capacity and electricity generation trends.

Data from [97].

5.2 Geothermal Power Plants

It is realized while digging a hole down the earth that the temperature keeps on rising with the depth and the reason is the heat inside the earth that is employed to extract heat or electricity in terms of geothermal energy. Geothermal power plants extract electricity by employing the hot steam to the turbine that generates the mechanical work, and the generator converts it into electrical power. The produced electricity can be employed to provide heating, cooling, and power along with other practices. In the meantime, geothermal heat pumps are employed for fluid circulation using underground pipes where heat is absorbed.

Geothermal energy can primarily be extracted using two different techniques:

- Geothermal power plants
- Geothermal heat pumps

To construct a geothermal power plant, 1 to 2 miles deep wells are drilled into the earth to pump hot water or steam to the surface depending upon the application. The geothermal power plants are feasible to be constructed in the areas having some geysers, hot springs, or volcanic activity as the earth is predominantly hot right underneath the earth's surface at these places. Geothermal power plants employ the heat from underneath the earth to produce steam that is used to extract electricity.

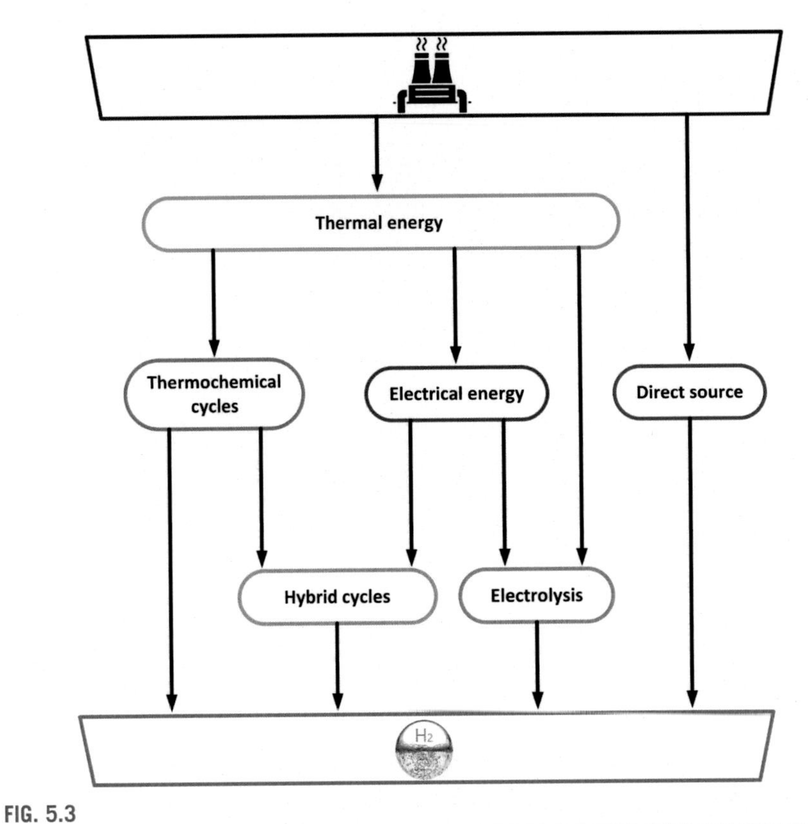

FIG. 5.3

Classification of geothermal energy utilization for hydrogen production.

Figs. 5.4a and 5.4b display the general demonstrations of the geothermal power plants with reinjection and without reinjection. Geothermal areas are also dependent upon the hot water reservoir and volume carried out that should be reinjected back to the earth that makes geothermal energy a sustainable resource. The sustainable geothermal power production employed for hydrogen production offers clean, renewable, sustainable, consistent, and environmentally benign solutions for hydrogen production. The geothermal energy-based hydrogen production system undergoes zero GHG emissions. Fig. 5.4a shows the general schematic of the geothermal power plant with reinjection for power generation while Fig. 5.4b exhibits the schematic of a geothermal power plant without reinjection. The generated power using geothermal power plant is employed to the electrolyzer (PEM/ALK/SOE) for water splitting, and the produced hydrogen is stored in the storage tank.

Geothermal power plants employ hydrothermal resources that consume both thermal energy (heat) and water (hydro). The steam temperature of higher than

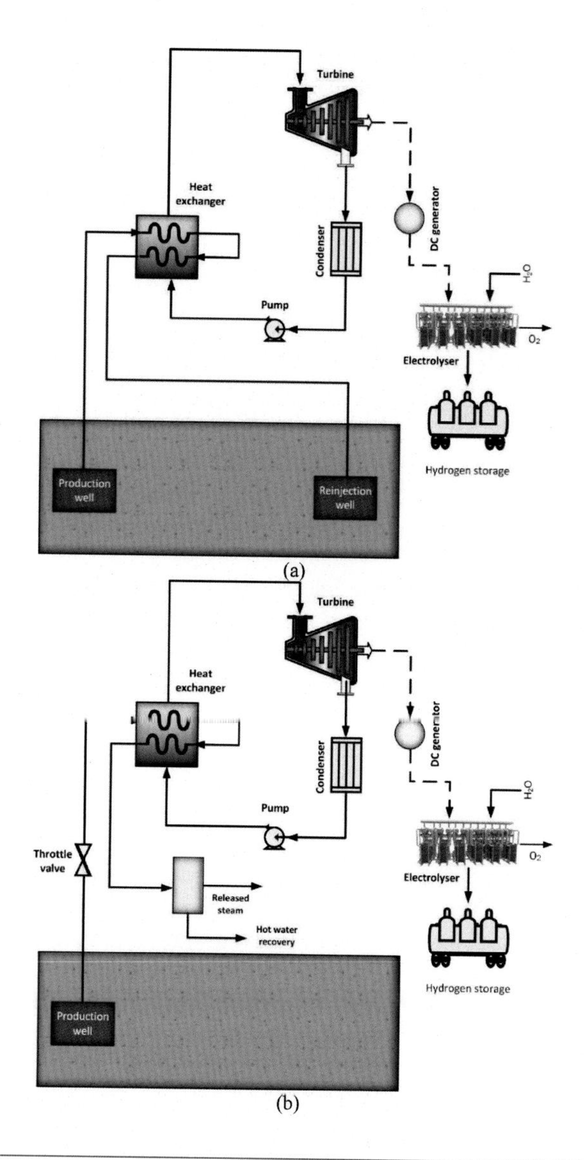

FIG. 5.4

Demonstration of geothermal power plants (a) with reinjection (b) without reinjection.

150°C or greater is required from hot water or dry steam wells to turn the turbines effectively and produce an efficient amount of electricity employing geothermal energy. The constructing undergoes deep digging to extract hot water or steam, and pressure is dropped when water approaches the surface that converts water into steam. The produced steam is employed to the turbine to generate mechanical work and the generator converts it into electricity. It is significant to enlist the steps followed to generate electrical power through geothermal power plants. Following are the significant steps that are followed to generate electricity using geothermal power plants:

- Hot water that is extracted by deep digging underneath the earth is pumped through a well under high pressure.
- The pressure of the hot water or steam is dropped when water approaches the surface that converts water into steam.
- The steam is employed to the turbine that rotates and produces mechanical work, and a generator produces electricity.
- A cooling tower is followed by the turbine that is employed to cool down the steam back into the water.
- The cooled water from the cooling tower is pumped back to continue the process.

5.3 Types of Geothermal Power Plants

Geothermal power plants generate electricity for the chosen facilities using geothermal energy. To get the benefit of the heat underneath the earth, the deep digging is carried out to extract hot water or steam and pressure is dropped when water approaches the surface that converts water into steam. This steam is employed to the turbine that produces mechanical work and coupled generator converts it to the electrical power. The majority of the power plants employ fossil fuels for boiling that causes numerous environmental emissions while geothermal power plants utilize the steam generated from hot water reservoirs that are found underneath the earth's surface.

A recent study[98] conducted the modeling and economic analysis of optimum pipe diameter using energy and exergy analysis of the geothermal district heating system. The system under consideration consisted of three significant cycles of the Bariskent region, transportation network and Danistay region in Turkey. The thermal capacities and supply and return temperatures of the mentioned regions were 7975 kW and 21,025 kW and 80°C and 50°C, respectively. Based on the transportation network assessment, minimized cost was found to be 561,856.9 USD/year for 300 DN nominal diameter. The exergy losses were revealed to be 1.94% of exergy input to the designed system for the distribution network. The overall exergy and energy efficiencies of the designed system were revealed to be 50.12% and 40.21%.

The development and analysis of a geothermal energy integrated solar humidification–dehumidification desalination system were proposed in a recent study.[99] They investigated both the economic and technical feasibility of employing a combination of solar and geothermal energy sources in a humidification–dehumidification desalination system. The novel humidification–dehumidification system was an improved solar and geothermal combined system. The geothermal water tank temperature ranged from 60 to 80°C that reproduces low-grade geothermal energy. The experimental results revealed the freshwater cost to be 0.003 USD/L.

Geothermal power plants may generally be classified into three significant types as follows:

- Dry steam power plants
- Flash steam power plants
- Binary cycle power plants

5.3.1 Dry Steam Power Plants

Dry steam power plants are drawn from underground steam resources. In dry steam power plants, steam passes straight through underground wells and is employed directly to the turbine coupled with a generator. Geothermal power plants employ hydrothermal resources that consume both thermal energy (heat) and water (hydro). The steam temperature of higher than 150°C or greater is required from hot water or dry steam wells to turn the turbines effectively and produce an efficient amount of electricity employing geothermal energy. The steam is employed to the turbine that drives the generator to generate electrical power. The steam extracted from the geothermal energy eradicates the fossil fuel burning to operate the turbine as well as avoids fuel transportation and storage requirements. These dry steam power plants produce excess steam and negligible quantities of gases that exist underneath the earth's surface.

The dry steam power plant system was the very first geothermal power production plant built in Italy at Lardarello in 1904. The world's largest geothermal power plant is effective at The Geysers in northern California.

Principally, dry steam power plant stations employ the steam flowing outside the geo-deposits that heats the secondary fluid to operate the turbines and produce power. The extracted steam appears at a temperature of approximately 150°C that is sufficient enough to operate the geothermal plant turbine.

5.3.2 Flash Steam Power Plants

Flash steam power plants are known as the most common geothermal power plants. They employ geothermal water reservoirs offering temperatures more than 182°C. This hot water moves up from the wells underneath the ground in terms of its pressure. As this hot water moves upward, the drop in pressure causes the hot water to turn into steam. The water is separated from steam and separated steam is employed to the turbine/generator to generate power. The condensed steam and leftover water are reinjected into a reservoir that helps in continuing the process and makes this a sustainable resource.

Flash steam power plant forces water into the injection well through a groundwater pump. The injection well is required to be sunk sufficiently deep to approach underground rocks at a higher temperature than water boiling point and filters through the rocks to extract heat and rises back through production well. This hot water reaches a flash tank where it flashes into vapor due to the reduced pressure.

The remaining liquid water in the flash tank is pumped back to the earth again. The water vapors are employed to drive the steam turbine that operates the electric generator to produce electrical power. The exit stream of the turbine then reaches to the condenser where it gets cooled. The condenser converts the water vapors into a liquid state and is pumped back to the earth again with the remaining liquid water in the flash tank. The condensed vapors can also be utilized for irrigation and drinking as it becomes distilled. The flash tank needs to be cleaned and flushed periodically to eliminate the buildup of mineral. If production well water contains high content of minerals, the flushing and cleaning of the flash tank need to be carried out more frequently.

5.3.3 Binary Cycle Power Plants

Binary cycle geothermal power plants are designed to operate at comparatively lower temperatures ranged from 107 to 182°C. These plants extract the water heater to boil an organic compound that offers a comparatively low boiling point. A heat exchanger is employed that vaporizes the organic compound that drives the turbine. The turbine operates to generate mechanical work and the generator converts mechanical energy into electrical. Water is injected back underneath the ground where it gets reheated again. Both, working fluid and water do not have a direct contact to avoid air emissions.

Binary cycle power plants are differently designed as compared with flash and dry steam power plants as steam or hot water from the geothermal reservoir does not come in direct contact with the turbine and is only used to evaporate the organic compound that is used as working fluid. The steam or hot water is employed to the heat exchanger that causes the organic compound to flash into vapors that are employed to the turbine and generator unit to produce electrical power.

These power plants undergo a closed-loop system and practically only water is released to the environment. Small-scale power plants can be employed for widespread applications, especially in rural areas. Energy distributed resources denote a number of small modular power-generating techniques that can collectively make the electricity generation system further efficient.

The recent methods take benefit of the boiling point differences that are evaluated using density. This type of geothermal power plant employs a working fluid that offers a considerably low boiling point as compared with water and steam or hot water is employed to vaporize the secondary fluid owing to the low boiling point, and this secondary fluid drives the turbines and generator to produce electricity.

5.4 Geothermal Heat Pumps

A geothermal or ground source heat pump is a central cooling and/or heating system that is employed to transfer heat to or from the ground. It employs and benefits from the earth without any intermittency. In summer, it works as a heat sink and in winter, it operates as a heat source. During the winter season, the refrigerant or water employed absorbs thermal energy from earth and the pump conveys this extracted heat to the facility above while during summer, some heat pumps operate in reverse and provide cooling to the facilities.

Geothermal energy is not only employed to generate power using geothermal power plants but geothermal heat pumps are also constructed that can be employed for various applications, such as heating and cooling small residential homes to the massive infrastructures. Geothermal heat pumps employ a specific fluid type for heat transfer using pipes that are installed just underneath the earth's surface where the temperature ranges from 10 to 15°C. Fig. 5.5 shows the basic demonstration of the geothermal heat pump for cooling and heating. During the winter season, the geothermal heat pump is used to provide heating while it operates in reverse during the summer season.

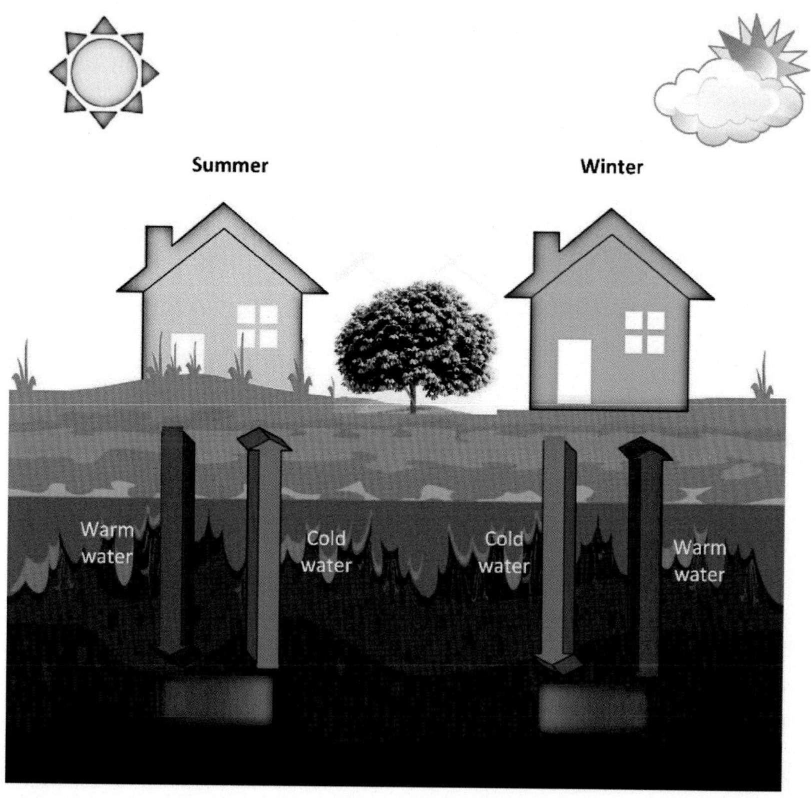

FIG. 5.5

Demonstration of the geothermal heat pump for cooling and heating.

It is also substantial to enlist the significant steps followed to extract cooling or heating through geothermal heat pumps. Following are the significant steps that are followed to produce cooling or heating:

- Refrigerant or water circulates through a pipe loop.
- In the winter season, the refrigerant or water gets heated up as it circulates through the pipe loop buried underground.
- Upon the arrival of the warmed refrigerant on water back above ground, the warmed fluid transfers heat to the facility.
- The refrigerant or water transfers the heat and cools down. After this, it is reinjected back inside the ground to continue the process again.
- In the summer season, the system operates in reverse. The refrigerant or water cools the facility and reinjected underground.

5.5 Types of Geothermal Heat Pumps

Geothermal or ground-source heat pumps have been functional since the 1940s. These systems benefit from the constant earth temperature rather than outside air-temperature as an exchange medium. The heat pumps are classified into different categories and the most suitable is picket up depending on the climate, available land, soil conditions, and local site installation costs. All of these approaches can be used for residential and commercial building applications. The two significant types of geothermal heat pumps that are closed-loop and open-loop heat pumps and subtypes are classified as follows:

- Closed-loop
  - Vertical
  - Horizontal
  - Lake/pond
- Open-loop
- Hybrid Systems

5.5.1 Closed-Loop Systems

Three different types of closed-loop systems are described in this section.

Horizontal

The installation of the horizontal heat pumps is usually the most cost-effective approach for residential installations and predominantly for new installation with sufficient land availability. The horizontal heat pumps need approximately 4 ft deep trenches. This type comprises the two most common designs. One design employs one pipe buried at 4 ft and other at 6 ft, while the second design employs the two pipes that are positioned at 5 ft and side by side in 2-ft wide trench.

Vertical

The schools and large commercial buildings mostly employ vertical heat pumps because of the prohibited land area requirement for horizontal heat pumps, and these specific types of heat pumps is also employed at the locations of shallow soil for trenching to minimize the existing landscaping disturbance. To construct a vertical heat pump system, around 100–400 ft deep holes are drilled approximately 20 ft apart and four inches in diameter. The two pipes are installed in these holes connecting through U-bend for looping at the bottom. The constructed vertical loops are attached with horizontal pipes that are positioned in channels and coupled with heat pumps in the facility.

Pond/lake

The locations or sites having a suitable water body can offer the cheapest option in terms of cost. A pipe supply line is route underground connecting the building to the water and looped coils into circles no less than 8 ft beneath the surface to prevent and avoid freezing. The looped coils must only be positioned in source water that satisfies the minimum depth, volume, and quality criteria.

5.5.2 Open-Loop System

The open-loop systems employ a surface body or well water as a fluid to exchange heat that circulates straight through the geothermal heat pump system. As soon as it is circulated through the heat pump system, the water moves back to the recharge well to return to the ground. The open-loop system options are practical and constructed only at the sites having moderately clean water supply, and all resident codes, limitations, and regulations are met concerning the groundwater discharge.

5.5.3 Hybrid Systems

The hybrid system employs a combined geothermal resource and outside air or more than a few different geothermal resources that are an alternative technological option. These hybrid systems are predominantly effective and efficient at the facilities requiring significant cooling as compared with heating requirements. Standing column well offers another alternative depending on if local geology permits. In such open-loop systems, water moves the through standing column bottom and returns to the top using deep vertical wells.

It is significant to explore the continuous growth in geothermal energy consumption. Fig. 5.6 displays the direct use of global geothermal energy and predictions from 2012 to 2024. This figure does not only exhibit the data of the global geothermal energy until today but also provides future predictions until 2024 provided by International Energy Agency (IEA). The geothermal energy usage is classified depending on the applications into three significant categories of building, industry and agriculture all these usages are drawn on the primary *y*-axis in exajoules while percentage geothermal heat is displayed on the secondary *y*-axis. It can be

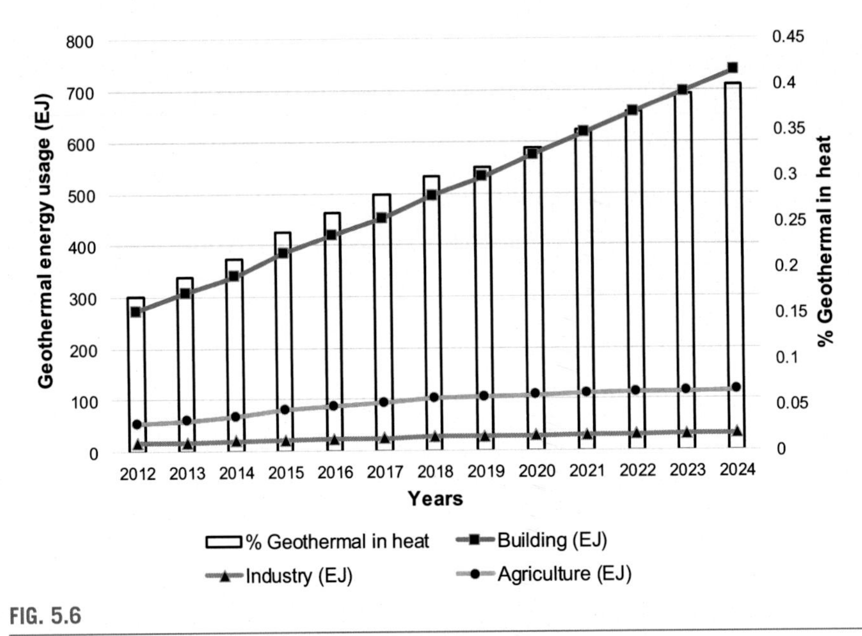

FIG. 5.6

Direct use of global geothermal energy and predictions from 2012 to 2024.

Data from [101].

depicted that from 2012 to 2024, the geothermal energy usage in buildings is expected to rise from 272 to 737 EJ, the geothermal energy usage in the industry is expected to rise from 16 to 33 EJ, the geothermal energy usage in agriculture is expected to rise from 53 to 116 EJ, and the percentage geothermal heat is expected to rise from 0.17% to 0.4%.

A recent study[100] conducted the investigation and performance assessment of a geothermal energy-assisted hydrogen production system employing chlor-alkali cell. In this research study, the significant components were investigated including steam turbine, separator, organic Rankine cycle, condenser, chlor-alkali cell, and NaCl solution reservoir. To low-grade heat from ORC was employed to heat NaCl solution to improve the cell performance of the cell. Consequently, the designed integrated system supplied electricity to the grid and three significant outputs of chlorine, hydrogen, and sodium hydroxide. The parametric studies revealed that with a rise in geothermal temperature from 140 to 155 °C, power generation rises from 2.5 to 3.9 MW, and production of hydrogen rises from 10.5 to 21.1 kg/h. Thus, the exergetic and energetic efficiencies were revealed as 22.4% and 6.2% at 155°C temperature.

A comprehensive review paper targeted the geothermal energy-driven hydrogen production system in a recent study.[47] Hydrogen production is currently carried out using many different methods using conventional and alternative energy resources, such as natural gas, coal, wind, biomass, solar and geothermal energy. Hydrogen has been gaining the industrial and research institution's attention over the last decade. Hydrogen production is a promising candidate for complete renewable energy utilization. In this review study, a summary of the current status and advancements in hydrogen production techniques was followed by the detailed research linked with geothermal energy-based hydrogen production, and the process description as well as technical, environmental, and economic aspects of geothermal energy-driven hydrogen production was addressed. Lastly, a detailed environmental and cost comparative evaluation was performed on diverse energy sources for hydrogen production. The results revealed that geothermal-assisted production of hydrogen-based electrolysis was found the most promising and cost-effective method in comparison with other conventional and alternative energy resources.

A geothermal heat pump gets benefits out of the constant high-temperature of the upper 3 m of earth surface to provide the heat to the facility in winter while in summer, it extracts heat from the facility and transfers it back to the comparatively cooler ground while geothermal power plant employs geothermal water reservoirs offering high-temperatures. This hot water moves up from the wells underneath the ground in terms of its pressure. As this hot water moves upward, the drop in pressure causes the hot water to turn into steam. The water is separated from steam, and separated steam is employed to the turbine/generator to generate power.

Fig. 5.7 displays the installed geothermal capacity around the globe in 2018. The installed geothermal capacity was found in the significant global countries and regions of Austria, Chile, China, Costa Rica, El Salvador, Ethiopia, France, Germany, Guadeloupe, Guatemala, Honduras, Hungary, Iceland, Indonesia, Italy, Japan, Kenya, Mexico, New Zealand, Nicaragua, Other Asia and Pacific, Papua New Guinea, Philippines, Portugal, Russia, Total Africa, Total Asia Pacific, Total CIS, Total Europe, Total North America, Total South and Central America, Turkey, and the United States.

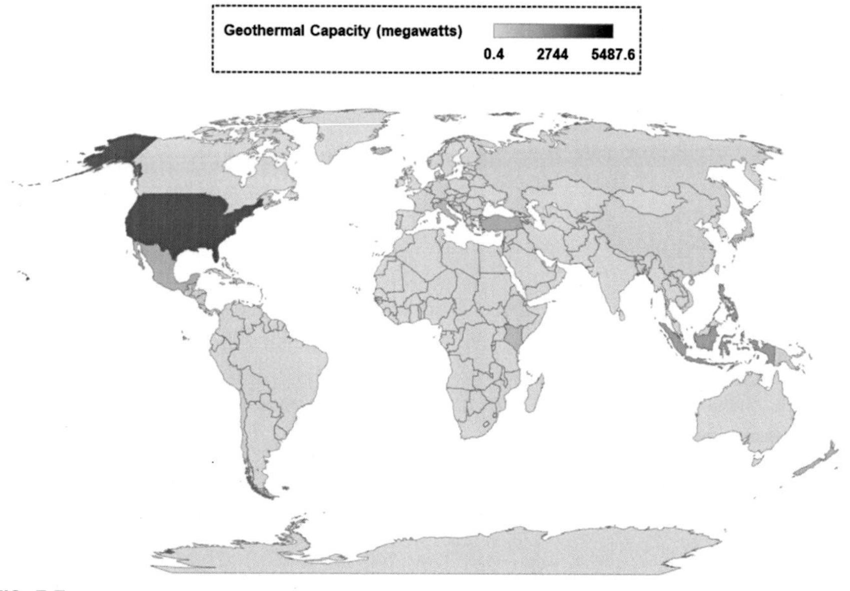

FIG. 5.7

Installed geothermal capacity around the globe in 2018.

Modified from [102].

5.6 Flashing Types of Geothermal-Assisted Hydrogen Production Plants with Reinjection

A recent study[103] conducted the thermodynamic analysis and assessment of geothermal energy-assisted production of hydrogen using proton exchange membrane (PEM) electrolyzer. The work is generated using a geothermal-driven organic Rankine cycle, and output power is employed to the water electrolysis process for hydrogen production and waste geothermal water is used to preheat the electrolysis input water. The geothermal water and water electrolysis temperature effects on the rates of hydrogen production were investigated and found proportional. A geothermal resource was considered at 160°C with a flow rate of 100 kg/s that was capable of generating 3810 kW power that is fed to the electrolyzer. The electrolyzer input water was preheated at 80°C using heat from waste geothermal water and the hydrogen production rate was found to be 0.034 kg/s. The energetic and exergetic efficiencies of the binary geothermal power plant and overall system were revealed as 11.4% and 45.1% and 6.7% and 23.8%, respectively.

In a recent study[104], geothermal energy-assisted production of hydrogen and liquefaction was investigated. This study proposed six different models for the production of hydrogen and liquefaction and investigated the performance of each designed model using a thermodynamic approach to maximize the rate of hydrogen production and minimize the geothermal energy usage, and geothermal water temperature effect on the performance of designed models was investigated. The first model employed geothermal output power to the water electrolysis process, the second model used partial geothermal heat to generate electrical work for electrolysis and remaining heat was employed to heat electrolysis input water, the third model employed the geothermal heat to absorption refrigeration cycle for precooling of gas before the liquefaction cycle, the fourth model employed partial geothermal heat to absorption refrigeration for precooling of hydrogen and remaining heat was employed to generate work for liquefaction cycle, the fifth model employed geothermal output power for liquefaction cycle, and the sixth model employed partial geothermal power for water electrolysis and remaining for liquefaction.

A recent study[105] conducted the exergy analysis on the varying flashing geothermal power plants to investigate the performance. Flash steam power plants evolve a substantial share of global geothermal power. In this research study, single to quadruple flashing steam geothermal power plants were investigated undergoing reinjection options as reinjecting the water back makes the geothermal energy resource sustainable. The optimum operating points were found explicitly using optimum flashing pressures along with turbine output power and energetic and exergetic efficiencies. The output power was found to be rising with the increase in flashing stages from single to double while turbine output power was dropped significantly with the rise in stages from double to quadruple. Both, exergetic and energetic efficiencies were found to decreases from 72.6% to 69.8% and 28% to 23.5% for single to quadruple flashing options.

The strategy to depict the optimum reinjection is dependent on the geothermal system type. The vapor-dominated systems that need continuous water supply must not run out of water and reinjection must be infield. Whereas for liquid-dominated two-phase and hot water systems of low and medium enthalpy, reinjection is probable to include a combination of the outfield and infield injection. The infield reinjection offers pressure support and decreases the subsidence potential, while outfield reinjection lessens the cold water returning risk to the production area. Deep reinjection can be carried out to reduce the hazard of ground surface inflation and groundwater contamination. The infield to outfield reinjection proportion and shallow or deep location varies case to case and characteristically the rate of infield reinjection varies with time as part of the steam field managing strategy.

5.6.1 Single-Flash Geothermal-Assisted Hydrogen Production Plant

The schematic displaying the single-flash geothermal-assisted hydrogen production is shown in Fig. 5.8. The geothermal fluid arrives the system from the production well in a saturated liquid state and flashed to a low pressure through a throttle valve. Even though the flashing reduces temperature and pressure, the quality of geothermal fluid rises. Consequently, the flashing output consists of a saturated vapor—liquid mixture that passes through a separator to separate vapors from the liquid state. The separated liquid from the separator is reinjected to the reinjection well while the saturated vapors exiting the separator are fed to the steam turbine to produce mechanical work, and the accompanied generator converts the mechanical work into electrical power. The electrical work generated using the steam turbine and generator is employed to the PEM water electrolysis process along with water, and electrolyzer splits water into its constituent using electrical work. The produced hydrogen is stored in the hydrogen storage tank. The steam turbine output stream leaves at lower temperature and pressure consisting of saturated vapor mixture and fed to the condenser. The condenser helps the saturated vapor mixture to release heat and converts saturated vapors to the saturated liquid. The resulting saturated liquid is then reinjected back to the reinjection well to reheat the water again. The reinjection makes the geothermal energy resource reliable, consistent, and sustainable.

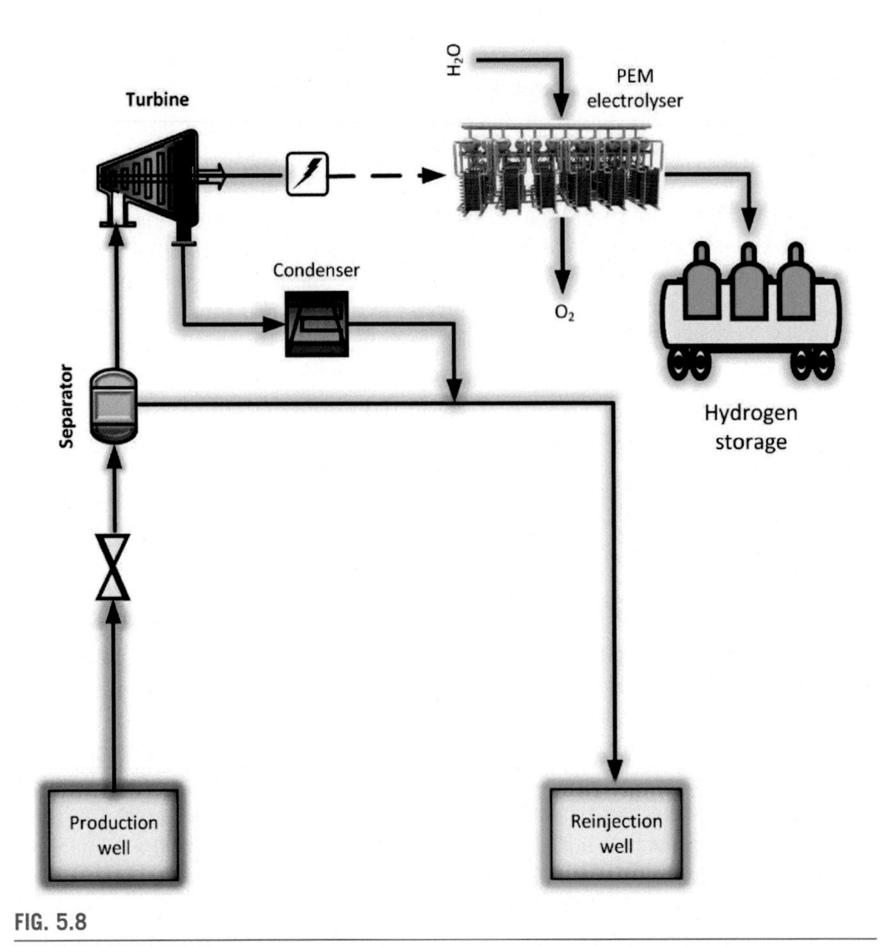

FIG. 5.8

Schematic displaying the single-flash geothermal-assisted hydrogen production.

5.6.2 Double-Flash Geothermal-Assisted Hydrogen Production Plant

Double-flash geothermal-assisted hydrogen production configuration is displayed in Fig. 5.9. In the double-flash type geothermal plant, the separated saturated liquid from separator 1 is employed again to the throttle valve 2 and flashed again to a lower temperature and pressure. This flashing converts the saturated liquid into a saturated vapor–liquid mixture, and this mixture is then fed to separator 2 to separate vapors from the liquid. The separated liquid from the separator 2 is reinjected back to the reinjection well while the saturated vapors exiting separator 2 are fed to steam turbine 2 to produce mechanical work and thus electrical work through generator. The electrical work generated using the steam turbine 2 is employed to the PEM water electrolysis process along with water, and PEM electrolyzer 2 splits water into its constituent using electrical work.

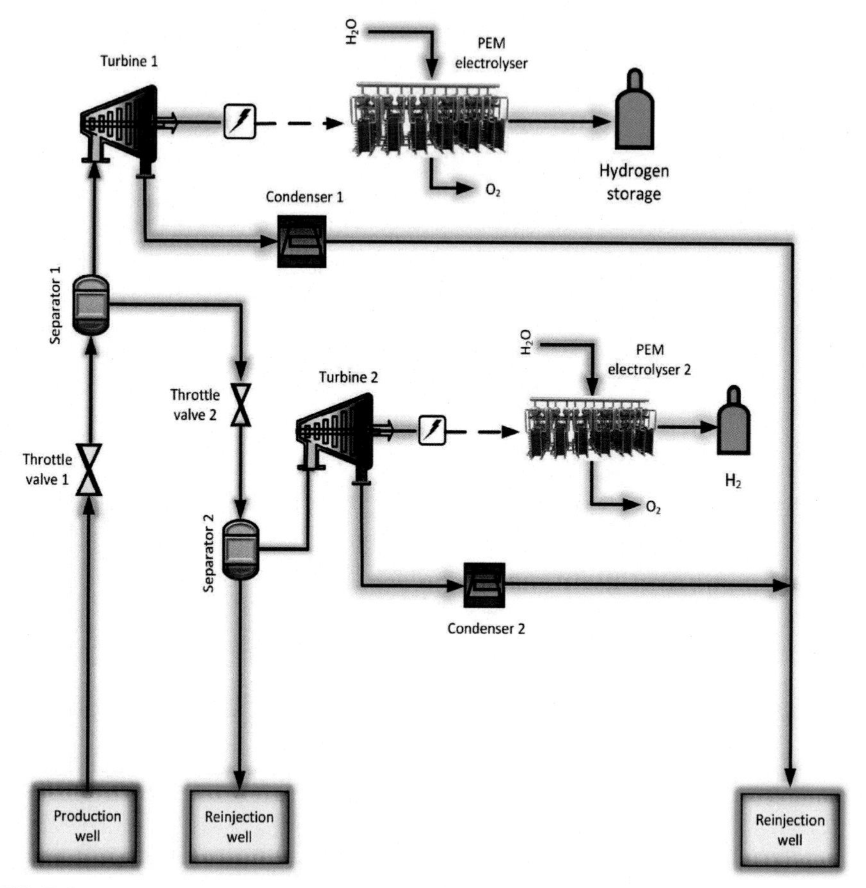

FIG. 5.9

Schematic displaying the double-flash geothermal-assisted hydrogen production.

The produced hydrogen is stored in the hydrogen storage tank. The output stream of steam turbine 2 leaves at lower temperature and pressure consisting of saturated vapor mixture and fed to the condenser 2. The condenser 2 helps the saturated vapor mixture to release heat and converts saturated vapors to the saturated liquid. The resulting saturated liquid is then mixed with the saturated liquid leaving the condenser 1 and reinjected back to the reinjection well to reheat the water again.

5.6.3 Triple-Flash Geothermal-Assisted Hydrogen Production Plant

The configuration displaying the triple-flash geothermal-assisted hydrogen production is shown in Fig. 5.10. In the triple-flash type geothermal plant, the separated saturated liquid from separator 2 is employed again to the throttle valve 3 and flashed

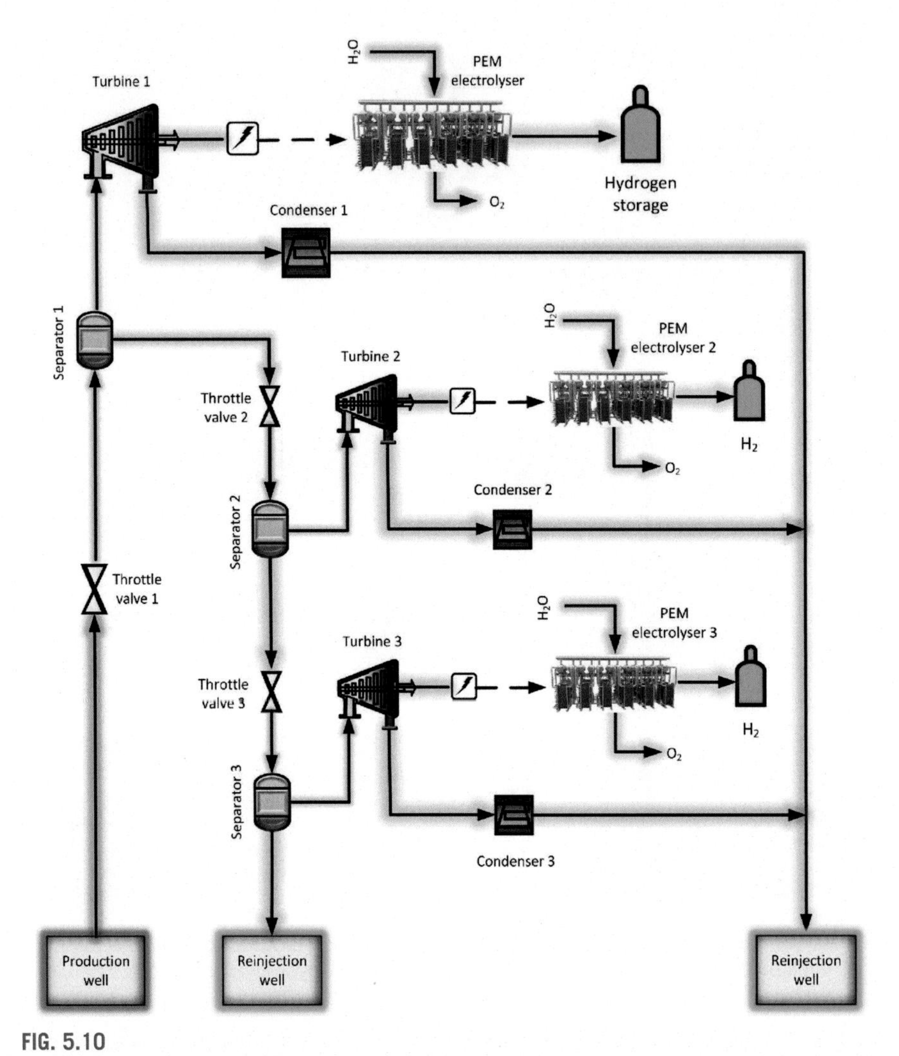

FIG. 5.10

Schematic displaying the triple-flash geothermal-assisted hydrogen production.

again to a lower temperature and pressure. This flashing converts the saturated liquid into a saturated vapor–liquid mixture, and this mixture is then fed to separator 3 to separate vapors from the liquid. The separated liquid from separator 3 is reinjected back to the reinjection well, while the saturated vapors exiting separator 3 are fed to the steam turbine 3 to produce additional mechanical work and thus electrical work through generator.

The electrical work generated using the steam turbine 3 is employed to the PEM water electrolysis process along with water and PEM electrolyzer 3 splits water into its constituent using electrical work. The produced hydrogen is stored in the hydrogen storage tank. The output stream of steam turbine 3 leaves at lower temperature and pressure consisting of saturated vapor mixture and fed to the condenser 3. The condenser 3 helps the saturated vapor mixture to release heat and converts saturated vapors to the saturated liquid. The resulting saturated liquid is then mixed with the saturated liquid leaving the condenser 1 and 2 and reinjected back to the reinjection well to reheat the water again.

5.7 Case Study 5

It is significant to design a case study to investigate the geothermal energy-assisted hydrogen production system. The global reinjection experience in geothermal power plants and heat pumps is reviewed. The information of 91 electric generating geothermal power plants showed that a reinjection strategy must be developed as soon as possible on the geothermal field with flexible options. The reinjection makes the geothermal energy a reliable, consistent, and sustainable energy source as water is pumped back to the reinjection well and water is heated again to continue the process.

5.7.1 Description

Fig. 5.11 displays the case study for geothermal-assisted hydrogen production with reinjection. The geothermal fluid arrives at the system from the production well in the saturated liquid state and flashed to a low pressure through a flashing chamber. The flashing reduces the water pressure and thus, converts water to the saturated vapor–liquid mixture. This mixture is fed to the separator to separate vapors from the liquid state. The separated liquid from the separator is reinjected to the reinjection well while the saturated vapors leaving the separator are employed to the steam turbine to generate electrical power. The electrical work harvested through the steam turbine is used to split water into its constituents using the PEM water electrolysis process, and the produced hydrogen is stored in the hydrogen storage tank.

The output of the steam turbine exits at low temperature and pressure and consists of a saturated vapor mixture. This mixture is then fed to the condenser that releases the heat from the saturated vapor mixture and converts the saturated vapors to the saturated liquid. The exiting saturated liquid from the condenser is then reinjected to the reinjection well where water gets reheated again. It is substantial to reinject the geothermal fluid back to the ground to reheat the water to continue the process, and reinjection makes the geothermal energy source sustainable and consistent.

Table 5.1 displays the input parameters and design constraints of the geothermal energy-assisted hydrogen production system. The significant parameters of the geothermal power plant, namely geothermal well temperature, geothermal fluid mass flow rate, geothermal fluid inlet condition, geothermal fluid flashing pressure,

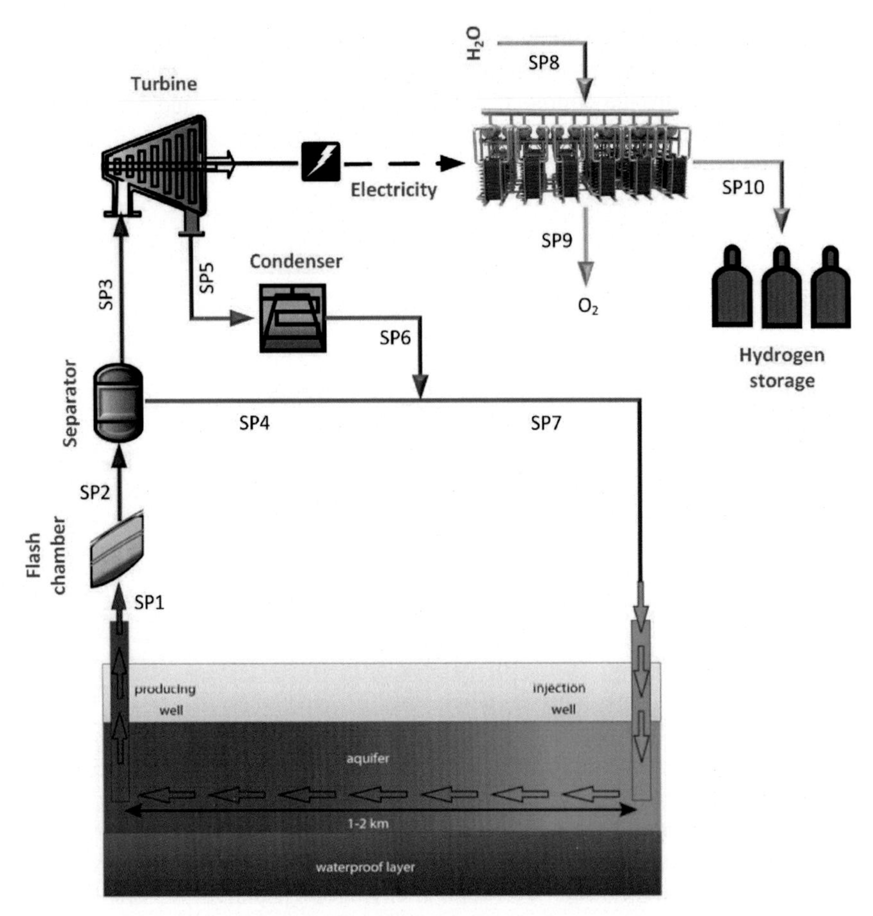

FIG. 5.11

Case study for geothermal-assisted hydrogen production with reinjection.

and geothermal fluid inlet pressure along with the significant parameters of the water electrolysis system, namely membrane thickness, anodic preexponential, cathodic preexponential, Faraday constant, electrolysis temperature, anodic activation energy, and cathodic activation energy are presented in the table.

5.7.2 Analysis

The design model equations of the PEM electrolyzer are described in the previous chapter in detail. The analysis of this case study includes the thermodynamic equations of each significant component of the designed case study. The thermodynamic equations for each component of the geothermal-assisted hydrogen case study are as follows:

Table 5.1 Design Constraints and Input Parameters.

Input Parameter	Value
Geothermal Energy Source	
Geothermal well temperature	230°C
Geothermal fluid mass flow rate	100 kg/s
Geothermal fluid inlet condition	Saturated liquid
Geothermal fluid flashing pressure	650 kPa
Geothermal fluid inlet pressure	3347 kPa
Ambient pressure	101 kPa
Ambient temperature	25°C
Isentropic efficiency of turbine	80%
Generator efficiency	80%
PEM Water Electrolysis	
Membrane thickness	100 μm
Anodic preexponential $\left(J_a^{\mathrm{ref}}\right)$	17×10^4 A/m^2
Cathodic preexponential $\left(J_c^{\mathrm{ref}}\right)$	46×10 A/m^2
Faraday number	96486 C/mol
Electrolysis temperature	80°C
Anodic activation cnergy E_{act_a}	76000 J/mol
Cathodic activation energy E_{act_c}	18000 J/mol

Flash Chamber

The mass, energy, entropy, and exergy balance equations for the flash chamber are as follows:

$$\dot{m}_1 = \dot{m}_2 \tag{5.1}$$

$$\dot{m}_1 h_1 = \dot{m}_2 h_2 \tag{5.2}$$

$$\dot{m}_1 s_1 + \dot{S}_{\mathrm{gen}} = \dot{m}_2 s_2 \tag{5.3}$$

$$\dot{m}_1 ex_1 = \dot{m}_2 ex_2 + \dot{E}x_{\mathrm{dest}} \tag{5.4}$$

Separator

The mass, energy, entropy and exergy balance equations for the separator are as follows:

$$\dot{m}_2 = \dot{m}_3 + \dot{m}_4 \tag{5.5}$$

$$\dot{m}_2 h_2 = \dot{m}_3 h_3 + \dot{m}_4 h_4 \tag{5.6}$$

$$\dot{m}_2 s_2 + \dot{S}_{\mathrm{gen}} = \dot{m}_3 s_3 + \dot{m}_4 s_4 \tag{5.7}$$

$$\dot{m}_2 ex_2 = \dot{m}_3 ex_3 + \dot{m}_4 ex_4 + \dot{E}x_{\mathrm{dest}} \tag{5.8}$$

Turbine

The mass, energy, entropy and exergy balance equations for the turbine are as follows:

$$\dot{m}_3 = \dot{m}_5 \tag{5.9}$$

$$\dot{m}_3 h_3 = \dot{m}_5 h_5 + \dot{W}_{\text{Turbine}} \tag{5.10}$$

$$\dot{m}_3 h_3 + \dot{S}_{\text{gen}} = \dot{m}_5 h_5 \tag{5.11}$$

$$\dot{m}_3 ex_3 = \dot{m}_5 ex_5 + \dot{W}_{\text{Turbine}} + \dot{E}x_{\text{dest}} \tag{5.12}$$

Generator

The isentropic efficiency of the turbine and generator efficiencies are assumed to be 80%. The turbine generates mechanical work that is converted to the electrical power using a generator. Therefore, the net electrical output from the generator can be determined using the product of generator efficiency and mechanical work.[106]

$$\dot{W}_{\text{el}} = \eta_{\text{generator}} \dot{W}_{\text{Turbine}} \tag{5.13}$$

Condenser

The mass, energy, entropy, and exergy balance equations for the condenser are as follows:

$$\dot{m}_5 = \dot{m}_6 \tag{5.14}$$

$$\dot{m}_5 h_5 = \dot{m}_6 h_6 + \dot{Q}_{\text{out}} \tag{5.15}$$

$$\dot{m}_5 h_5 + \dot{S}_{\text{gen}} = \dot{m}_6 h_6 \tag{5.16}$$

$$\dot{m}_5 ex_5 = \dot{m}_6 ex_6 + \dot{Q}_{\text{out}} \left(1 - \frac{T_o}{T}\right) + \dot{E}x_{\text{dest}} \tag{5.17}$$

Performance Assessment

The performance of the geothermal energy-assisted hydrogen production system is analyzed thermodynamically using energetic and exergetic efficiencies and following are the expressions of the efficiencies:

$$\eta_{\text{Geothermal}} = \left(\frac{\dot{m}_{H_2} \text{LHV}_{H_2}}{\dot{m}_1 h_1 - \left(\dot{m}_4 h_4 + \dot{m}_6 h_6\right)} \right) \tag{5.18}$$

$$\psi_{\text{Geothermal}} = \left(\frac{\dot{m}_{H_2} ex_{H_2}}{\dot{m}_1 ex_1 - \left(\dot{m}_4 ex_4 + \dot{m}_6 ex_6\right)} \right) \tag{5.19}$$

5.7.3 Results and Discussion

The geothermal energy-assisted hydrogen production system is analyzed using a thermodynamic approach. The performance of each component of the designed case study is important to be investigated. Flashing pressure affects the geothermal-assisted hydrogen production system significantly. Thus, it is substantial to investigate the effect of flashing pressure on different significant parameters, such as exergy destruction rate of the flash chamber, flash temperature, vapor quality, specific enthalpy of steam, turbine output power at different inlet source pressures, work and exergy destruction rates of turbine, and exergy destruction rate of flashing chamber.

Fig. 5.12 displays the flashing pressure effect on exergy destruction rate of the flash chamber and flash temperature. The system is designed to operate at 650 kPa flashing pressure. Thus, the range of flashing pressure is taken from 400 to 800 kPa to investigate the system operation under varying flashing pressures. Figure depicts that the increase in flashing pressure from 400 to 800 kPa causes the flashing temperature to rise from 143.6 to 170.4°C and exergy destruction rate of the flashing chamber drops from 2724 to 1274 kW.

The turbine specific power outputs for the geothermal energy-assisted hydrogen production plant for diverse combinations of flashing pressures and inlet source are shown in Fig. 5.13. The inlet source pressures are ranged from 100 to 2000 kPa, and this sensitivity analysis also helps in evaluating the optimum turbine output power

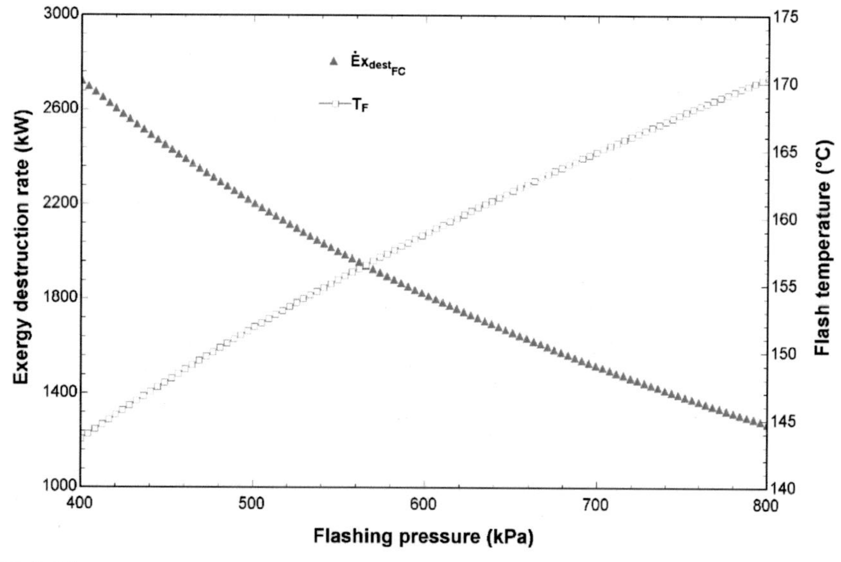

FIG. 5.12

Flashing pressure effect on exergy destruction rate of flash chamber and flash temperature.

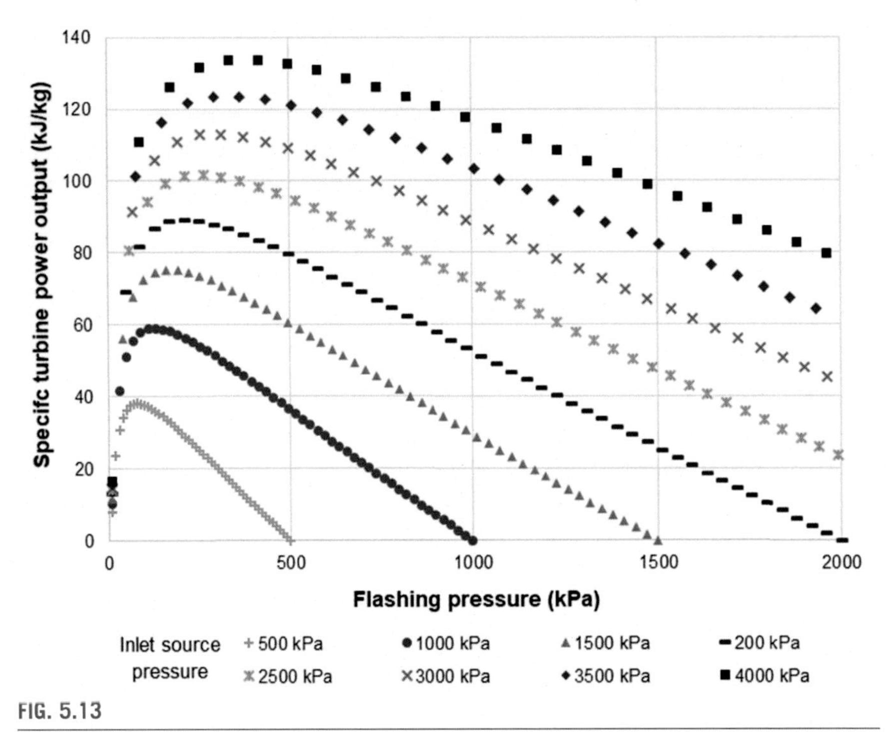

FIG. 5.13

Flashing pressure effect on turbine output power at different inlet source pressures.

Modified from [107].

for each case. The optimal flashing pressure can be depicted from the figure for each inlet source pressure value. For example, the optimal flashing pressure occurs nearly at 500 kPa for 3000 kPa inlet source pressure and the turbine work decrease with the rise or drop in the flashing pressures. Thus, using the optimal flashing pressure is recommended to generate the maximum steam turbine output power and to undergo minimum exergy destruction rates. As can be depicted from Fig. 5.13, an optimal flashing pressure exists for each inlet source pressure that offers the maximum turbine output power. At 4000 kPa inlet source pressure, the optimal value of flashing pressure is shown as 417.1 kPa that offers the maximum steam turbine specific power of 133.5 kW while at 3500 kPa inlet source pressure, the optimal value of flashing pressure is shown as 336.1 kPa that offers the maximum steam turbine specific power of 123.5 kW. Likewise, such analysis must be conducted to achieve the optimal flashing point for any geothermal well if the pressures are known.

The geothermal fluid flow rate affects the significant system parameters, such as steam turbine work rate and exergy destruction rate and hydrogen production capacity. Any parameter that causes the steam turbine work rate to vary varies the hydrogen production capacity as well because the water electrolysis process employs the power supplied by the geothermal energy-assisted steam turbine.

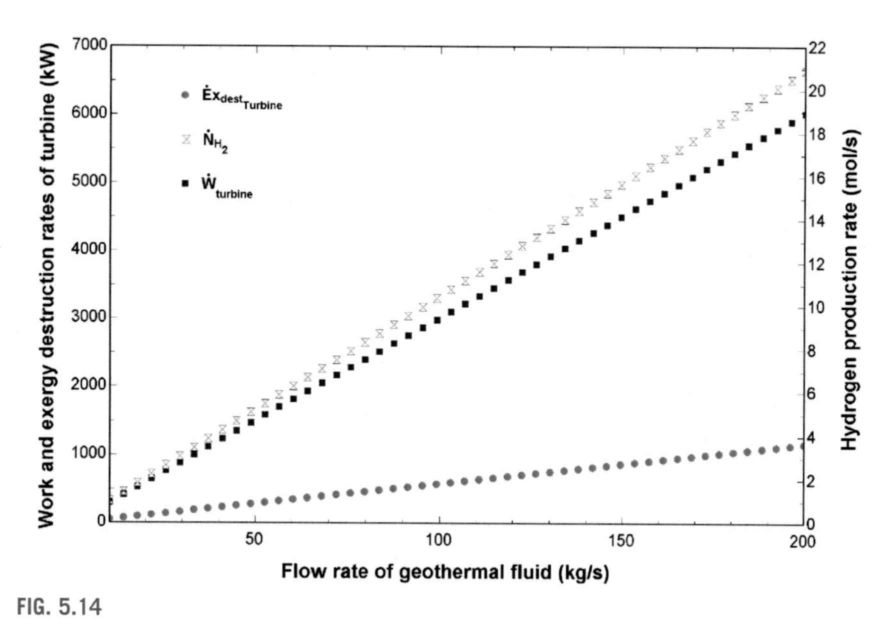

FIG. 5.14

Geothermal fluid flowrate effect on turbine work and exergy destruction rates and hydrogen production.

Fig. 5.14 exhibits the geothermal fluid flowrate effect on turbine work and exergy destruction rates and hydrogen production capacity. The system is designed to operate at 100 kg/s of geothermal fluid flowrate and the considered geothermal fluid flowrate range for sensitivity analysis is 10–200 kg/s to explore the system functionality under dissimilar conditions. The increase in geothermal fluid flowrate causes the work and exergy destruction rates of the turbine to rise from 301.1 to 6021 kW and 58.33 to 1167 kW, respectively, and hydrogen production capacity using PEM water electrolysis rises from 1.045 to 20.91 mol/s.

The effect of isentropic efficiency on the substantial parameters, such as turbine work and exergy destruction rates and hydrogen production. As isentropic efficiency causes the steam turbine work rate to vary, it affects the hydrogen production capacity as well. Fig. 5.15 displays the effect of isentropic efficiency on steam turbine work rate and exergy destruction rate and hydrogen production molar flow rate. The turbine is designed to operate at 0.8 isentropic efficiencies, and the considered isentropic efficiency range for sensitivity analysis is 0.75–0.90 to explore the system functionality under different operating efficiencies. The increase in isentropic efficiency from 0.75 to 0.90 causes the steam turbine work to rise from 2822 to 3387 kW and exergy destruction rate to drop from 729.2 to 291.7 kW and hydrogen production flow rate using PEM water electrolysis increases from 9.8 to 11.76 mol/s, respectively.

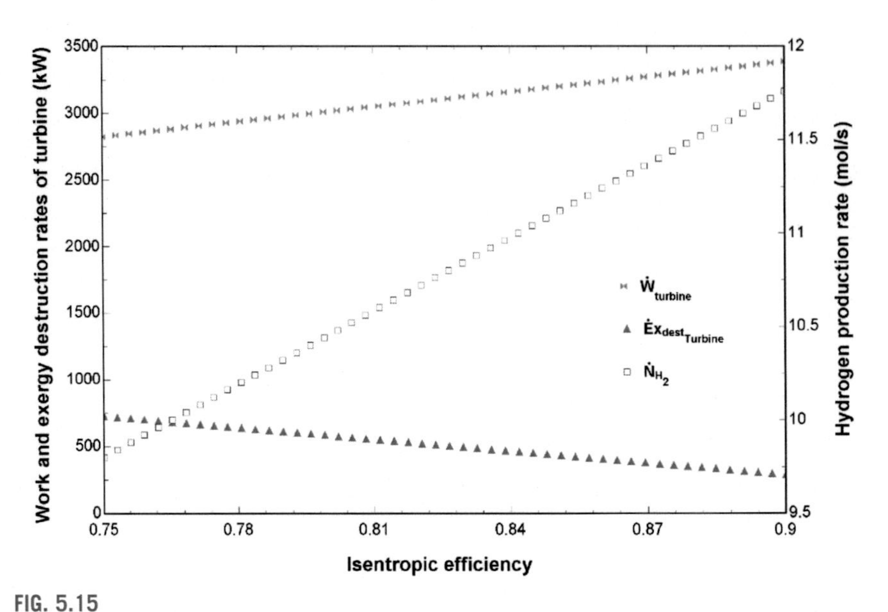

FIG. 5.15

Effect of isentropic efficiency on turbine work and exergy destruction rates and hydrogen production.

5.8 Closing Remarks

Geothermal energy is known as a heat driving source within the earth's subsurface. Contingent on the applications, geothermal energy can be employed for cooling and heating, power and also can be employed to produce clean and sustainable hydrogen. Geothermal energy is among the promising global energy resources to produce clean and environmentally benign hydrogen, and this promising route reduces GHG emissions as well. Geothermal energy is extracted in the form of clean, consistent, and sustainable heat from the Earth. Scientists, experts, and professionals consider geothermal energy as a promising, cleaner, consistent, efficient, environmentally benign, sustainable, and cost-effective solution for clean and sustainable hydrogen production as compared with fossil fuel burning that creates numerous harmful emissions causing global warming. Geothermal energy resource offers the cheapest, clean and sustainable hydrogen production solution followed by the wind energy source and gradually becoming cheaper, optimum, and cost-effective as compared with fossil fuel-fired energy.

Unlike solar and wind energy sources, the geothermal energy source is not intermittent by nature and thus, offers reliable, consistent, and sustainable energy solutions. Other than the amount of GHGs exist underneath the earth's surface that are mitigated to the earth's surface and atmosphere, the geothermal power plants or heat pumps undergo zero carbon emissions. The geothermal energy-assisted hydrogen production technology can play a significant role to make the global transition to 100% renewable energy possible. Geothermal-to-hydrogen stands as a promising route to produce clean and environmentally benign hydrogen and undergoing zero carbon emissions.

A case study is designed in this chapter to investigate the geothermal energy-assisted hydrogen production system. Numerous sensitivity analyses are conducted to explore the system performance under variable operating conditions. Geothermal power is employed in the water electrolysis process to produce clean hydrogen. A sensitivity analysis is conducted to evaluate optimum turbine output power at different steam pressures and flashing pressures. Thus, using the optimal flashing pressure is recommended to generate the maximum steam turbine output power that results in increased hydrogen produced capacities and minimized exergy destruction rates. The stored hydrogen can be employed to the fuel cells to generate power or can be used for diverse applications.

CHAPTER

Hydro Energy-Based Hydrogen Production

6

Hydroelectricity or hydropower is recognized as the energy conversion from flowing water to electricity. In 1970s hydropower was fully considered clean and renewable by nature. Hydropower was therefore treated as a renewable energy resource as a cycle of water is continuously renewed through the sun. Mechanical milling was the first use of hydropower historically as grinding grains. One of the most significant applications of hydropower is hydroelectricity that is employed to produce electricity that is the primary hydropower use today. Hydroelectric power plants usually consist of a water reservoir that is employed to extract the falling water energy in terms of the kinetic energy of water. During the past decade hydropower, particularly large hydro, has been treated as non-environmentally-benign (no longer fully renewable) due to some ecological and environmental impacts caused by huge reservoirs. At present, there is a common consensus and understanding that micro and small hydropower plants are renewable.

Hydropower is considered among the oldest planet power sources that generate electricity using the flowing water that rotates a turbine or spins a wheel. The ancient Greek farmers employed this technique to perform different mechanical tasks, such as grinding grain. This hydropower energy source generates electricity and undergoes no toxic byproducts or air pollution. Water occupies mass, and it always falls and flows downside because gravity and hydroelectric technique harnesses the kinetic energy of water for power generation.

6.1 Working Principle

Hydropower plants extract the falling water energy for electricity generation. A turbine is employed that utilizes the falling water kinetic energy and converts it into mechanical energy. This mechanical energy spins the generator connected with a turbine that converts mechanical into electrical work.

The construction of a dam is recommended in a large river having a large elevation drop. The dam is constructed in a way to have water intake near the bottom and enable the storage of large quantities of water. Due to gravity, waterfalls inside the dam through the penstock, and a turbine is positioned at the penstock end that is rotated through moving water. The turbine shaft feeds the mechanical work to the accompanying generator that generates electrical power. The power lines from the generator are connected to the grid to supply electricity upon demand. This

Renewable Hydrogen Production. https://doi.org/10.1016/B978-0-323-85176-3.00012-3

electricity can be employed in the water electrolysis process to produce clean and sustainable hydrogen by splitting water into its constituents.

A hydraulic turbine employs the flowing water energy and converts the kinetic energy into mechanical energy, and an accompanied hydroelectric generator converts mechanical into electrical work. The generator operation was principally discovered by Faraday who depicted that electricity flows when the magnet moves past the conductor. In a large generator, electromagnets are constructed using wire loops of direct circulating current that are wounded around magnetic steel stacks that are known as field poles and stranded on rotor perimeter. This rotor is connected to the turbine shaft that rotates at static speed. On the rotor turning, the field poles move past the conductors that are mounted in the stator that initiates and develops the flow of voltage and electricity at generator output terminals.

The electricity demand does not remain constant and varies throughout the day, and less electricity is required during the night times for households, businesses, and other amenities. Hydroelectric plants offer a more effective substitute to meet peak electricity demands for short periods as compared with nuclear power plants and fossil fuels. The most common route is employing pumped water storage that reuses the water over again.

The pumped-storage route signifies the pumped water reserve to meet the peak period of electricity demands. The same water that ran through turbines can be pumped and stored in the pools overhead the power plant during the intervals of low electricity demand, for instance, at midnight. The stored water is allowed to move through the turbine accompanied by a generator during the high demand times.

The water reservoir consisting of the pumped water behaves like a battery as it stores power in terms of water during low demands and generates maximum output power for the seasonal peak intervals. A significant pumped-storage advantage is that such hydroelectric units can be started quickly and rapid adjustments can be made in output power and generate efficient results if operates for several hours or a single hour. The construction costs of the pumped water storage reservoirs are comparatively less as they are smaller in size in comparison with conventional hydropower plants.

Fig. 6.1 displays the total primary energy supply by source from 1990 to 2015. A gradual rise in the total primary energy supply in each source can be depicted from the figure. The global power generation can be distributed among the significant sectors of coal, natural gas, nuclear, hydro, wind, solar, etc., biofuels and waste, and oil. From the years 1990 to 2015, the electricity production increased from 2,220,466 to 3,852,538 ktoe in the coal sector, increased from 1,663,608 to 2,949,909 ktoe in the natural gas sector, increased from 525,520 to 670,298 ktoe in the nuclear sector, increased from 184,102 to 335,519 ktoe in the hydro sector, increased from 36,560 to 204,190 ktoe in the wind, solar, etc., sector, increased from 902,367 to 1,286,064 ktoe in the biofuels and waste sector, and increased from 3,232,737 to 4,329,220 ktoe in the oil sector.

A water turbine principally works on the rotational motion like a rotary machine that converts the kinetic energy of water into mechanical energy, and the associated generator converts mechanical work into electrical output. The development of water turbines was established in the 19th century and targeted the industrial power former to the electrical grids. Currently, these turbines are mostly employed to

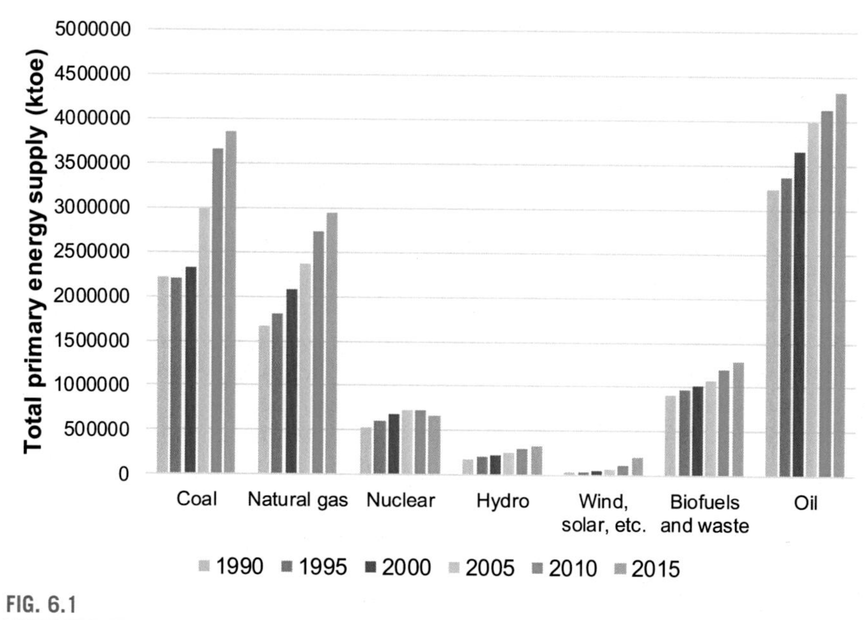

FIG. 6.1

Total primary energy supply by source from 1990 to 2005.

Data from [108]

supply generated power to the grid. Such water turbines are generally constructed in dams to produce power using the kinetic energy of water.

Even though other renewable energy sources namely wind and solar are catching up quickly with hydroelectric power evaporation, it still grasps the largest global electricity share among other renewables. The hydropower source offers more efficient and consistent results in comparison with wind and solar sources as it does not have the intermittent nature of the other two renewable energy sources. The hydroelectric power source was so huge in the 20th century that it was termed white coal for its abundance and power. Hydroelectric power generation is the first and simplest technology for electricity generation.

The primary production employed granite block, timber, and rock construction for the low dam to store water into the reservoir from rainfall and surface runoff. From the water reservoir, water was directed toward the turbine or water wheel. The guided water that fell on the turbine blades resulted in spinning the shaft which was further connected with an accompanying generator that converted mechanical work from the turbine to electrical power. The technology has been fully fledged grown significantly, and water is generally collected from flowing rivers and falling. The technologies that are being employed today with the advancements contain diversion, impoundment, and pumped water storage hydropower. Table A-6 in the appendix displays the installed hydro capacity around the globe from 1990 to 2018 among the significant countries and regions of Africa, Asia Pacific, CIS, Canada, Central America, China, Europe, Japan, Middle East, North America, South and Central America, the United Kingdom, and the United States.

Harnessing or extracting the flowing water energy and producing electricity is called hydroelectric power generation. The water that is employed for the hydroelectric power generation flows from higher lower to elevation. The hydropower plant is typically adjacent that employs the turbine connected to a generator to generate power.

Hydroelectric power is categorized as renewable energy because water is employed as the working fluid that vaporizes into clouds using the sun and recycles back to the earth's surface as precipitation. The cycle of water is constantly energized; thus, it can be employed to generate electrical power over and over again. Hydroelectric power means power generation using falling or flow of water. Dams are built on rivers to generate power using hydro energy. The continuously flowing water is employed to rotate the turbine that utilizes the moving water kinetic energy, and water flow rotates the turbine blades and the rotating shaft turns the generator that converts the mechanical work from the turbine to electrical power. The water exiting the turbine is returned to the dam through the output stream. Thus, flowing water is an influential energy source that can enlighten not only households but also cities and countries. This energy source is also employed to meet the peak power demand in some countries.

Even though energy production using hydropower does not evolve any greenhouse gas (GHG) emissions and environmental issues, constructing huge dams and blocking can cause some social and environmental effects in terms of varying the normal river flow, migratory fish passage blockage, earthquakes, unexpected floods, and displacement of local communities.

Fig. 6.2 exhibits the global electricity generation by renewable energy sources from 1990 to 2015. A gradual rise in the power generation using each source can

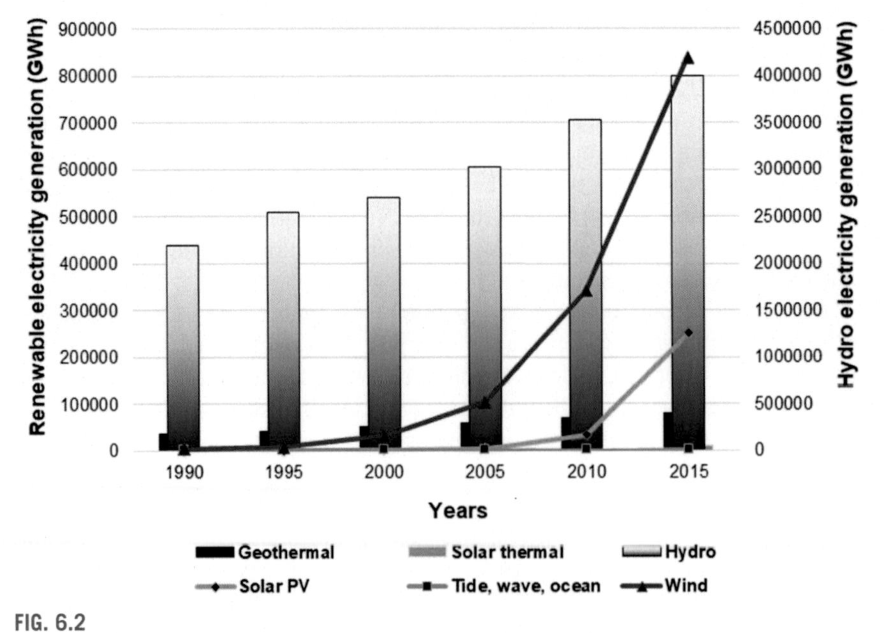

FIG. 6.2

Global renewable electricity generation from 1990 to 2015.

Data from [108].

be depicted from the figure. The primary y-axis is used for all the renewable energy sources excluding hydro source-based power generation, which is drawn using the secondary y-axis. It can be depicted that the hydro energy source is dominant among all other renewable energy sources globally for power generation, and solar photovoltaic and wind energy sources are globally emerging renewable energy sources for power generation. The global power generation can be distributed among the different renewable energy sources of geothermal, solar thermal, hydro, solar PV, tide, wave, ocean, and wind. From the years 1990 to 2015, the global electricity generation using the mentioned sources increased significantly. The electricity generation increased from 36,426 to 80,562 GWh using a geothermal energy source, increased from 663 to 9605 GWh using a solar thermal energy source, increased from 2,191,675 to 3,989,825 GWh using a hydro energy source, increased from 91 to 250,574 GWh using solar photovoltaic source, increased from 536 to 1006 GWh using tidal, wave, and ocean energy sources, and increased from 3880 to 838,314 GWh using a wind energy source.

6.2 Advantages and Disadvantages of Hydro Energy

Hydroelectric power has lately been in headlines as a large number of companies have started recognizing the flowing water potential to generate electricity. That being said, hydroelectric power accompanies some pros and cons that should be considered while looking at hydroelectric power and its global impacts. This section contains both advantages and disadvantages associated with hydroelectric power.

6.2.1 Advantages of Hydro Energy

The following are treated as the significant advantages associated with hydro energy:

Renewable energy source

The hydroelectric energy source is considered a renewable energy source as it employs earth to water to generate electrical power. The earth's water is evaporated using the heat from the sun, and the formed clouds reach the earth again in rain and snow form. Lakes and rivers that are characteristically leveraged to produce hydroelectric power cannot be disappeared.

This indicates that this source cannot be used out and one should not be worried about its shortage or insufficiency. Many suitable locations for the repositories can be found to install hydroelectric power plants to generate electrical power.

Contribution in remote community development

Hydroelectric power plants provide electrical power to remote communities to attract highway constructions as well as industrial and commercial constructions. Such kind of activities assists in to rise of the economy of remote areas, improve access and admittance to healthcare and education, and improve the life quality of the residents. Hydroelectricity has been quantified, computed, and harvested for more than a century.

Clean energy source

As it is well known, the hydroelectric source of power generation is a clean and green alternative energy source that employs earth water. Generating electrical power using hydroelectric energy does not contaminate itself. The electrical power generated using hydroelectric power plants does not yield any greenhouse or toxic gases that can cause environmental pollution. The contamination can be caused while building the power plants.

The hydroelectric power plants release a negligible amount of greenhouse gas emissions as compared with the power production sources using fossil-based power sources that assist to alleviate climate change, smog, and acid rains. Hydroelectric power plants do not only improve the quality of air but also do not produce any toxic byproducts.

Sustainable development

Energy technologies associated with hydroelectric power generation are established and operate in economically viable, environmentally sensible, and communally responsible models that represent the chief sustainable development concepts. This represents the development models that are being used today without disturbing the competence of approaching generations to meet their own needs to address the needs of individuals.

Cost competitive

Hydroelectric power generation plants can be among the most cost-competitive energy sources although the upfront costs for the building can be high. The river or lake water offers an infinite water resource that is employed for power generation does not get affected, however, by market volatility while conventional methods of power generation using fossil fuels are severely affected by market volatility that drives their prices.

Hydroelectric power plants offer an average lifetime of approximately 50–100 years that shows the potential of strategic investments to support future generations. These hydroelectric power plants can be upgraded easily to be fallen in line with the advanced technological requirements and require substantially lower maintenance and operational costs.

Recreational opportunities

The lake that is formed overdue the dam can also be utilized for recreational opportunities, such as boating, fishing, and swimming. The lake water can be employed for irrigation purposes as well. The large dams also become an attraction hot spot for tourists.

Hydroelectric power plants offer massive water storage capacity that can also be employed for irrigation purposes during the intervals of no rainfall and also to meet shortage consumption. The capability to store high water capacities is beneficial as it protects water tables from collapse and reduces the vulnerability to floods and droughts.

6.2.2 Disadvantages of Hydropower

The following are considered some disadvantages associated with hydro energy:

Environmental impact

Natural water flow interruptions can play a great role in the environment and river ecosystem. Some species of fish and other creatures usually migrate during the breeding season or due to the food shortage. The dam constructions may lead to reproduction shortage or fish deaths under certain conditions.

The hydropower natural results are recognized with nature interventions due to the water damming, the altered flow of water, and the advancements in power lines and streets. Hydroelectric power plants may affect the fishing department as they can cause the migration of fishes and other creatures; however, this process is complicated to research, and determination cannot be made on this one factor.

Flood risks

Communities residing on the lower elevations are susceptible to flooding due to the strong water currents from dams. In due course, the livelihood and the source of income of the people residing in such areas can be affected.

High upfront capital costs

Hydropower plants can be extremely expensive to construct in terms of cost but such power plants offer an average lifetime of approximately 50–100 years that shows the potential of strategic investments to support future generations. Hydroelectric power plants require high investment for construction because of the logistical challenges, such as topography, putting underwater foundations, and construction materials. The advantage is that such power plants require less maintenance after completion. The hydroelectric power plants will still need to be operated for a long period to recoup the investment.

Methane and carbon dioxide emissions

The reservoirs constructed to operate the hydroelectric power plants emit high quantities of methane and carbon dioxide gases although they are very minor in comparison with traditional fuels, such as coal, oil, and natural gas. The areas nearby the dam always remain full of water; thus, plants beneath the water start to decompose. This type of plant decomposition without the presence of oxygen emits high amounts of methane and carbon dioxide gases that cause the rise in environmental pollution.

Conflicts

Countries having rich hydroelectric power sources generally construct dams in the rivers to generate electricity using water. While this action is creditable, it may cause an interruption in the flow of natural water in different directions. When one specific location does not need a large amount of water, the water is directed toward other locations to accommodate those needing to construct dams in specific locations. Nevertheless, if water shortage hits those particular areas needing water supply, this shortage can cause conflict that means water guided to the dams need to be ceased.

Droughts

One of the significant disadvantages of employing hydroelectric power plants for electricity generation is the local droughts risk. The overall costs of power and energy are determined using the water accessibility that can be pointedly affected by dry spells that leads to the individuals not retrieving the electricity need.

People living in the lower elevation areas are naturally in danger of flood as nearby areas can be destroyed when dam water is released. Building the dams in explicit areas offers the aptitude to produce massive power amounts that can undergo several challenges. The righteous action is to systematically evaluate the local statistics in detail former to the construction of the state-of-the-art structure in the selected area. Among the numerous key considerations, evaluating the effect of constructing a dam on the environment and community and local safety requirements needs to be evaluated before building the structure.

Fig. 6.3 displays the installed hydropower generation and installed capacity trends from 2010 to 2017. A steady rise in both hydropower electricity generation and installed capacity can be depicted from the figure. Fig. 6.3 exhibits the installed

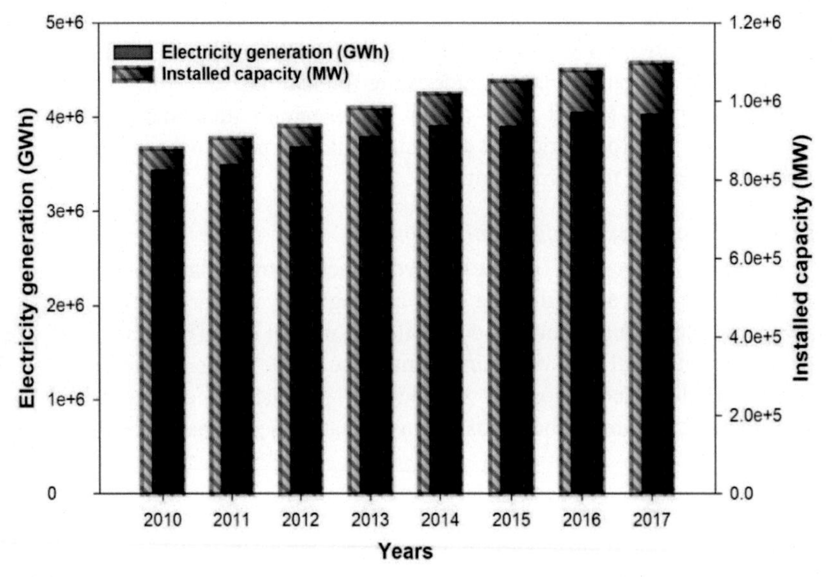

FIG. 6.3

Installed hydropower generation and installed capacity trends from 2010 to 2017.

Data from [111].

hydropower electricity generation and installed capacity trends. A steady increase in electricity generation can be depicted from 3,437,371 to 4,030,628 GWh from 2010 to 2017 and meanwhile, installed capacity increased from 880,606 to 1,099,007 MW.

A comprehensive review[109] of the exergy comparison of renewable energy-driven hydrogen production approaches. The objective of the review study was to offer a comparative study of various hydrogen production methods using an exergetic efficiency approach. This study employed the renewable sources of solar, wind, hydroelectric, and biomass-based methods. It was revealed that the hydropower-based hydrogen production method offered the maximum exergy efficiency of 5.6%, and photovoltaics energy-driven hydrogen production using the water electrolysis process offered the minimum exergy efficiency. A recent study[110] published the global water footprint evaluation of hydropower. Hydropower is among the most significant renewable energy sources; however, it consumes large water portions that can contribute to the scarcity of water. Some previous literature employed a gross evaporation method for water footprint evaluation in which entire evaporation is accredited to hydropower. Such studies failed to consider two significant factors of evapotranspiration former to the construction of the dam, which needs to be abstracted from water footprint and seasonal dynamics storage of water. Such factors are critical to assess reservoir influences on the scarcity of water employing temporally specific indices of water. This study filled this gap by calculating the 1500 hydropower plant water footprints that covered approximately 43% of global yearly hydroelectricity production. They also analyzed the environmental flow necessities as variations in flow regimes can unfavorably affect ecosystems. This study revealed that previous studies in the literature overrated the hydropower influences on the scarcity of water as reservoirs can store and release water during the periods of low and high scarcity and release water during months of high water scarcity. On the contrary, alterations in the water flow normally disturb the environment more as compared with water consumption. Subsequently, effects differ largely among plants and plant-specific assessments are necessary.

A recent study[112] investigated the reliability-constrained hydropower evaluation. To exploit the long-term hydropower production value entails supervision on ambiguous reservoir inflows, potentially flexible parameters on outflows, and revelation to probably wildly changing power prices. They described a stochastic dynamic program design method to quantify the reliability of the reservoir, for instance, measurements of overtopping or failing downstream reservoirs for risk assessment and a correlated approach to determine the flow strategy of the reservoir that maximizes the expected revenue subjected to the distinct levels of target reliability. The modeling and hydropower plant controlling was investigated thoroughly in a recent study.[113] This chapter discussed the historical perspective of hydropower. The boom in the construction of the hydropower dam globally was explored in a recent study.[114] Growth in the human population, climate change, economic development, and the requirement to meet the gap in electricity have encouraged researchers to explore new renewable energy sources. Due to such reasons, significant new initiatives were underway to be explored in hydropower development. At that point in time, approximately 3700 dams were either under planning or construction in emerging economies countries each offering 1 MW capacity. These planning and constructions were projected to raise the global hydropower capacity by 73%. Even this dramatic expansion was not sufficient to meet the growing electricity demand. Moreover, they indicated that this may partially meet the electricity gap but might not significantly reduce GHG emissions (primarily methane and carbon dioxide) and may not eradicate social conflicts and interdependencies. Simultaneously, it may lessen the free-flowing large rivers. The recommendations were to evaluate the economic, social, and ecological complications in the dam construction to make hydropower a feasible and consistent source to generate power.

The probability and potential of employing hydrokinetic turbines in existing tailwaters of traditional hydropower stations were investigated in a recent study[115] using a hydropower system (combined cycle). This study aimed at debating the feasibility and potential to install hydrokinetic turbines behind the working traditional hydropower stations to increase the hydropower generation using a hydropower system (combined cycle). This system was proposed to harvest supplementary power from the remaining energy in water exiting dams. This study proposed the two different modes of hydrokinetic turbines that can be installed directly behind the existing conventional turbine or it can be installed at the sites in powerhouse locality and discussed the advantages as well as challenges associated with the installation of hydrokinetic turbines. It was revealed that the installation of a hydropower system (combined cycle) can be promising and possesses the substantial potential to generate supplementary clean energy, and this might become promising clean energy generating a route to mitigate climate change.

A comprehensive review[116] was published on the operation of hydropower plants. Energy plays a significant contribution to the growth of a country. The traditional energy sources are employed to meet the significant energy demand that is being replaced with a renewable energy source, and hydropower source is one of the most valuable sources among other renewables. Hydropower offers an insignificant environmental and social impacts in comparison with conventional energy sources. Appropriate operation of hydropower plants is very imperative for employing the obtainable potential to generate maximum energy. They conducted a detailed review study on the operative features of hydropower plants that contribute to generate energy undergoing minimal environmental effects and offering minimal operational cost. Every single component and parameter were explored that contribute to hydropower plant operation to achieve maximum power generation and minimum operational cost. The advanced mathematical models and techniques employed in the recent literature to investigate the hydropower plant operation to maximize the power generation and minimize the operational cost were presented to benefit the researchers. They recommended that further research studies need to be conducted for the optimal operation of small river hydropower plants.

A research article[117] was presented in 1985 for hydropower plant-based hydrogen production but hydrogen was not accepted as a clean fuel and energy carrier at that point of time. Hydroelectric power offers the benefit of being an eternally renewable and clean resource. Alternate energy sources, such as fossil fuel-driven thermal plants need to be replaced with renewable and clean energy resources because of environmental problems. Numerous research studies are actively exploring the opportunities to increase the energy output of already existing hydropower plants by substituting or accompanying existing production plants; however, hydroelectric power plants still utilize a minor portion of hydraulic energy from the river because of economic and hydrologic restraints. The possibility of making the hydropower plants more efficient by converting the water to hydrogen using water electrolysis to extract the wasted or excess energy was explored in this study that can be employed to generate electricity through fuel cells or gas turbines. It was proposed that the excess electricity generated during the off-peak or massive river flow periods can be stored in hydrogen form that can later be employed to meet the high-energy demand. The significant advantage was that the site is accompanied by the required source of water for hydrogen production as it is required to install the electrolyzer near the hydropower plant. It was revealed that the electricity conversion to hydrogen at hydropower plants and employing this hydrogen in the gas turbine can be economically and technically feasible. They also predicted that the introduced idea can improve the projected economy by generating additional revenues.

Fig. 6.4 displays the installed hydropower capacity around the globe in 2018. The installed hydropower capacity was found in the significant global countries and

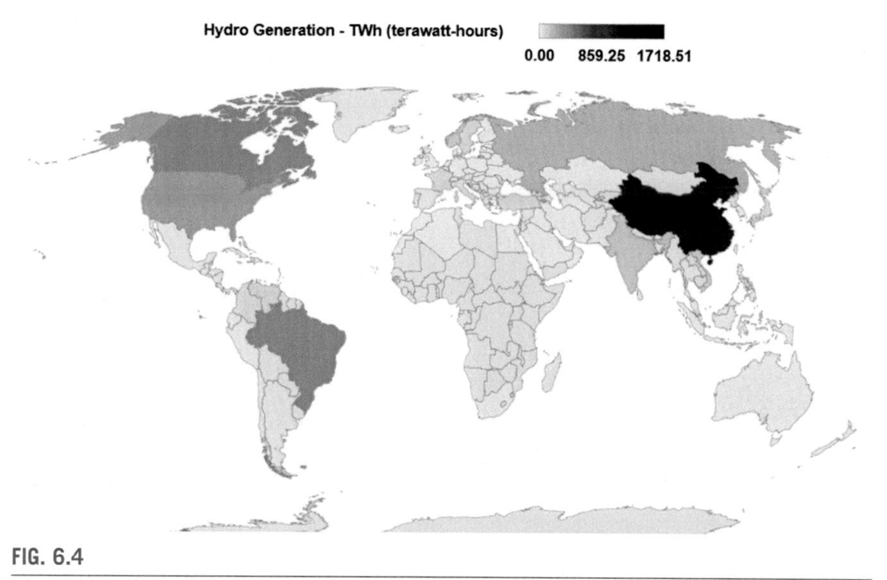

FIG. 6.4

Installed hydropower capacity around the globe in 2018.

Modified from [102].

regions of Austria, Chile, Costa Rica, Portugal, China, El Salvador, Russia, Philippines, Ethiopia, Germany, France, Total Africa, Guadeloupe, Honduras, Guatemala, Hungary, Iceland, Indonesia, Italy, Japan, Kenya, Mexico, New Zealand, Nicaragua, Papua New Guinea, Total Asia Pacific, Total CIS, other Asia and Pacific, Total Europe, Total S. and Cent. America, Total North America, the United States, and Turkey.

6.3 Classification of Hydropower Plants

The classification of hydropower plants can be depicted from Fig. 6.5. Hydropower plants can be classified into different categories as follows:

- According to capacity
  - Pico
  - Micro
  - Mini
  - Small
  - Medium
  - large
- According to head
  - High
  - Medium
  - Low
- According to purpose
  - High
  - Medium
  - Low
- According to facility type
  - Single purpose
  - Multipurpose
- According to hydrological relations
  - Reservoir
  - River
  - Pumped storage
  - In-stream
- According to the transmission system
  - Connected to grid
  - Isolated

The main hydropower principle is simple as the solar heat causes large water amounts to vaporize, consequently increasing the potential energy. The vaporized water forms clouds that assist to concentrate vapors into droplets that fall on the ground as rain that becomes more concentrated by directing into rivers naturally

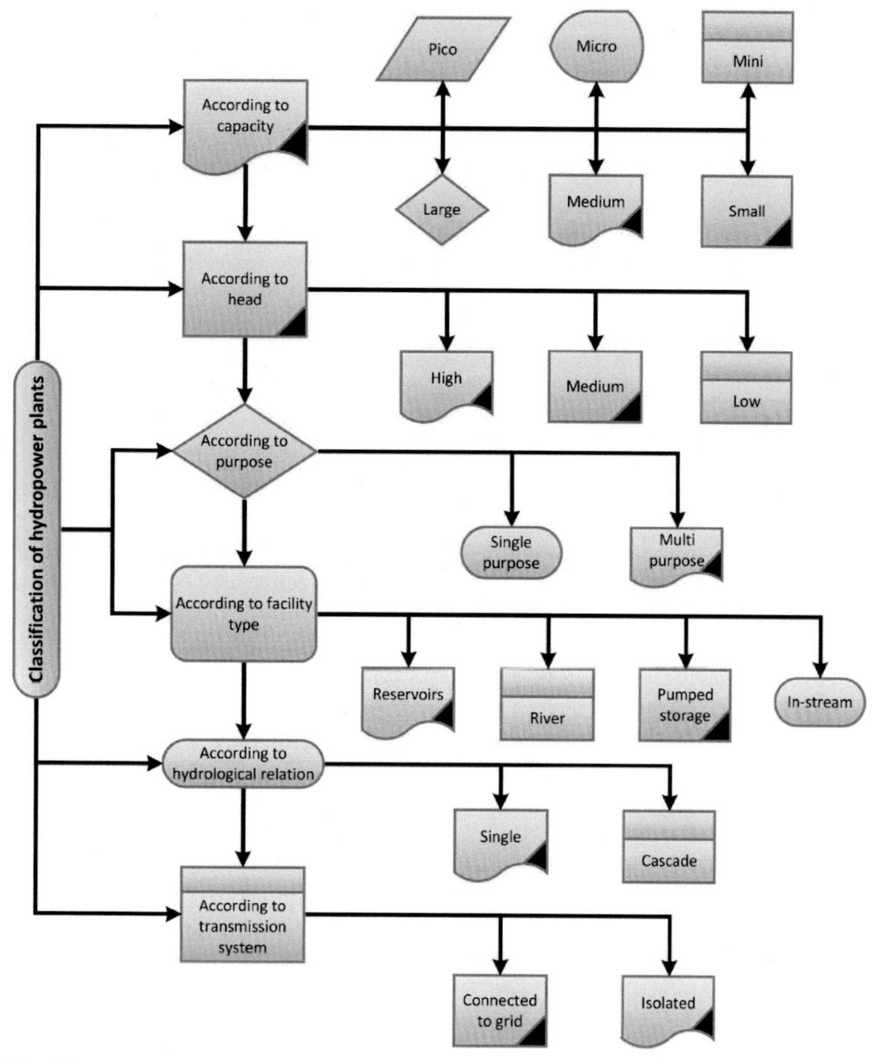

FIG. 6.5

Classification of hydropower plants.

and provides a convenient and free renewable fuel. Extracting water energy comprises several steps that are as follows:

- Entertaining some water into artificial channels
- Offering storage of water to level the head and flow
- Employing a waterwheel or turbine that converts water energy into mechanical
- Employing the sluice gate and gearbox to control the system inlet flow and mechanical output.

The described critical features occur in the earliest historical hydropower descriptions and can be still recognized in modern-day efficient installations.

A comprehensive review[109] of the exergy comparison of renewable energy-driven hydrogen production approaches. The objective of their review study was to offer a comparative study of various hydrogen production methods using an exergetic efficiency approach. This study employed the renewable sources of solar, wind, hydroelectric, and biomass-based methods. It was revealed that the hydropower-based hydrogen production method offered the maximum exergy efficiency of 5.6% and photovoltaics energy-driven hydrogen production using the water electrolysis process offered the minimum exergy efficiency.

Even though renewable energy sources like wind and solar are growing rapidly but hydropower still grasps the largest global electricity share. In the 20th century, hydropower was so huge that it was named as white coal because of its abundance and power. Hydropower is the simplest and foremost electrical power generation technology.

The preliminary production employed granite block, timber, and rock construction to gather and store water from rainfall and surface runoff into the reservoir. From the reservoir, water was channeled toward the pipe leading to the turbine or water wheel. The directed controlled water moved through the turbine blades and started rotating the connected shaft that was further attached to a generator to generate power. The technology has been growing rapidly and water is generally collected from flowing and falling rivers. The advanced technologies that are being employed nowadays include diversion, impoundment, and pumped storage.

Even though energy production using hydropower does not release GHG emissions, however, constructing massive dams on rivers can cause serious social and environmental effects in terms of changing the normal river flow, unexpected flood incidences, upsurge in earthquakes, and relocating the local communities.

6.4 Hydroelectric Turbine and Generator

A hydraulic turbine principally converts flowing water energy into mechanical energy, and an integrated hydroelectric generator converts the mechanical energy from the turbine into electricity. The hydraulic turbine operation is based on the Faraday principles as he discovered that the movement of a magnet past a conductor causes the flow of electricity. In a generator, electromagnets are circulated through direct current wire loops around magnetic steel stacks termed as field poles and mounted on the rotor perimeter. The rotor is connected to the turbine shaft that is rotated at a fixed speed. When the rotor starts turning, it moves the field poles past the conductors that are mounted in the stator that causes the flow of electricity, and voltage to be developed at the output terminals of the generator. Fig. 6.6 displays the schematic of a hydroelectric turbine and generator.

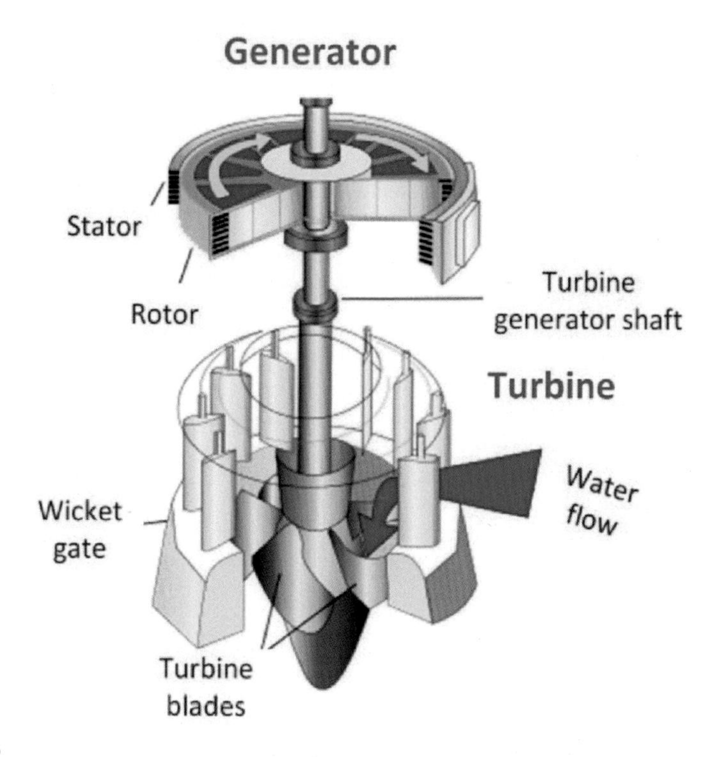

FIG. 6.6

Schematic of an hydroelectric turbine and generator.

Modified from [118].

The turbine blade is a distinct component that makes the turbine unit of water, steam, or gas turbine. The turbine blades are accountable for extracting energy using the working fluid. Turbine blades and wicket gates are the significant components of a hydroelectric turbine while the rotor and stator are the significant components of a hydroelectric generator. A turbine-generator shaft connects the hydroelectric turbine with a hydroelectric generator.

Hydropower may be classified by capacity as displayed in Fig. 6.7. In terms of capacity, pico-hydro-type hydroelectric power ranges from 0 to 5 kW, micro-hydro-type hydroelectric power ranges from 5 to 100 kW, mini-hydro-type hydroelectric power ranges from 100 kW to 1 MW, small hydro-type hydroelectric power ranges from 1 to 15 MW, medium hydro-type hydroelectric power ranges from 15 to 100 MW, and large hydro-type hydroelectric power is more than 100 MW.

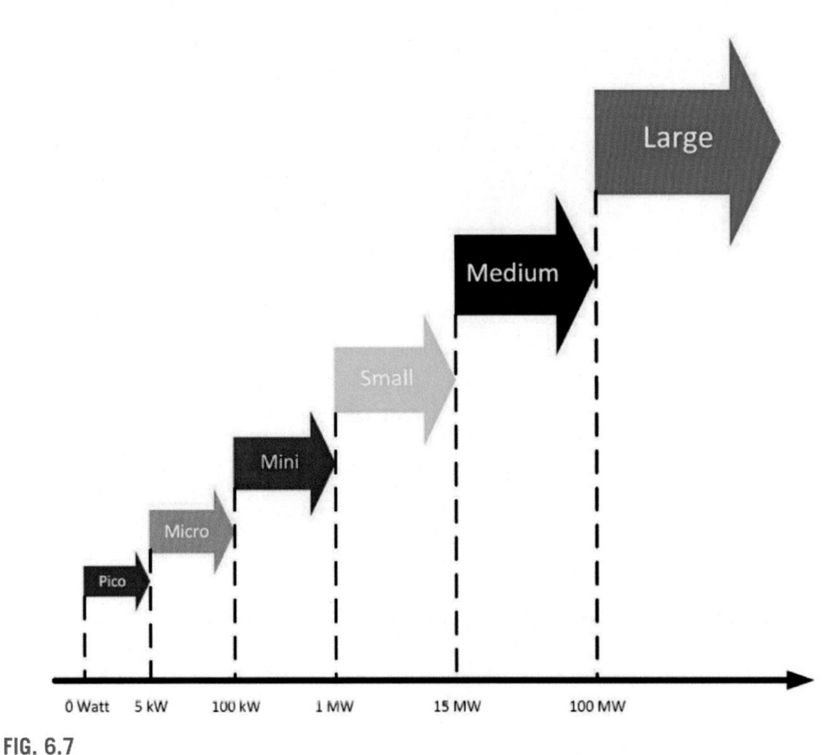

FIG. 6.7

Hydropower classification in terms of capacity.

6.4.1 Hydroelectric Power Plant and Pumped Storage

The schematic of the hydropower plant generating power is displayed in Fig. 6.8. The dams are constructed on the rivers to store large quantities of water in the reservoirs. The large capacities of water are stored in the upper reservoir so that it can be employed to harvest electricity. The stored water can also be utilized to generate power during high-demand periods. The water from the upper reservoir is employed in the turbine or waterwheel that extracts the energy from water to generate mechanical energy. This produced energy is employed to the generator connected with the turbine that converts mechanical into electrical energy. The output water stream from the turbine output is fed to the lower reservoir, and the generated electricity is supplied to the electricity grid. This way, water is fed back to the river, and the cycle continues to generate electricity that can be supplied to the community or grid, depending upon demand.

The most common pumped-storage-type hydropower system is founded on a dam and reservoir power plant, but a lower reservoir is also constructed at the tailrace end that collects the water exiting the powerhouse as shown in Fig. 6.8. The plant operates as a conventional hydropower plant under normal scenarios

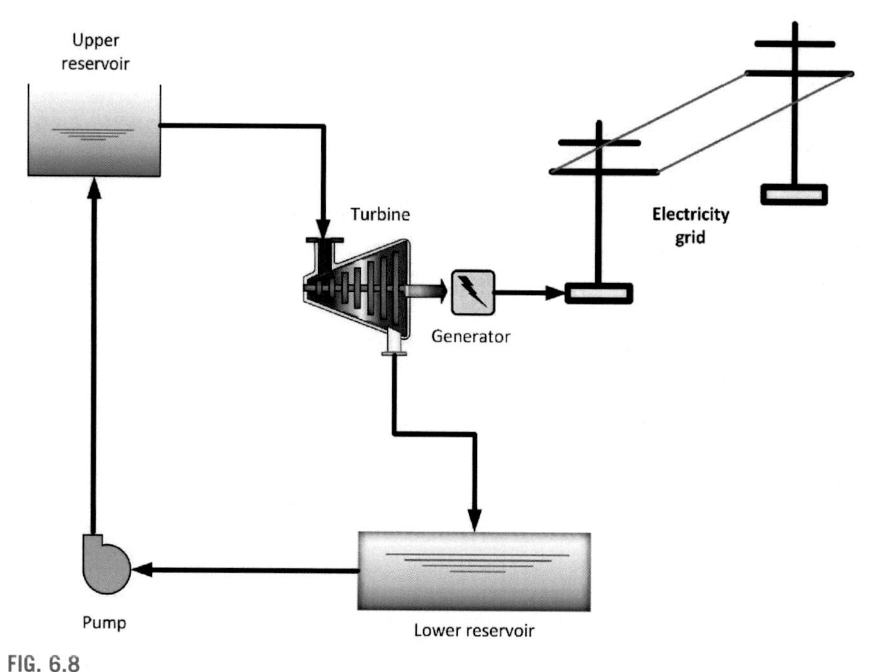

FIG. 6.8

Schematic of the power generation using hydropower plant.

nevertheless with regulated output to supply the electrical power during peak-demand periods instead of functioning as a base-load plant. To be suitable as pumped storage, the pump also needs to be incorporated in the plant that pumps water back to the main reservoir to meet the surplus power during the intervals of low water levels. This type of plant is termed an on-stream pumped-storage plant.

It is frequently appropriate to utilize the existing dam and reservoir power plant to construct a pumped-storage hydropower plant as one reservoir that already exists. Numerous plants that are constructed using this design can function as both generating plants and pumped storage. The United States has the largest pumped-storage hydropower plant at Bath County that offers 3000 MW of power generation capacity. The levels of water in both upper and lower reservoirs fluctuate during the operation, and the maximum head between reservoirs is 400 m.

Every so often, it is more appropriate to build two man-made reservoirs not positioned on any waterway; however, a water source will be desirable for reservoirs filling and to offer them refilling due to the continuous losses. Old mines and diggings have been employed to deliver a reservoir for such a system. This type of plant basically circulates the water from the upper to lower reservoir and lower to the upper reservoir, and it can operate as a storage plant and is generally termed closed-cycle pumped-storage hydropower plant.

Specifying the desired location for two reservoirs to be constructed or exploited is among the most challenging tasks as reservoirs need to be apart from each other to offer sufficient heads required for an efficient and effective pumped-storage plant. A high-head is desirable between the two reservoirs as it provides more energy compared to the low-head system. Another option that is considered but hardly ever employed is for the sea to deliver a lower reservoir of the pumped-storage system. High cliffs must be next to the shore, and a site appropriate for forming a lake must be at the cliff top to utilize this technique effectively.

Constructing the second reservoir underground is another alternative. This alternative is attractive as it carries a very negligible environmental impact as compared with constructing a new reservoir overhead the ground; however, it is restricted by the appropriate underground availability. The underground mining sites are most attractive and exist in abundance. Though the site must be watertight otherwise, water can leak away before pumping back to the upper reservoir but none of these systems have been built.

The size of the reservoir is an additional factor that should be considered during the system design, predominantly if reservoirs are to be constructed. The capacity of the water storage plant is dependent on as much upper reservoir can hold water. The larger the capacity of the reservoir, the more power for more hours can be stored. The pumped energy storage plants offer round-trip efficiency of 70%–80% when functioned on a daily cycle.

6.5 Types of Hydropower Turbines

The hydropower turbines can be classified into two significant categories of impulse turbine and reaction turbine that are further classified as follows:

- Impulse turbine
  - Pelton
  - Cross-flow
- Reaction turbine
  - Kaplan
  - Francis

6.5.1 Impulse Turbine

The impulse turbine typically employs the water velocity to operate the runner and is released at ambient pressure. In these turbines, all the obtainable potential energy is converted into velocity-head or kinetic energy by circulating water over contracting nozzle before its striking the buckets. The wheel rotates in the air, and a specific part of the wheel remains in contact with water.

A casing is installed to guide water and avoid splashing from buckets. An impulse turbine is principally a low-speed wheel that is employed for comparatively high-heads. Pelton turbines and cross-flow turbines are significant types of impulse turbines.

Pelton

Pelton wheel carries one or more free jets discharging water into imposing on the runner buckets and aerated space. For impulse turbines, the draft tubes are not needed as the runner is obligatory to be situated overhead the maximum tailwater to allow atmospheric pressure operation. Water strikes the runner tangentially in the Pelton wheel.

Cross-flow

Cross-flow turbine is usually drum-shaped and employs a rectangular-section, elongated nozzle positioned counter to the curved vanes on the runner with a cylindrical shape. This type of turbine permits the water to flow twice through blades. In the first pass, water flows from outside to the inside of the blades, and in the second pass, from inside to outside. The guide vane positioned near the turbine inlet guides the water flow to a restricted runner portion. This cross-flow allows and accommodates the larger flows of water and lower heads as compared with Pelton.

6.5.2 Reaction Turbine

The reaction turbine produces power employing the combined effort of moving water and pressure. The runner is directly positioned in the flowing stream of water through the blades instead of striking each runner individually. Such turbines are usually designed to be installed on sites with high water flow and lower heads where impulse turbines cannot be feasible as they require high-head to generate maximum power.

In reaction turbines, a part of obtainable potential energy is utilized and converted into a velocity-head at the runner entrance. The inlet turbine pressure is considerably higher as compared with the outlet pressure. It fluctuates all over the water passage through the turbine. Typically, power is established through the difference in acting pressure on the front and back of the runner blades. A small power portion originates from the velocity dynamic action. After the pressurized system, the complete flow from head to tail race occurs in the closed system. Kaplan and Francis turbines are the two imperative kinds of reaction turbines.

Kaplan

Kaplan turbine is principally a propeller having adaptable blades inside the tube. Both the wicket gates and blades are adjustable that allows a wider operation range. Kaplan is principally an axial-flow turbine that means the direction of flow does not vary while crossing the rotor. In Kaplan and propeller turbines, the flow is parallel or axial to the turbine shaft axis. The selection criteria to evaluate a suitable turbine type primarily depend upon the head availability and the quantification of required waste.

The guide-vanes allowing the inlet water can be closed or opened to regulate and control the flow volume that can be passed through the turbine. During the intervals of zero electricity requirement or demand, the flow of water is completely stopped and the turbine is brought to the rest. Contingent to the inlet guide-vanes position, they introduce variable swirl amounts to the water flow and confirm that water is hitting the rotor at the maximum efficient angle to achieve maximum efficiency.

The pitch of the rotor blade is adjustable similarly from flat profile for low water flows to heavily pitched profiles with high water flows. The adjustability of both rotor blades and inlet guide vanes means that the operating range of flow is very extensive, the efficiency curve is flat, and the turbine is highly efficient. The efficiency curve is the adjustable rotor blades feature that allows the optimal alignment of the blade against the oncoming flow.

Francis

Francis turbine employs a runner through fixed buckets generally nine or more. The water is introduced overhead the runner and all over it and then falls over initiating the spin. Other significant components are wicket gates, scroll cases, and draft tubes beside the runner. Francis turbine supports the inward radial water flow. In the advanced Francis turbines, the flow enters inward radially but exits in a parallel direction and is termed as mixed flow.

Francis turbines are among the most commonly used water turbines today. These turbines operate at the water head ranged from 40 to 600 m and are primarily employed for electrical power generation. The electric generator that is employed in such turbine types usually undergo power output ranging from a few kW to 800 MW; however, installations of mini-hydro may be lower. The diameter of the penstock ranges from 3 to 33 ft, and the turbine speed range lies between 75 and 1000 rpm. A wicket gate around the rotating runner turbine outside is installed to control the water flow rate over the turbine to achieve different rates of power production. Such turbines are practically at all times straddled with the vertical shaft to separate water from the generator. This phenomenon facilitates maintenance and installation as well.

6.6 Hydropower-Based Hydrogen Production

Classification of hydroelectric power plants for hydrogen production can be depicted in Fig. 6.9, and power generated by any of these classified hydroelectric power plants can be employed directly in the water electrolysis process for hydrogen production. The classification of hydroelectric power plants is as follows:

- Waterfall availability
  - Reservoir power plants
  - Runoff river power plants without pondage
  - Run of river power plant with pondage
- Water heat availability
  - High-head
  - Medium-head
  - Low-head
- Loading type
  - Pumped-storage plant for pick load
  - Base load
  - Peak load

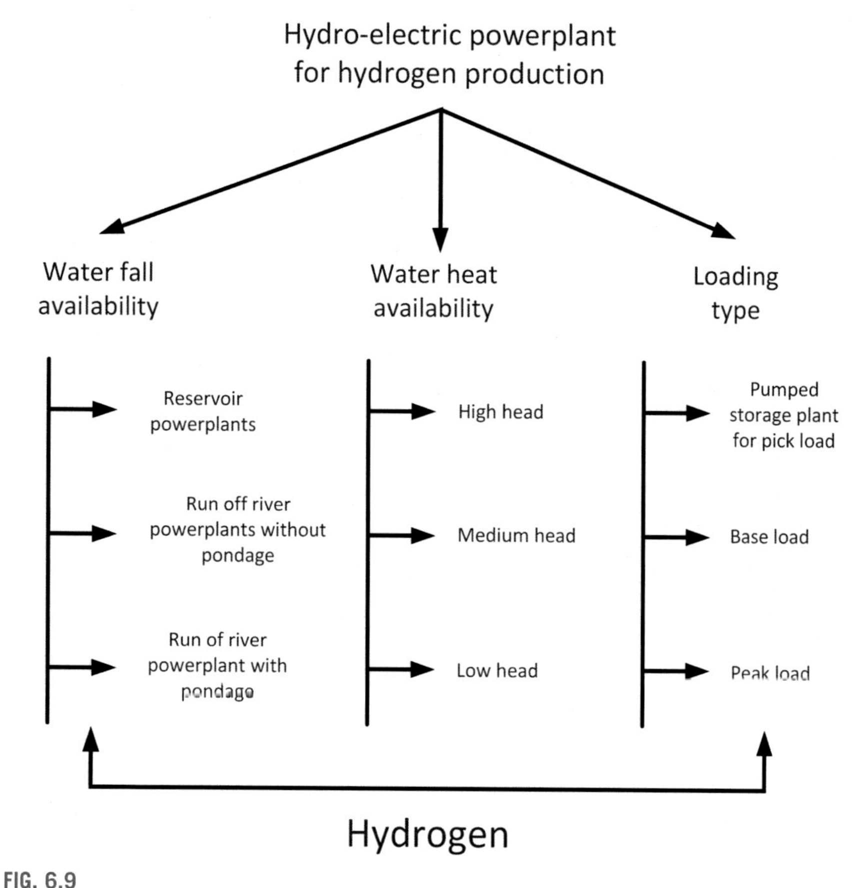

FIG. 6.9

Classification of hydroelectric power plants for hydrogen production.

The schematic of the hydropower-based hydrogen production system is displayed in Fig. 6.10. The dams are constructed larger to increase the storage capacity in the reservoirs. The large capacities of water stored in the upper reservoir are directly employed in the turbine or waterwheel to generate electricity. The water in the dams can also be stored to meet the periods of high-energy demand that is usually done through pumped storage or flywheels.

Utilizing the existing dam and reservoir power plant to construct pumped-storage hydropower plant as an already existing reservoir is more appropriate. The plants following such a constructed design can be operated as generating plants as well as pumped storage. The selection of the location desired to construct reservoirs is among the most challenging tasks as reservoirs need to be apart from each other to offer sufficient heads required for an efficient and effective pumped-storage plant. A high-head is desirable between the two constructed reservoirs as it offers higher energy generation as compared with the low-head system.

Water flows from the top through the reservoir and is employed to the turbine that extracts the energy from water to generate mechanical energy. Generator is

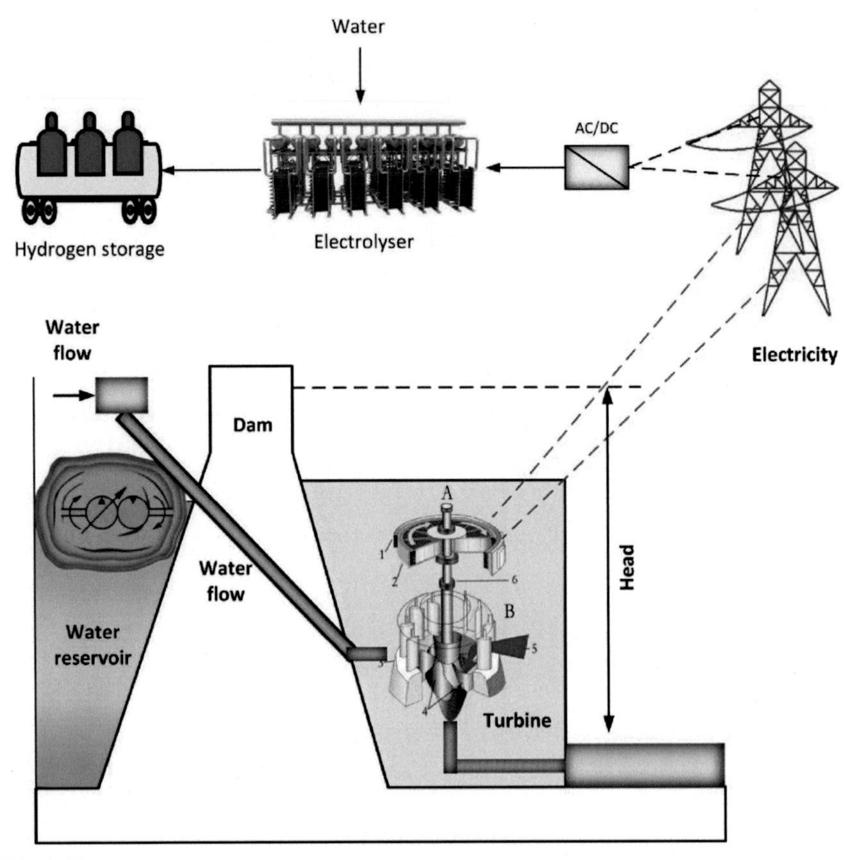

FIG. 6.10

Hydropower-based hydrogen production.

accompanied by a turbine that converts mechanical energy into electrical. The output water from the turbine output moves to the lower reservoir, and the generated electricity is supplied to the electricity grid. This way, water is fed back to the river and the cycle continues to generate electricity that can be employed to produce hydrogen. The electrical power generated by the hydropower plant is converted to the DC through AC/DC converter, and the electricity is fed to the electrolysis process with water to produce hydrogen. This type of system produces clean and environmentally benign hydrogen undergoing zero carbon emissions.

The mechanical power produced by the hydraulic turbine and represented as (P_T) can be calculated using hydraulic head available denoted as (h), turbine flow expressed as (q), water density denotes as (ρ), gravitation acceleration expressed as (g), turbine head symbolized as (h), and turbine efficiency denoted by (η) through Eq. (6.1). The efficiency factor reduces the total mechanical power generated by the turbine and accounts for the power losses as well. Turbine torque at evaluated head and speed is about linearly related to the position of guide vane from no-load to evaluated load. The model of the turbine is based on a steady-state operational equation to correlate the output power with the flow of water and turbine head[113]:

$$P_T = \eta q \rho g h \tag{6.1}$$

The efficiency factor of the turbine must consider that turbine cannot be 100% efficient; thus, no-load flow (q_{nl}) is subtracted from net flow to evaluate the actual flow that is multiplied by the turbine head to determine the mechanical power production. The turbine damping effect that is a function of guide vane opening also needs to be considered in the mechanical power calculation. Thus, the turbine power per unit can be represented as follows:

$$\overline{P_T} = A_T \overline{h}\left(\overline{q} - \overline{q_{nl}}\right) - D_n \overline{G} \Delta \overline{n} \tag{6.2}$$

The turbine MW evaluation is used as base power, q_{base} is the turbine flowrate, with fully open guide vanes (position of guide vane = 1), and h_{base} water is static head of column h_0. The D_n parameter ranged from $0.5 \leq D_n \leq 2.0$ that accounts for the speed variation Δn effect on turbine efficiency. The turbine gain is represented as A_t and expressed in the following equation as the ratio of effective to actual gate position:

$$A_t = \left(\frac{1}{G_{fl} - G_{nl}} \times \frac{\text{Turbine MW rating}}{\text{Generator MW rating}}\right) \tag{6.3}$$

where G_{fl} denotes full-load guide vane position and G_{nl} signifies no-load guide vane position, and both of these factors are determined at the head and rated speed.

The turbine features explain base flow using the relationship among flow (q), head (h), and guide vane position (G). The per-unit flowrate in the turbine employing base flowrate q_{base} and assuming the based head h_{base} equivalent to static head h_0 is represented by the characteristic of the valve and expressed as follows:

$$\overline{q} = \overline{G}\sqrt{\overline{h}} \tag{6.4}$$

The turbine can be signified using the following linearized Taylor series approximations for small differences in operating point concerning the torque and flow to head, guide vane position and speed:

$$\Delta q = a_{11}\Delta h + a_{12}\Delta n + a_{13}\Delta g \tag{6.5}$$

$$\Delta m = a_{21}\Delta h + a_{22}\Delta n + a_{23}\Delta g \tag{6.6}$$

The parameters a_{ij} are representing the partial derivatives of torque and flow concerning guide vane position and head speed. For an ideal turbine effective at rated head and speed, the partial derivatives against load are presumed as follows:

$$a_{11} = 0.5,\ a_{12} = 0,\ a_{13} = 1,\ a_{21} = 1.5,\ a_{22} = -1,\ a_{23} = \frac{dm}{dg} \tag{6.7}$$

Note that the representation of the turbine mainly depends on the coefficient a_{23}.

6.6.1 Modeling of Single Penstock

A general hydropower plant comprises a solitary channel contributing to the turbine-generating unit, and the primary model development is constrained to the inelastic water column case. The characteristics of the turbine and penstock are evaluated through three basic equations concerning the water velocity in penstock, water column acceleration under the gravity influence, and mechanical power production in the turbine. At first, a nonlinear appropriate expression is established to consider the great changes in power and speed.

The elementary water column model characterizes a single penstock with no surge or large tank. Water is assumed to be an incompressible fluid for penstock modeling to neglect the water hammer impact. Consider a rigid conduit of cross-section area (A) and length (l), where penstock head losses as a result of water friction h_f against penstock wall are directly proportional to the square of flow (q) and expressed as follows:

$$h_f = f_p q^2 \tag{6.8}$$

where f_p denotes the head loss coefficient due to friction in the penstock.

Assuming penstock water a solid mass, the change in flow rate can be correlated to the water head employing Newton's second law of motion. The force on water mass can be expressed as follows:

$$\left(h_0 - h - h_f\right) \rho g_a A = \rho A l \frac{dv}{dt} \tag{6.9}$$

where h_0 denotes water column static head, h signifies the head at turbine admission, h_f symbolizes frictional head loss, f_p represents head loss coefficient, and v signifies water velocity.

The rate of flow change in penstock can be expressed as follows:

$$\frac{dv}{dt} = \left(h_0 - h - h_f\right) \frac{g_a A}{l} \tag{6.10}$$

This equation can also be expressed in per-unit form to standardize the system representation. The per-unit representation is as follows:

$$\frac{d\bar{q}}{dt} = \left(\bar{1} - \bar{h} - \bar{h}_f\right) \frac{h_{base} g_a A}{l \, q_{base}} \tag{6.11}$$

$$\frac{d\bar{q}}{dt} = \frac{\left(\bar{1} - \bar{h} - \bar{h}_f\right)}{T_w} \tag{6.12}$$

where $T_w = \frac{l q_{base}}{h_{base} g_a A} = \frac{l v_{base}}{h_{base} g_a}$ represents water inertia time constant.

The starting time of water or inertia time constant represents the required time for the base head (h_{base}) to accelerate water in penstock since standstill at base velocity (v_{base}) that is determined between the surge tank or forebay and turbine inlet depending upon the existence of large surge tank. Considering a simple penstock provided by an open reservoir being discharged into the atmosphere, guide vane opening in time Δt reasons the water velocity in penstock to rise by Δv and drop in the turbine head by Δh.

The water acceleration due to change in the turbine head that is characterized using the second law of Newton can be expressed as follows:

$$\rho A l \frac{d\Delta v}{dt} = -\rho g_a A \Delta h \tag{6.13}$$

The equation for the acceleration can be represented in per-unit form by dividing with base velocity v_{base} and based head h_{base}:

$$\left(\frac{l v_{base}}{g_a h_{base}}\right) \frac{d\Delta \overline{v}}{dt} = -\Delta \overline{h} \tag{6.14}$$

$$T_w \frac{d\Delta \overline{v}}{dt} = -\Delta \overline{h} \tag{6.15}$$

Eq. (6.15) indicates a significant hydraulic plant characteristic. The examination of this equation demonstrates that backpressure arises and causes water to decelerate if the guide vane is closed. Explicitly, a negative acceleration change will occur in the case of a positive pressure change. Likewise, a positive acceleration change occurs in the case of a negative pressure change. The maximum acceleration arises right after the opening of the guide vane as the entire pressure difference is offered for water acceleration. The water inertia time constant for nonuniform penstock with dissimilar cross-sectional areas can be expressed as follows:

$$T_w = \frac{\sum lv}{g_a h} \tag{6.16}$$

where $\sum lv$ signifies length summation and v denotes the velocity.

6.6.2 Surge Tank Modeling

The hydraulic pressure changes and transients are measured and controlled by surge tank usage. The surge tank is arranged as an open reservoir and positioned right above the high-pressure shaft. Even though it is larger in size but still for modeling purposes, treated as a conduit. Therefore, the system significantly is composed of three-way compound conduits, namely low-pressure tunnel, high-pressure tunnel, and surge tank together with throttling orifice.

The model of the surge tank is derived from the modeling and controlling hydropower plants book[113]:

Flow down the upper tunnel is expressed as:

$$q_t = q_{st}(\text{flow into surge tank}) + q_t(\text{flow to turbines}) \tag{6.17}$$

Flow into surge tank is dependent on the area of surge tank denoted as (A_s), and rate of change of tank level represented as (h_s):

$$q_s = A_s \frac{dh_s}{dt} \tag{6.18}$$

The per-unit water level in the surge tank is expressed as follows:

$$h_s = \frac{q_s}{C_s s} \tag{6.19}$$

here C_s denotes the storage constant of surge tank that is expressed as follows:

$$C_s = \frac{A_s h_{base}}{q_{base}} s \tag{6.20}$$

The surge tank orifice head losses are directly proportional to the flowrate times coefficient f_0 times the absolute flowrate value that supports to maintain the head loss direction. Henceforth, the lower penstock head is calculated by subtracting the orifice head losses from the surge tank level head. The surge tank level describes the head across the lower penstock. The surge tank effect inclusion is acceptable in cases when dynamic performance is simulated using a few minutes of real time.

The surge tank addition gives an upsurge to the poorly damped oscillation among the reservoir and tank. These oscillations are normally relatively slow and can be neglected to govern the load frequency control for few minutes per cycle order. In a rigid column lumped system theory, if the head loss is neglected in both orifice and tunnel, the surge tank oscillation consumes a period (T_{st}), which is the interval of time between maximum surge occurrence and turbine load change. The calculation expression can be expressed as follows:

$$T_{st} = 2\pi\sqrt{\frac{lA_s}{gA}} \tag{6.21}$$

where l represents the tunnel length between reservoir and surge tank, A_s signifies surge tank cross-sectional area, and A denotes the tunnel cross-sectional area.

6.6.3 Wave Travel Time

Wave travel time T_w depends on the tunnel length and sound velocity in water that can be determined using the following correlation:

$$T_w = \frac{1}{a} \tag{6.22}$$

Bergant et al.[119] presented the correlation to determine the pressure wave velocity propagation in a pipe expressed as follows:

$$a = \sqrt{\frac{1}{\rho\left(\frac{1}{k} - \frac{d}{eE}\right)}} \tag{6.23}$$

where ρ represents the density of water, d denotes penstock diameter, k signifies water bulk modulus, e indicates penstock wall thickness, and E denotes elastic modulus.

For an impeccably rigid pipe, the wave propagation velocity related to liquid is identical to the sound velocity in an infinite liquid expanse. Consequently, the correlation for a rigid pipe can be expressed as follows:

$$a = \sqrt{\frac{k}{\rho}} \tag{6.24}$$

Water bulk modulus is $2.05 \times 10^9 \mathrm{N/m^2}$, and rigid pipe wave velocity can be expressed as follows:

$$a = \sqrt{\frac{2.05 \times 10^9 \mathrm{N/m^2}}{10^3 \mathrm{kg/m^3}}} = 1432 \mathrm{\ m/s} \tag{6.25}$$

The tunnels are perfectly lined in the rock, and the tunnel wall thickness is much larger than the tunnel diameter. Therefore, the wave velocity rigid pipe approximation can be applied to estimate the wave travel time for the hydraulic system.

6.6.4 Head Loss Coefficient

The flow in penstock is generally turbulent and hereafter highly complex. Water flowing over a penstock causes a drop in pressure in the flow direction.

The pressure drop through the penstock length can be represented concerning the head loss (h_f) as follows:

$$h_f = f_r \left(\frac{1}{d}\right)\left(\frac{v^2}{2g_a}\right) \tag{6.26}$$

where f_r denotes friction factor and d signifies penstock diameter.

This correlation presented in Eq. (6.26) is termed as D'Arcy—Weisbach form, and the ($v^2/2g$) factor is known as velocity head. The head loss is dependent on water velocity in penstock, and therefore, on several operational units. The coefficient of head loss (f_p) employed former in water column modeling is determined using Eq. (6.8).

Table 6.1 displays the head losses for the Dinorwig hydraulic system that characterizes the plant hydraulic pressure losses once a single unit is operating at full load undergoes 65 (m^3/s) water flow. Frictional factor (f) is the function of relative roughness of pipe and Reynolds number that is determined graphically.

Table 6.1 Head Losses for Dinorwig Hydraulic System.

Type	Diameter (m)	Length (m)	H_f (m)	f_r	$f_p \left(\frac{m}{\left(\frac{m^3}{s}\right)^2}\right)$
L.P. tunnel	10.5	1695	0.05	0.0151	$1.65 \times e^{-5}$
H.P. tunnel	9.5	446	0.031	0.0154	$7.33 \times e^{-6}$
H.P. shaft	10	412	0.0277	0.0147	$4.9 \times e^{-6}$
Concrete penstock	3.8	50	0.416	0.0189	$9.85 \times e^{-5}$
Manifold	9.5	77	0.0053	0.0154	$1.25 \times e^{-6}$

Adapted from [113].

6.7 Closing Remarks

Hydropower states the generation of electricity employing the flowing water energy. This power production sector is classified as a renewable energy resource as the water cycle is continuously renewed through the sun. Hydropower is among the promising global energy resources to produce clean, sustainable, consistent, and environmentally benign hydrogen, and this promising route does not emit GHG emissions to the environment. The hydropower-driven hydrogen production technology can play a vital role in providing the global transition to 100% renewable energy. Hydroelectricity is the most significant application of hydropower that is employed to generate electricity. Hydroelectric power plants typically involve a water reservoir that is employed to extract the falling water energy in terms of kinetic energy associated with water. Hydropower is among the oldest planet sources that produce electrical power using the flowing water that is employed to spin a wheel or rotate a turbine. This technique was historically employed by the ancient Greek farmers to accomplish diverse mechanical tasks, for instance, grinding grain. Hydropower energy source extracts electricity that can be employed directly in the water electrolysis process and undertakes no toxic byproducts or air pollution. Water occupies mass and continuously falls and flows downside because of gravity, and the hydropower technique extracts the kinetic energy of water for electricity generation. A water turbine principally operates on the rotational motion like a rotary machine, such as a wheel or turbine that undergoes the energy conversion from kinetic to mechanical energy, and the accompanying generator converts mechanical work into electrical output. This promising and renewable power generation source can be employed to produce clean, sustainable, and environmentally benign hydrogen.

Although other renewable energy sources, such as wind and solar are attracting the significant research scientific community, however, hydropower still holds the largest global electrical power generation share among other renewables and it offers more consistent and efficient results as it does not have an intermittent nature in comparison with other renewable energy sources. The power production by the hydropower resource can be supplied directly to the water electrolyzer process that breaks water into its constituents and produces clean hydrogen that can be used for numerous applications, such as hybrid and fuel cell vehicles, combustion, ammonia synthesis, heating, as energy carriers, ammonia and methanol synthesis, and hydrogen storage system can employ the stored hydrogen to the described applications on demand. This promising route to produce clean, consistent, sustainable, and environmentally benign hydrogen undergoing zero carbon emissions can play a significant role to reduce and even replace the conventional power production methods employing fossil fuels and hydrogen production methods undergoing natural gas reforming. Hydropower applications may also cover energy storage options (using flywheels) to meet the peak demands, and additional methods can be employed to produce clean hydrogen that can be utilized for numerous applications on demand.

CHAPTER

7 Ocean Energy-Based Hydrogen Production

The ocean occupies one of the largest, but slightest explored sources of renewable energy on earth. Ocean energy source has the potential to provide a significant amount of renewable and consistent energy around the globe. The first patent was published in 1799 by Girard in Paris to utilize ocean wave energy. A primary wave-power application device was constructed in 1910 by Bochaux-Praceique, and the extracted power was employed to light the house.

The ocean energy source can generate two types of energy, namely thermal energy through solar heat and mechanical energy through waves and tides. Closed-cycle systems employ the warm surface water of the ocean to vaporize, the working fluid, and the working fluid offering a low-boiling point is selected, such as ammonia. The vapor expansion turns a turbine. Ocean energy includes all renewable energy forms derived from the ocean. There are three primary ocean energy technology types as follows:

- Tidal energy
- Wave energy
- Ocean thermal energy

All ocean energy forms have been attracting attention for commercialization and implementation. Tidal energy and ocean thermal energy conversion are determined to be cost-competitive as compared with wave-energy technology. Ocean energy offers consistent and renewable energy sources as well as high energy density as compared with other forms of renewable energy. Tidal turbines offer approximately 80% efficiency that is comparatively much higher than wind and solar energy sources. Barrages condense the high tidal surges damage on land.

Wave energy is principally power drawn using waves. The blowing wind across the surface of the sea transfers energy to the waves. The stronger the waves, the more power generation capability is carried out. The energy extracted can then be employed for generating electricity, powering power plants, or water pumping.

Ocean waves contain immense energy before they break onshore. The energy that is extracted using a single wave can be employed to drive the electric car for hundreds of miles. The research community is focusing on developing the techniques to extract and convert this energy into reliable and cost-effective electricity. America can stand the frontline in developing the new energy industry that can meet a huge demand for energy with consistent and forecastable clean energy as it has a 50% population residing inside the 50 miles of coastlines.

Renewable Hydrogen Production. https://doi.org/10.1016/B978-0-323-85176-3.00014-7

7.1 Ocean Energy Productions Steps

The process of ocean waves energy conversion to electricity comprises different stages as follows:

Wind Blows Create Waves

The solar heat from the sun heats the air at different places that produce wind that blows through ocean surfaces. Wind produces surface waves that vary in size from ripples to approximately 100 ft high and can travel hundreds and thousands of miles before reaching the land with nearly zero energy loss.

Waves Approach Land

Unlike the regular light or radio waves, ocean waves do not offer constant amplitude and frequency but interact with the environment, each other and weather. When the wave reaches the land, it is different from any other usage which, research is intended to harvest the energy.

Waves Encounter Machines

Wave-energy converter devices are employed to converts ocean waves into electricity. Researchers are expecting the distinctive full-scale wave-energy converter devices to be attached in deep water miles offshore that offers the strongest wave energy. As wave-energy converter extracts energy through all sized waves moving in different directions, recognizing the machine type that can generate electricity most effectively is a significant challenge.

AquaHarmonics won the wave-energy prize competition that advanced the wave-energy converter energy capture potential and highlighted a variability of capable promising devices. They designed a point absorber, a two-part device that established a fivefold rise in the energy capture potential. One part moves statically underneath the surface and the other rides the surface waves. The surface part moves quicker as compared with the submerged part, and the wave-energy converter converts relative motion into electricity.

Machines Converting Waves into Electricity

Ocean waves move a wave-energy converter that motion drives a generator that generates electricity. The machines employing low-speed ocean waves motion having high energy content and converting these into high-speed motion that is essential for the generator is yet to be explored in terms of economics and reliability as it needs to survive severe ocean conditions.

Electricity Provided to the Grid

Wave energy can power the swaths of seaside homes and other facilities. Wave energy is extremely probable and can be established nearby load centers to condense the transmission needs and comfort the integration to the grid. Moreover, wave energy can power disseminated applications, such as desalination plants that desalinate the saltwater to assist the water-insecure societies and military bases.

Electricity Used for Hydrogen Production

Wave energy is enormously predictable and can be advanced nearby load centers. To eliminate the transmission needs and comfort the grid integration, the produced energy can directly be employed in the water electrolysis process to produce clean hydrogen. The produced hydrogen can be stored through the storage unit and can be employed for numerous purposes, such as electricity generation using fuel cells, electric and hybrid vehicles, combustion purposes, heating, as an energy carrier, and for the formation of ammonia, methanol, and other compounds.

Ocean water offers a significant source of renewable energy. Dissimilar technologies employ diverse approaches to harvesting energy. Some potential ocean energy sources can specifically be listed as follows:

- Tidal streams
- Tidal range (rise and fall)
- Ocean currents
- Ocean thermal energy
- Waves
- Salinity gradients

Ocean water involves a substantial amount of energy. Fig. 7.1 exhibits the potential of ocean energy resources. The theoretical potentials of tidal power, wave power, ocean thermal energy conversion, and salinity gradient were explored and quantified as 26,280 TWh/year, 32,000 TWh/year, 44,000 TWh/year, and 1650 TWh/year,

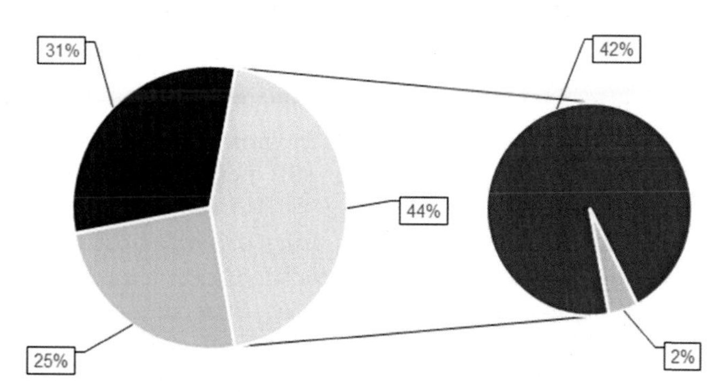

FIG. 7.1

Potential of ocean energy resources.

Data from [120].

respectively. The percentage share of the ocean energy potential classified into different categories of tidal energy, wave energy, ocean thermal energy conversion, and salinity gradient was explored and found to be 25%, 31%, 42%, and 2%, respectively.

Ocean currents and tidal streams offer reliable, environmentally benign, and locally generated renewable energy. They offer quite a lot of substantial advantages over other sources of renewable energy.

- Ocean energy offers a predictable and consistent electricity generation source. Dissimilar to the other renewable energy sources, namely wind and solar, ocean energy sources, namely wave energy, tidal energy, and ocean thermal energy are practically 100% predictable. The interminable flows generate consistency in the availability of future energy.
- Ocean currents and tidal streams are globally available on all the continents.
- Ocean water offers energy-intensive solutions as moving ocean water is 832 times denser as compared with moving air and creates efficient energy conversion conditions.
- In countless regions, the land is a threatened resource. Consequently, onshore solutions like solar and wind strive with other users, but ocean energy source requires minimal land use and has no graphical impact. Ocean energy technologies are hidden in the ocean depth and do not strive for land space.
- The large ocean currents and tidal streams resource can be exploited with comparatively slight environmental interaction, thus, contributing and presenting one of the most lenient large-scale electricity production methods.

There is a growing prerequisite for the development of new technology with the rising clean energy source requirement to complement solar and wind energy sources. One example is an ocean energy technology that can be installed in conjunction with solar and wind to enhance additional renewables onto the grid or environmentally benign hydrogen production system and reduce greenhouse gas (GHG) emissions.

Wave power offers an enormous resource potential but system development that can be established to withstand severe ocean environments and can efficiently generate electricity employing ocean energy is still a challenging task that needs to be further explored. Not only the technology needs to be engineered, explored, and established to harvest energy efficiently using dissimilar ocean conditions, nevertheless, technological development also needs to establish a design to endure and function reliably and consistently in extreme ocean conditions. The advanced systems for harvesting the wave energy to meet these challenges can be predictable to offer a more consistent energy supply in comparison with wind or solar energy.

Credit needs to be given to the recent advances in engineering simulation software and computing power wave technology companies; namely, Oscilla Power is continuously investigating and now is able to establish more precise model simulations and devices that can respond in actual ocean waves together with extreme waves. These technological advancements help them to analyze, evaluate, improve, and optimize dissimilar designs rapidly under dissimilar ocean conditions, develop efficient and innovative wave-energy technological solutions. After long research, analysis, testing, and modeling, Oscilla Power is certain that it has congregated on an efficient and affordable solution termed as a Triton system, which is a wave-energy converter exclusively designed for optimum energy conversion under dissimilar wave conditions together with survivability in exciting weather conditions.

To understand the Triton system, it is significant to comprehend a little regarding ocean waves. On the ocean surface, the concentrated ocean energy is found that reduces expressively with the rise in water column depth.

These implications carry a twofold impact on wave technology:

- Wave technology devices produce maximum energy on the ocean surface but become vulnerable under extreme conditions.
- If the operational floating devices will submerge beneath the surface, they will become harmless from extreme waves.

Oscilla constructed and established the shape, architecture as well as power conversion system of Triton to benefit from these factors. The consequential design comprises a large floating surface designed to move under all wave conditions and a submerged big concrete ring stay under the surface. The qualified movement of a highly active floating surface and the stable concrete ring is brought about to generate efficient energy capture under diverse conditions. Nevertheless, when the floating surface turns out to be susceptible to get harmed in extreme conditions, such as storms or tornados, Oscilla can fully or partially submerge the whole system remotely, to resist the floating surface on reacting to the waves. Even in such survival conditions, the system still generates the power; however, it is just underneath the surface.

Fig. 7.2 displays tidal-stream development. Fig. 7.2a exhibits the device types breakdown into four different technological categories of the horizontal axis (axial flow), crossflow, reciprocating systems, and others. The horizontal axis (axial flow) category involves 76% of technologies, crossflow involves 12% of technologies, reciprocating systems involve 8% of technologies, and 4% of technologies are classified as others. In total, 68% of the concepts employ a single turbine while the remaining 32% of technologies utilize the multiple rotors on each floating platform.

Fig. 7.2b exhibits the support structures breakdown into four different technological categories of rigid connection, tether/mooring, monopile, and other. The rigidly seabed connected category involves 56% of technologies that employ nonmonopile foundation system, tether/mooring category involves 36% floating systems technologies, monopile foundation systems involve 4% of technologies, and 4% of technologies are unspecified. In total, 48% of technologies employ gearbox and generator, 44% employ direct-drive perpetual magnet generator, and 8% are unspecified.

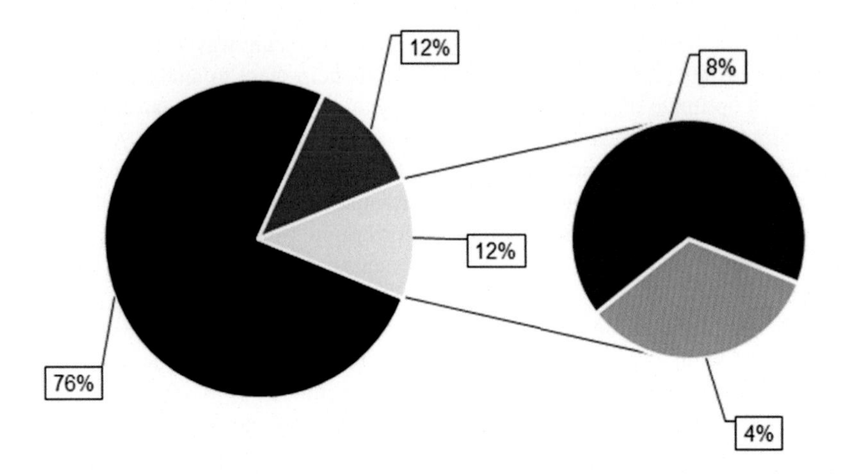

(a)

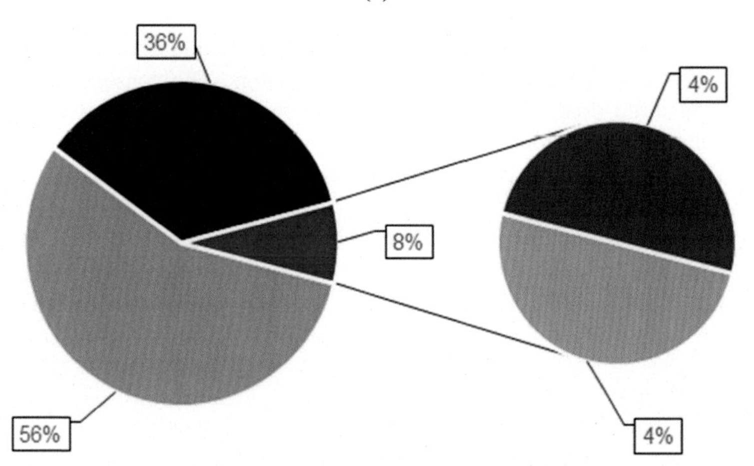

(b)

FIG. 7.2

Tidal-stream development: (a) device types breakdown; (b) support structures breakdown.

Data from [121].

Ocean renewable energy extracts the ocean water power to generate electricity that can be done using several ways; nevertheless, the resources having the greatest instantaneous potential concerning energy production are the following:

Waves: employing wave-energy converters to produce electricity
Tides: employing tidal barrages and turbines to produce electricity

More than a few wave-energy converters, tidal-stream devices, and tidal turbines are at development stages globally. Tides, currents, and waves can be employed to generate electrical power. Promising ocean technologies are as follows:

- Wave energy: converters extract the energy confined in ocean waves and employ it to produce electricity. Converters consist of oscillating columns of water that trap the air pockets to operate a turbine, wave motion is employed by oscillating body converters, and height differences are employed by overtopping converters.
- Tidal energy: formed either by tidal-stream technologies, tidal-range technologies, or hybrid applications employing a barrage using a dam to produce electrical power among high and low tides.
- Salinity gradient energy arises from different salt concentrations and occurs anywhere a river drains into an ocean. Demonstration projects use a water desalination system to process flowing water through a membrane to upsurge the saltwater tank pressure and reverse electrodialysis having salt ions passing through salt- and freshwater alternating tanks.
- Ocean thermal energy conversion: that produces electrical power using the temperature difference among the cold seawater in depths and warm surface seawater.

7.2 Ocean Energy Conversion

Oceans cover above 70% of the earth's surface that makes them the largest global solar collectors. The solar heat warms the water surface much more as compared with deep ocean water, and temperature difference produces thermal energy. Just a tiny share of heat trapped in the ocean can power the world.

7.2.1 Types of Ocean Thermal Energy Conversion Systems

Ocean thermal energy conversion can be employed for numerous applications, such as power generation. In general, there are three kinds of electricity conversion systems as follows:

1. Closed cycle
2. Open cycle
3. Hybrid

Closed-cycle systems employ the warm ocean surface water to extract the heat and vaporize the low-boiling point working fluid like ammonia. The vapor expansion rotates the turbine, and the accompanied generator generates electricity. Open-cycle systems boil seawater by functioning at low pressures that form steam that drives the turbine and generator while the hybrid systems are formed with the combination of both open- and closed-cycle systems.

Ocean mechanical energy is dissimilar from ocean thermal energy. Although the sun affects ocean activities, waves are primarily driven by the winds and tides are primarily driven by gravitational pull. Therefore, waves and tides are intermittent energy sources whereas, ocean thermal energy is comparatively consistent. Likewise, dissimilar to thermal energy, the wave and tidal energy-based electricity conversion includes mechanical devices.

A barrage is characteristically employed to convert tidal energy into electrical power by passing water through turbines connecting to a generator. There are three elementary systems for wave-energy conversion: floating systems that drive hydraulic pumps, channel systems that flue waves into reservoirs, and oscillating water column systems. The mechanical power produced from such systems either stimulates the generator directly or transmits to a working fluid, such as water or air that operates a turbine and generator.

Ocean energy could be the future wave, and wave-power systems that are state of the art in clean, renewable energy. Wave-energy systems employ the movement of water to make electrical power. Some of such types of devices extract the power of the breaking wave while others utilize swells. Most of the devices employ wave pressure near the ocean floor thus far, and all carry the same objective of converting wave energy into electricity that can either be employed to produce clean hydrogen or power the electric grid. The electric grid is the cable network that transfers electricity to buildings and homes, or it can directly be employed in the water electrolysis process to generate hydrogen.

7.2.2 Wave Power Generation

Wave power is constrained to sites nearby oceans. With all wave-power potential, the research community is testing different generator types that can convert ocean energy into electrical power efficiently. Meanwhile, they are also trying to make certain advancements so that sea life will not be harmed during this process. Numerous steps are trailed to generate wave power as follows:

Figuring out best sites to install energy converters

Not all coastal sites are suitable for producing wave power. The land shape underneath the sea faces fluctuations in the shape and size of waves. Wave-energy converters are expensive as well; thus, the best sites must have plenty of acting waves nevertheless, not much that storms can damage the converters. To understand the best suitable sites, research scientists use computer models. The scientists are conducting research to discover suitable places to install wave-energy converters. The oceans occupy immense natural energy amounts, but the challenging part is to extract and efficiently convert it to the electrical power. A part of this challenge and encounter is the ocean itself due to the waves under severe conditions. The hardware can also practice under extreme weather conditions. Huge storm waves can harm the converters, and salty seawater can corrode or break down metal parts.

Sea carpet

Scientists communities and engineers are conducting numerous research to meet such challenges. Research works have resulted in numerous design types. A newly designed converter type floats on the water surface, tied on the ocean floor with wave generators. Another type has one end attached with the bottom of the sea by the other end that is free to flip side by side since waves flow over it while others employ water or air pressure to produce electricity.

One of the state-of-the-art systems looks like a flat carpet. A converter is designed at the University of California to mimic the muddy seafloor. Sites having lots of mud are suitable to absorb incoming waves. Fishermen search for the muddy sites in shallow seas during rough weather conditions.

The carpet part of this designed converter is constructed using a smooth rubber sheet. It exists close to the seafloor wherever it can flex and bend along the waves. As it travels up and down, it thrusts the posts in and out of the piston pump. The pump produces the electrical power using the movement of the piston and travels along the cable to the electric grid or the water electrolysis process for hydrogen production.

Environment friendly

Finding new renewable energy sources is suitable for the environment as it causes less environmental pollution and less GHG emissions to the surroundings. Wind energy can cause birds, migration; for instance, some estimations indicate that thousands of birds might die from colliding with enormous spinning blades each year. The minor heights of wave-energy converters reduce the interference probably with migrating animals; nevertheless, the interaction with the ocean environment is needed to be considered.

One alarming factor is regarding the ecological impacts to absorb all the energy after incoming waves. Energy appointed from waves can reduce the remaining energy as ocean waves stay near the shore. Smaller-sized waves might cause fewer nutrients to mix inside the water column that is the water between the ocean bottom and overhead surface. This can affect the living species but can also be advantageous as wave-energy converters can provide coastal protection support by reducing erosion.

The electric generators can also affect wildlife interaction. Numerous birds and marine animals chase fishes in the sites that are ideal for wave converters. Converters can likely attract fishes if smaller critters seek refuge there which can attract hungry predators. This can assist in boosting marine life in nearby areas. However, fish types and other animals can get scrambled up in extended cables that can bind surface-floating energy converters. Therefore, researchers must investigate where they are suitable to mount and install such converters to validate that they will not be harmful to the local ecosystems.

The noise from the converters is another concern that can cause trouble for fishes and other animals that count on the sound to search for food and to communicate. The bottomless boat growls and loud sonar sound cause problems for ocean species. These critters might struggle to search for food or becoming disoriented. Though, wave converters are improbable to generate high noise levels. The noisiest part occurs once converters are primarily mounted at some site, but they become quiet once they start operating.

From the enormous ocean resources, wave energy carries the immense potential to make a massive contribution to the future clean energy requirements. However, this must be completed but keeping harmony with the marine environment in a sustainable way.

Ocean energy is an enormous clean, consistent, and predictable source of renewable energy that is still unexploited. The question is if this energy can assist in providing the global transition toward clean and sustainable energy solutions. Fig. 7.3 displays the installed ocean power generation and installed capacity trends from 2010 to 2017. A steady rise in both ocean power generation and installed capacity can be depicted from the figure. Fig. 7.3 exhibits the installed ocean power generation and installed capacity trends. A steady increase in electricity

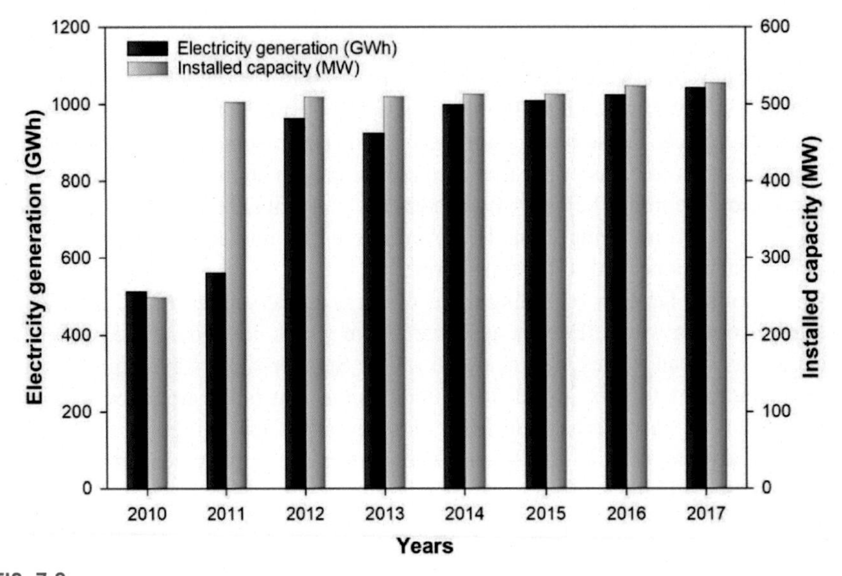

FIG. 7.3

Installed ocean power generation and installed capacity trends from 2010 to 2017.

Data from [122].

generation can be depicted from 514.6 to 1041.4 GWh from 2010 to 2017 and meanwhile, installed capacity increased from 249.6 to 527.6 MW.

A recent study[123] was published on the extraction of tidal energy using three-dimensional ocean models. High-performance computing access has made 3-D modeling for the assessments of tidal energy resources. Advancements in numerical modeling codes and computing resources have aided high-resolution 3-D ocean model applications at basin scales, although, at comparatively higher computational cost as compared with a conventional 2-D modeling approach. A comparative analysis between 2-D and 3-D modeling techniques for tidal energy extraction was commenced and methods differences were inspected through both impact assessment and resource viewpoints was conducted in this study. It was established that 3-D modeling tidal energy harvesting can be effectively incorporated in Pentland Firth regional ocean that is among the top global regions for the development of tidal-stream energy by conducting several numerical experiments. They demonstrated that 3-D flow resolving is significant to reduce environmental resource assessment uncertainty. Furthermore, they showed that 2-D methods of tidal energy extraction may lead to the velocity profile misrepresentation once functional to 3-D models, representing the significance of determining 3-D flows in the tidal array vicinity.

A low-cost hydrokinetic turbine farm to harvest the oceanic surface waves energy without employing plenum chambers and provided a feasibility study was presented in a recent study.[124] The plenum chamber eradication along with the complicated valve system reduced the wave-energy conversion expenses by 23%. The feasibility study was conducted for low- and high-frequency conditions. For low-frequency waves, 300 rpm angular velocity was offered as an optimal selection parameter for further studies while a consistent analytical approach founded using validated methods for high-frequency conditions was developed. Analytical results indicated that a wells, turbine employing a diameter of 60 cm could produce around 1600 W power in high-frequency waves. The power coefficient and efficiency values under high-frequency conditions are comparatively less in comparison with the low-frequency scenario; however, the immense utilization of small-scale and plenum chambers elimination reduced the overall conversion cost. The high-frequency surface wave abundance is a development motivation for the hydrokinetic wells turbine system to extract clean energy at a lower cost.

A recent study[125] has conducted the initial design and performance evaluation of a radial-inflow turbine for the ocean thermal energy conversion (OTEC) cycle. The OTEC cycle employs the temperature difference in the surface and deep seawater to generate power. The ocean thermal energy potential is substantial that offers an environmentally benign power system, even though, it offers very low thermal efficiency followed by the lower temperature difference in surface and deep seawater. Thus, the development of a high-efficiency turbine is significant to enhance the thermal efficiency of the OTEC cycle. A novel approach was proposed in this paper for the suitable selection of loading and flow coefficients. The three-dimensional viscous simulations and meanline analysis for designed turbines were directed to validate the projected method for off-design and in-design conditions. The outcomes established that the proposed model can be employed to offer the optimal radial-inflow turbine for design conditions.

The large-scale integration of ocean wave energy was considered in a recent study.[126] Their paper assessed the effective impacts of the development of large-scale ocean wave energy in the US Pacific Northwest. The forecasting and high-resolution wave-energy data was created for the wave-energy arrays that are spatially distributed along the coastal region. Geographic divergence is originated to bound the production variability scales rate with capacity installed over the timescales oscillating from minutes to hours. The condensed variability supports the forecasting of short-term wave generation precisely. Once modeled inside the operational structure of the primary balancing area authority of the region, the large-scale wave energy was originated to offer a comparatively high capacity rate and reduced integration costs as compared with wind energy.

People have been struggling to extract ocean energy for hundreds of years starting with Pierre-Simon Girard a French engineer in 1799. The electric power research institution approximates the waves contravening along the coastline can produce 2640 TWh of electricity each year. However, since shipping, naval operations, fishing, or environmental apprehensions take primacy in some areas, the recoverable power quantity is estimated to be 1170 TWh each year, which is practically one-third electricity amount of the United States consumption per year.

Energy is intrinsic in ocean waves movement, in the temperature difference among cold deep water and warm surface water in the disproportion in salinity among salt and freshwater and in ocean tides and currents. International Energy Agency estimates that the wave power can generate 8000−80,000 TWh power per year, ocean thermal energy conversion can generate 10,000 TWh, osmotic power can generate 2000 TWh using salinity difference and marine, and tidal currents can generate 1100 TWh of power. Ocean thermal energy conversion, marine currents, osmotic energy, and some other wave-energy types can be used to meet the base-load power or be employed to produce hydrogen.

7.3 Ocean Energy Devices and Designs

Alaska and the Pacific Northwest are the regions with the greatest wave-energy potential are in the United States. Tidally driven waves passing through the coasts of Korea, China, and European parts are the finest places to achieve ocean thermal energy. Researchers, engineers, and scientists have developed diverse devices and designs to employ ocean energy. The following are some of the significant examples.

Point Absorber Buoy

A floating buoy is attached to an alleviating base on the seafloor. As the buoy moves up and down through waves, it rotates the generator shaft that attaches the buoy with the base and creates electrical power. The produced electrical power is supplied to the onshore grid.

Surface Attenuator

The surface attenuator device involves multiple arms that hover on the water surface. The flutter of the waves produces a flex motion at the junctions coupled with hydraulic pumps that operate a generator and produces electricity. The electrical power is supplied to the shore through a cable passing under the sea or employed in the water electrolysis process for hydrogen production.

Oscillating Water Column

A tangible concrete structure is built including an enclosed chamber with below a sea level opening. As waves arrive and pass over the opening, the level of water increases in the chamber, obliging the overhead air to pass through the turbine coupled with the upper chamber opening. The airflow rotates the turbine that produces electricity using a generator. During the wave ebb tides, the air is lapped back inside the lower chamber, rotating the turbine in the opposite direction that is connected to the generator and producing electricity.

Overtopping Device

This structure and assembly can be either offshore floating or onshore. It extracts waves that fall apart into the storage reservoir. After that, waves pass through the turbine that produces electrical power as waves pass over it and reach back to the ocean.

Wave Carpet

A flexible membrane outspreads sideways the floor of the ocean and waves move over it that cause the up and down movement, which pushes seawater to the discharge pipe over vertical pump that is connected to the underneath of membrane. The high-pressure water coming from the pipe operates the onshore turbine that generates electrical power. The carpet extracts approximately 90% energy that is comparatively very efficient.

Oscillating Wave Surge Converter

Oscillating wave surge converter device comprises a fixed end to an assembly, and the other is strapped by waves and moves. The movement pushes the piston and supports in and out movement and drives pressurized water over piping toward an onshore turbine that generates electrical power.

7.4 Types of Ocean Energy

The significant types of ocean energy are as follows:

- Ocean thermal energy
- Osmotic power
- Tides and Currents

7.4.1 Ocean Thermal Energy

Around 70% of solar energy reaches the earth on the oceans, and the majority is captured in ocean surface layers as heat. The OTEC employs the temperature difference between cold deep water and warm water surface that is generally at a minimum of 20°C that is used to capture electricity.

Working principle

The temperature difference between cold deep water and the warm water surface is employed to generate power in the ocean thermal energy conversion system. Fig. 7.4 exhibits the ocean thermal energy conversion system for power generation. Both cold and warm seawater is fed to the heat exchanger that keeps the different fluids separated. Ammonia is usually employed as the working fluid as it offers a low-boiling point. The hot water from the surface is fed to the heat exchanger adjacent to ammonia, and warm seawater boils the ammonia and generates vapors using a heat exchanger. These pressurized vapors then drive the turbine that is further accompanied by the generator to generate electrical power. Once the vapors of ammonia leave the turbine, they are brought down using a pipe positioned in the chamber that is surrounded by cold seawater tubes. The vapors of ammonia are cooled and converted to the liquid again and recycled back to the heat exchanger to continue the cycle.

7.4.2 Osmotic Power

This technology is founded on the natural phenomenon that low-concentration saltwater searches for high-concentration saltwater. Once different salinity waters combine and meet at the river mouth to grasp uniform salinity, energy is released. Osmotic power stations imitate this existence by employing osmosis, separation of a freshwater tank from the saltwater tank using a semipermeable membrane that permits freshwater to pass resists the saltwater. Freshwater is strained into the side of saltwater that equalizes the water salinity and raises the saltwater tank pressure. This pressure is employed to rotate the turbine that generates electricity.

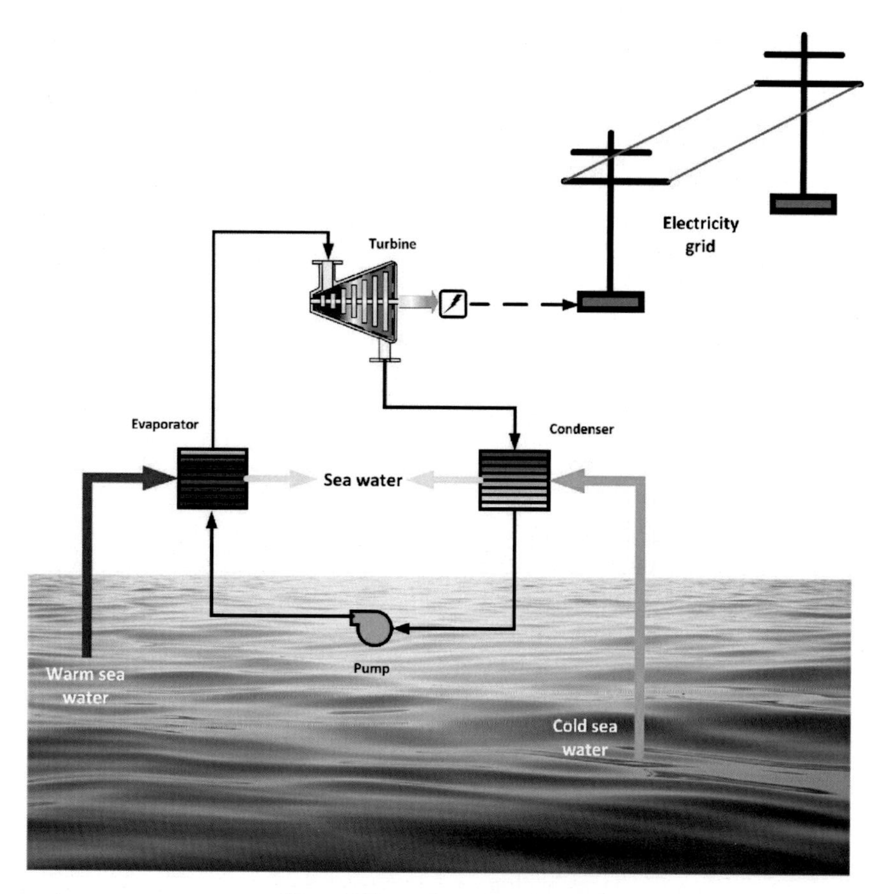

FIG. 7.4

Ocean thermal energy conversion system for power generation.

7.4.3 Tides and Currents

Tidal barrage

A structure similar to the dam is constructed by a lagoon, bay, or river. At high tide, the dam clutches the water out. Once channel gates are opened to permit the flow of water, the turbine at the channel gates produces electrical power as water travels through. When the bay fills up, the process is reversed that generates further electricity.

Dynamic tidal power

A 30 km or longer dam is constructed to the coast perpendicular and tailored with numerous turbines. A fence parallel to the coastline at the dam far end supports strengthening the pressure of water at both dam sides. The high-pressure water is employed in the turbines positioned in the dam. A 40-km dam can hold approximately 2000 turbines offering 5 MW power each and 10 GW net electrical power that can either be employed to power the millions of households or produce clean hydrogen.

Tidal current turbine

A tall turbine similar to the wind turbine attached to the base is positioned on the seafloor. The tidal currents pass through the rotors and generate electrical power. When tides move in the opposite direction and move out, the rotor reverses the direction and continues to produce power. The produced electricity is supplied to the onshore grid through cable or can be employed in the water electrolysis process for hydrogen production.

As a renewable resource of energy, ocean power consumes significant potential; however, it is in arrears as compared with other renewable energy sources as it experiences many challenges. Steel or concrete structure devices are required to stand up to continuous waves pounding and saltwater corrosion.

To scale up the ocean power industry, the construction of devices should not employ too complex or hard-to-obtain materials; in fact, materials should be readily available, economical, and easy to maintain. The significant technological challenges are not only associated with electrical power generation but mechanical systems, survivability and reliability, mooring and anchoring, predictability termed as wave forecasting, and integrating the generated power to the existing electricity grid. If generated power is employed in the water electrolysis process for hydrogen production, the challenge of integrating the generated power into the existing electricity grid can be eliminated.

Fig. 7.5 displays a visual illustration of the qualified measures for each type of ocean energy technology and their technological maturity level using a technology readiness level scale. The technology readiness level evaluates the maturity of growing technologies throughout their advancement and primary operations. The point to be noted is that even though technology readiness level scaling undergoes some boundaries, it is valuable however to provide an indicative maturity value for numerous technologies.

Some other significant parameters are noteworthy that impact the short-term development projections, for instance, manufacturing and economic performance to explore the other aspects outside the technology readiness level of different ocean energy technologies. Significant ocean energy technologies are as follows:

- Deep ocean current
- Tidal stream
- Wave-energy converters
- Tidal range
- Ocean thermal energy conversion
- Salinity gradient

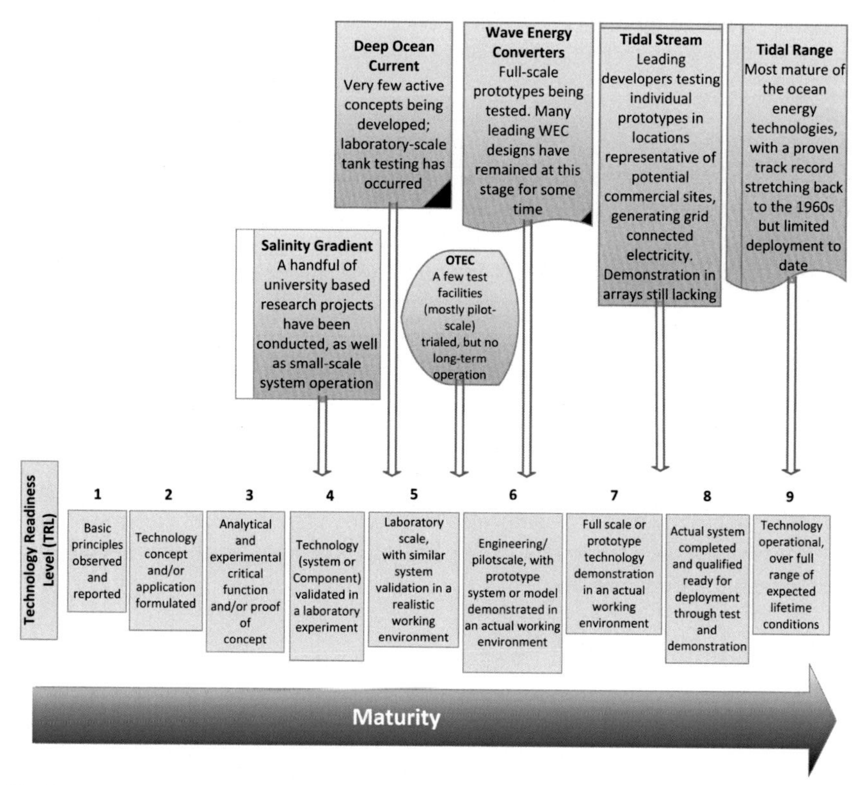

FIG. 7.5

Readiness of ocean energy technology.

Modified from [121].

7.5 Advantages and Disadvantages

Ocean energy is a massive clean, consistent, and probable source of renewable energy, and power is extracted using this source along the coastal regions. The energy extracted from the ocean looks like a harmless and unlimited supply. The question is if this energy has some drawbacks along with significant advantages. Thus, the advantages and disadvantages of ocean energy are explored in detail.

7.5.1 Advantages of Ocean Energy

Following are the significant advantages of harnessing ocean energy:

Renewable

The finest advantage regarding ocean energy is that it is continuous and consistent. The ocean waves will always be crashing the shores near the coastal regions. Even though waves flow back, they always return. The ocean water uses solar energy to absorb the heat, and the temperature difference is also employed to generate power in some ocean energy technologies. Ocean energy does not suffer from the fear of running out, unlike fossil fuels.

Environment friendly

Likewise, generating power using ocean energy produces zero harmful byproducts, for example, waste, gas, and pollution, unlike fossil fuels. Ocean energy can be employed directly to the turbine using the waves and to power the accompanied generators to convert mechanical energy into electrical. In the current energy-powered world with continuously growing energy demand, a clean energy source is hard to find.

Abundant and extensively available

One more advantage of employing this energy is the abundance and extensive availability. A large number of harbors and big cities exist subsequent to the ocean, and energy extracted using the ocean energy can either be supplied to the electricity grid to meet the electricity demand or can also be employed to the water electrolysis process for hydrogen production that can be stored and employed to the fuel cells for power generation upon demand.

Variety of methods to extract

Another advantage is that a variety of technological methods exist to harness ocean energy. Current energy extraction techniques range from installed power plants using hydro turbines to oceangoing vessels consisting of enormous structures that are positioned in the sea to extract the ocean energy and generate power.

Predictable

The principal advantage of ocean power in comparison with all other alternative sources of energy is that the amount of power that can be generated is predicted without difficulty. The ocean energy sources are reliable and consistent and demonstrate considerably better performance as compared with other sources depending on sun exposure or wind.

Less dependence on foreign oil

The dependence on foreign companies to employ fossil fuels is predicted to be condensed if ocean energy is extracted efficiently using wave power. It will not only assist in restricting air pollution but will also deliver green jobs.

No land damage

Dissimilar to the fossil fuels that cause immense land damage as digging and mining are processed to obtain fossil fuels to extract energy, ocean and wave power do not cause any land damage. It is one of the safest, cleanest, and preferred methods and approaches to use ocean energy for power extraction.

Reliable

Ocean energy is a very reliable and consistent energy source as water waves practically always keep moving. Even though there are tides and ebbs, the regular motion always continues; consequently, energy can be extracted continuously. Even though the energy quantification that is transported and produced through water waves differs from season to season, nevertheless, production of energy is continuous.

Huge energy amounts can be generated

An enormous amount of power can be generated from ocean energy using waves. It is so massive that along the shore, approximately 30—40 kW/m is the power density of a wave. While going further deeper inside the ocean, power density rises approximately to 100 kW, which is truly massive.

Offshore wave-power harnessing

Wave power can also be harvested offshore. The power plants harvesting electrical power can be constructed offshore. Such options can assist in resolving the power plant problems of being close to the land. When the power plants are positioned offshore, the potential energy of waves rises as well.

Subsequently to the flexibility in positioning offshore plants, the harmful environmental effects reduce as well. The only probable issue employing offshore power plants is the economical factor as they are expensive to construct. However, this option can be suitable for the environment to produce clean power.

7.5.2 Disadvantages of Ocean Energy

Following are the drawbacks and challenges of ocean energy:

Locations suitability

The major drawback to extract energy using water waves is location suitability. Only nearby cities and power plants can benefit straight from it. Landlocked nations, noncoastal, and cities distant from the water source need to explore the substitute power sources, or power generated using wave energy is supplied to the electrical grid.

Effect on ecosystem

Even though ocean or wave power is a clean energy source, it still produces some hazards for nearby creatures. Huge machines need to be positioned nearby and inside the water to extract energy from waves. These devices disrupt the seafloor, variation in the near-shore habitat creatures, such as starfish and crabs, and produce noise that interrupts the nearby sea life. Similarly, there is a toxic chemical danger that is employed to the wave-energy platforms polluting and spilling the nearby water.

Source of disturbance

Another drawback is that it disrupts private and commercial vessels. Power plants that harness the wave energy need to be positioned nearby the coastline to perform their job and they need to be around populated areas and cities to be utilized by the community.

Wavelength

Wind power is extremely wavelength-dependent, for instance, wavelength, wave speed, and water density. They involve a consistent powerful wave flow to produce a substantial wave power. Some zones involve inconsistent wave behavior, and forecasting the precise wave power becomes unpredictable.

Weak rough weather performance

The wave-power performance drops substantially through severe weather, but they must endure rough weather.

Visual and noise pollution

Wave-energy producers might be unpleasant for the community living very close to the coastal regions. They look similar to the large devices employed in the central ocean and abolish ocean beauty. They also make noise pollution; nevertheless, the noise is normally shielded by the noise of the waves that are much greater than wave generators.

Production costs

Even though wave energy has many advantages; however, the enormous production cost is one of the critical side effects. Energy generation using wave-energy entails an enormous setup. Likewise, the lifetime of technological usage is quite indeterminate in such cases as the waves are quite unpredictable.

Fig. 7.6 exhibits the device types breakdown of being followed by wave-energy technology developers. Fig. 7.6a displays the wave-energy applications are classified into nearshore, offshore, nearshore, and offshore and onshore categories. In total, 64% of designed technologies are precisely for offshore use, 19% designed technologies are categorized in nearshore use, 11% of designed technologies are used for both offshore and nearshore applications, and 6% are intended for shoreline structure usage. Fig. 7.6b shows the wave-energy installation that is classified into floating, fully submerged, and bottom-standing categories. In total, 67% of designed technologies are floating, 19% of designed technologies are bottom-standing, and 14% of designed technologies are fully submerged.

Fig. 7.6c represents the wave-energy orientation that is classified into point absorber, attenuator, and terminator categories. In total, 53% of designed technologies are absorbers, 33% of designed technologies are attenuators, and 14% of designed technologies are attenuators.

Fig. 7.6d displays the wave-energy power take-off that is classified into hydraulic, direct-drive, pneumatic, hydro, and unspecified categories. In total, 42% of designed technologies use hydraulic, 30% of designed technologies use direct-drive, 11% of designed technologies use pneumatic, another 11% of designed technologies use hydro machinery, and 5.6% are unspecified.

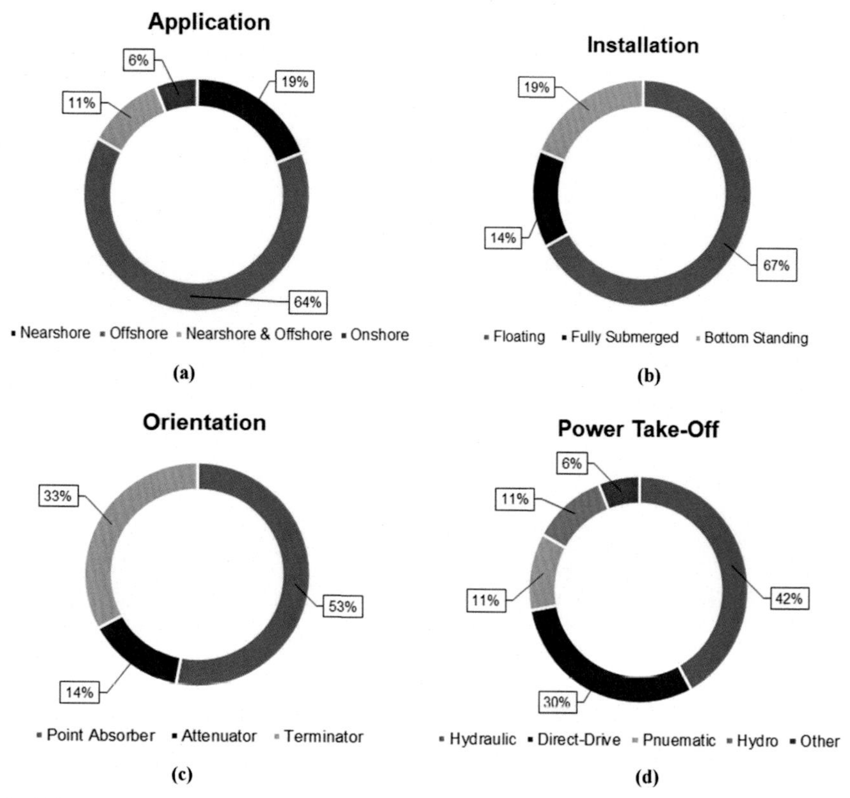

FIG. 7.6

Device types breakdown being followed by wave-energy technology developers: (a) application, (b) installation, (c) orientation, and (d) power take-off.

Data from [121].

7.6 Case Study 6

A case study is designed for the ocean thermal energy conversion-based hydrogen production shown in Fig. 7.7. Around 70% of solar energy reaches the earth on the oceans, and the mainstream is captured on the surface of the ocean as heat. The ocean thermal energy conversion employs the difference in temperature between cold deep water and warm water surface that is usually as lowest as 20°C that is employed to capture electricity.

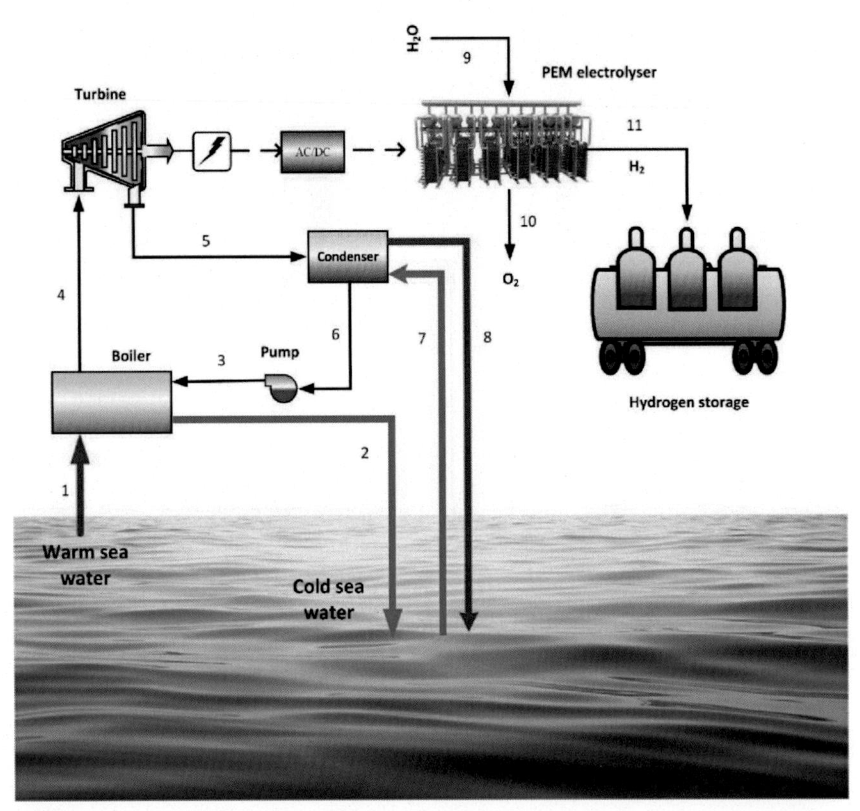

FIG. 7.7

Ocean thermal energy conversion system for hydrogen production.

7.6.1 System Description

The warm sea surface water and the working fluid pass through the boiler that works as a heat exchanger. Ammonia is usually employed as the working fluid as it offers a low-boiling point. The hot water from the surface is fed to the boiler adjacent to ammonia, and warm seawater boils the ammonia and produces vapors by extracting heat from the hot seawater. This boiler also keeps both the fluids, warm seawater and ammonia, separate from each other. The warm seawater transfers the heat to the working fluid and is fed back to the sea. A pump is installed before the boiler that raises the working fluid pressure before reaching the boiler. Thus, the pressurized vapors are then employed to drive the turbine that is further attached to the generator to produce electrical power. Once the ammonia vapors leave the turbine, they are brought down and fed to the condenser.

The cold deep-sea water is employed in the condenser that operates as the heat exchanger, and heat is transferred from ammonia to the cold deep-sea water that gets heated up. The vapors of ammonia are cooled and converted to the liquid again in the condenser and employed in the pump that increases the pressure and recycles back to the boiler to continue the cycle. The case study is investigated using two different scenarios. In the first scenario, the ocean thermal energy conversion is employed for power generation while in the second scenario, the produced power is employed in the water electrolysis process for hydrogen production. The efficiencies and performance of both scenarios are defined separately.

7.6.2 Analysis

The ocean thermal energy conversion cycle includes a pump, boiler, turbine, and condenser along with the proton exchange membrane (PEM) electrolyzer for clean hydrogen production. The model equations employed to investigate the performance of individual components and the overall system are presented. The ammonia is converted from saturated liquid to saturated vapor utilizing the heat of the warm seawater in the boiler. The pressurized ammonia vapors are employed to drive the turbine that expands the ammonia vapors to lower pressure and temperature and generates electrical energy using a generator. The deep seawater is employed in the condenser to extract the additional heat, and liquid is pressurized to feed the boiler using a pump that also circulates the working fluid through the cycle. Table 7.1 displays the design constraints of cold water temperature and warm water temperature used for the case study to produce power and hydrogen.

Table 7.1 Design Constraints of the designed Case Study for Hydrogen and Power Production [127].

Design Parameters	Value
Cold water temperature	5°C
Warm water temperature	28°C

Boiler

The mass, energy, entropy, and exergy balance equations of the boiler are as follows:

$$\dot{m}_1 = \dot{m}_2, \dot{m}_3 = \dot{m}_4 \tag{7.1}$$

$$\dot{m}_1 h_1 + \dot{m}_3 h_3 = \dot{m}_2 h_2 + \dot{m}_4 h_4 \tag{7.2}$$

$$\dot{m}_1 s_1 + \dot{m}_3 s_3 + \dot{S}_{gen} = \dot{m}_2 s_2 + \dot{m}_4 s_4 \tag{7.3}$$

$$\dot{m}_1 ex_1 + \dot{m}_3 ex_3 = \dot{m}_2 ex_2 + \dot{m}_4 ex_4 + \dot{E}x_d \tag{7.4}$$

Turbine

The mass, energy, entropy, and exergy balance equations of the turbine are as follows:

$$\dot{m}_4 = \dot{m}_5 \tag{7.5}$$

$$\dot{m}_4 h_4 = \dot{m}_5 h_5 + \dot{W}_{out} \tag{7.6}$$

$$\dot{m}_4 s_4 + \dot{S}_{gen} = \dot{m}_5 s_5 \tag{7.7}$$

$$\dot{m}_4 ex_4 = \dot{m}_5 ex_5 + \dot{W}_{out} + \dot{E}x_d \tag{7.8}$$

Condenser

The mass, energy, entropy, and exergy balance equations of the condenser are as follows:

$$\dot{m}_5 = \dot{m}_6,\ \dot{m}_7 = \dot{m}_8 \tag{7.9}$$

$$\dot{m}_5 h_5 + \dot{m}_7 h_7 = \dot{m}_6 h_6 + \dot{m}_8 h_8 \tag{7.10}$$

$$\dot{m}_5 s_5 + \dot{m}_7 s_7 + \dot{S}_{gen} = \dot{m}_6 s_6 + \dot{m}_8 s_8 \tag{7.11}$$

$$\dot{m}_5 ex_5 + \dot{m}_7 ex_7 = \dot{m}_6 ex_6 + \dot{m}_8 ex_8 + \dot{E}x_d \tag{7.12}$$

Pump

The mass, energy, entropy, and exergy balance equations of the pump are as follows:

$$\dot{m}_6 = \dot{m}_3 \tag{7.13}$$

$$\dot{m}_6 h_6 + \dot{W}_{in} = \dot{m}_3 h_3 \tag{7.14}$$

$$\dot{m}_6 s_6 + \dot{S}_{gen} = \dot{m}_3 s_3 \tag{7.15}$$

$$\dot{m}_6 ex_6 + \dot{W}_{in} = \dot{m}_3 ex_3 + \dot{E}x_d \tag{7.16}$$

PEM electrolyzer

The PEM electrolyzer splits the water into its constituents using electrical power. Conductivity shows the capability of a material to conduct electric current. To relate the resistance of a conductor with conductivity, the following correlation is used:

$$R_c = \frac{l}{\sigma A} \tag{7.17}$$

here R_c signifies the conductor resistance, σ denotes the conductivity, A indicates area, and l represents the conductor length. Ohmic losses can be calculated using the specific membrane electrical conductivity equations and determined using the following correlation presented by Ni et al.[66]:

$$\sigma(c,T) = (0.5139c - 0.326)\exp\left(1268\left(\frac{1}{303} - \frac{1}{T}\right)\right) \tag{7.18}$$

The electrical conductivity is well defined as $\sigma = \frac{dx}{dR}$ where R represents electrical resistance and differential equation can be represented as $dR = \sigma^{-1}dx$:

The correlations for the concentration overpotential, activation overpotential, and exchange current density are as follows:

$$\Delta E_{\text{conc}} = J^2 \left(\beta \left(\frac{J}{J_{\text{lim}}} \right)^2 \right) \tag{7.19}$$

$$\Delta E_{\text{act},i} = \frac{\text{RT}}{F} \ln \left(\frac{J}{J_{0,i}} + \sqrt{\left(0.5 \frac{J}{J_{0,i}} \right)^2 + 1} \right) \tag{7.20}$$

$$J_{0,i} = J_{\text{ref},i} \exp \left(-\frac{\Delta E_{\text{act},i}}{\text{RT}} \right) \tag{7.21}$$

where $J_{0,i}$ is cathodic and anodic exchange current density.

Performance assessment

The case study is investigated employing two different scenarios. In the first scenario, the ocean thermal energy conversion is employed to generate power while in the second scenario, the generated power is employed in the water electrolysis process for hydrogen production. The efficiencies and performance of both scenarios are defined separately in this section. Eq. (7.22) displays the net power output of the ocean thermal energy conversion system.

$$\dot{W}_{\text{net}} = \dot{W}_{\text{turb}} - \dot{W}_{\text{pump}} \tag{7.22}$$

The energetic and exergetic efficiencies of the first case scenario producing electrical power can be expressed as follows:

$$\eta_{en_{\dot{W}}} = \frac{\dot{W}_{\text{net}}}{\dot{m}_1 h_1 - \dot{m}_2 h_2} \tag{7.23}$$

$$\psi_{ex_{\dot{W}}} = \frac{\dot{W}_{\text{net}}}{\dot{m}_1 ex_1 - \dot{m}_2 ex_2} \tag{7.24}$$

The energetic and exergetic efficiencies of the second case scenario producing hydrogen can be expressed as follows:

$$\eta_{en_{H_2}} = \frac{\dot{m}_{H_2} \text{LHV}_{H_2}}{\dot{m}_1 h_1 + \dot{m}_2 h_2} \tag{7.25}$$

$$\psi_{ex_{H_2}} = \frac{\dot{m}_{H_2} ex_{H_2}}{\dot{m}_1 ex_1 - \dot{m}_2 ex_2} \tag{7.26}$$

7.6.3 Results and Discussion

The designed case study is analyzed and investigated in detail to explore the performance of individual components and the overall system. The efficiencies of both designed scenarios of power and hydrogen production are evaluated using parametric studies. The significant input parameters of the turbine input pressure, seawater flowrate, pump efficiency, and ambient temperature are employed to investigate the system performance.

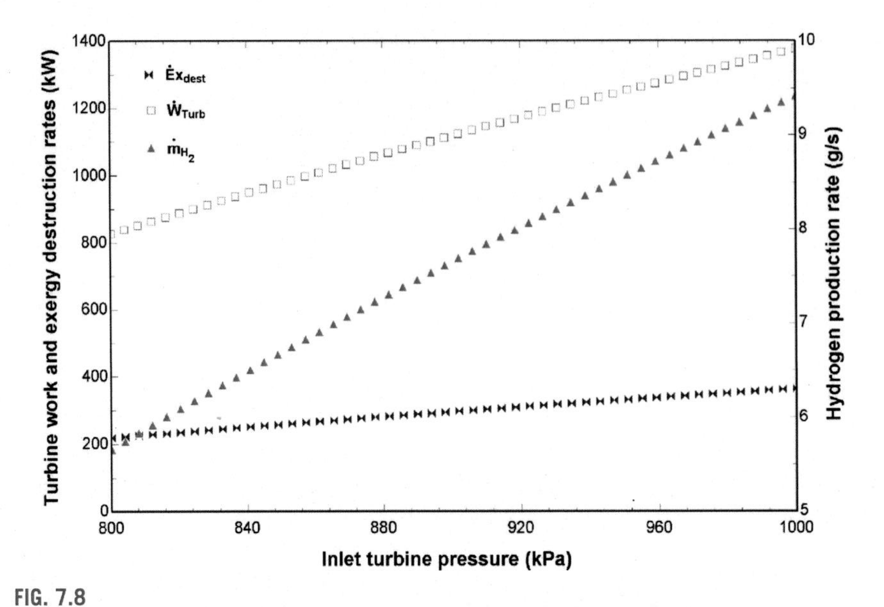

FIG. 7.8

Effect of inlet turbine pressure on work and exergy destruction rates of turbine and hydrogen production.

Fig. 7.8 displays the effect of inlet turbine pressure on work and exergy destruction rates of turbine and hydrogen production. The inlet turbine pressure ranged from 800 to 1000 kPa for the parametric study. The rise in inlet turbine pressure causes the turbine work rate to rise from 826.3 to 1376 kW, exergy destruction rate to rise from 218.9 to 364.7 kW, and hydrogen production rate to rise from 5.661 to 9.412 g/s.

It is important to investigate the effect of inlet turbine pressure on the significant system parameters to explore the variability in the designed parameters. Fig. 7.9 exhibits the effect of inlet turbine pressure on the turbine, pump, and overall work rates. The rise in inlet turbine pressure causes the pump work rate to rise from 11.2 to 21.16 kW, turbine work rate to rise from 826.3 to 1376 kW, and overall work rate to rise from 815.1 to 1355 kW.

Investigating the system performance in terms of energetic and exergetic efficiencies of two different case scenarios is significant to explore the variability in the designed parameters. Fig. 7.10 exhibits the effect of inlet turbine pressure on the efficiencies of power and hydrogen production case scenarios. The rise in inlet turbine pressure causes the energetic efficiencies of the power and hydrogen production systems to rise from 3.45% to 5.714% and 2.92% to 4.841%, respectively, and the exergetic efficiencies of the power and hydrogen production systems to rise from 8.355% to 13.84% and 6.847% to 11.34%, respectively.

The effect of pump efficiency on the significant parameters of the designed case study is also imperative to be investigated. Fig. 7.11 displays the effect of pump

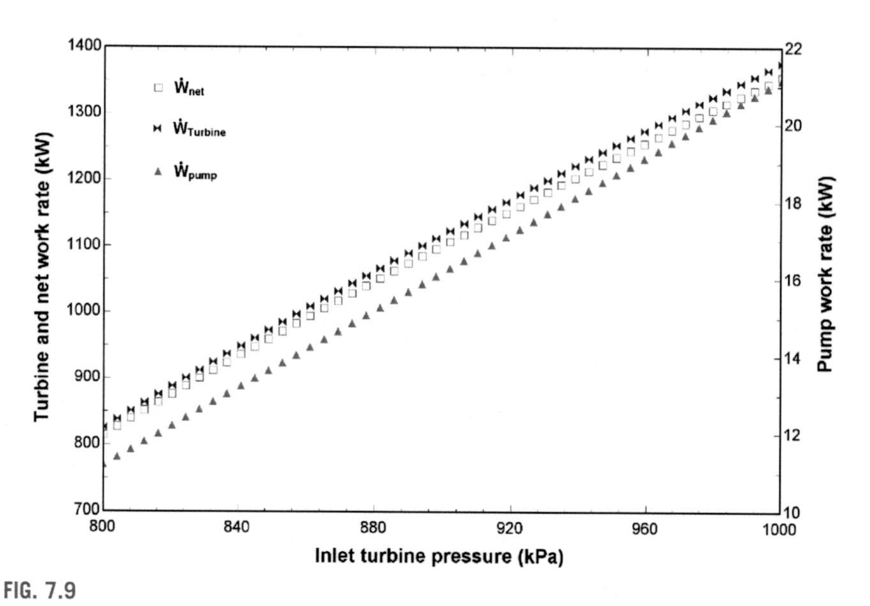

FIG. 7.9

Effect of inlet turbine pressure on the turbine, pump, and overall work rates.

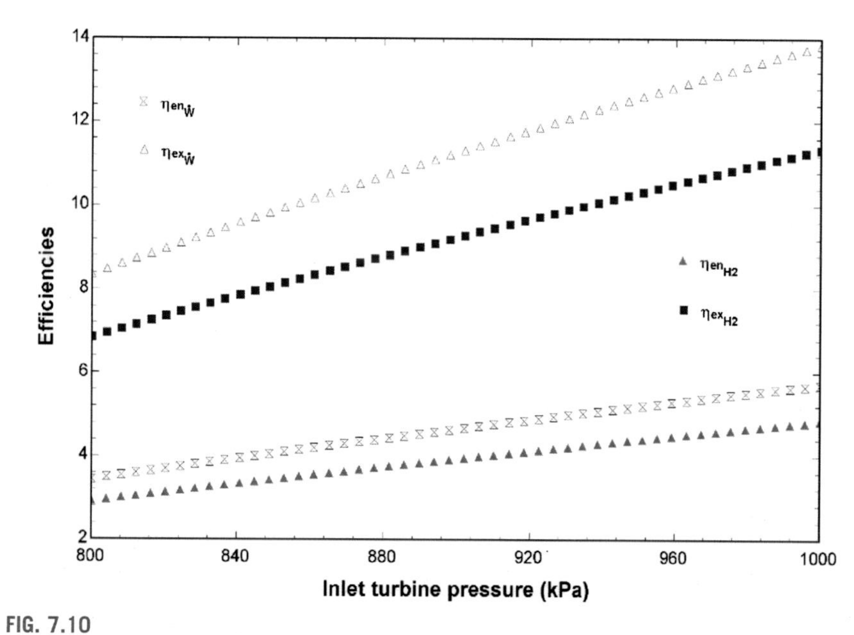

FIG. 7.10

Effect of inlet turbine pressure on the efficiencies of power and hydrogen production case scenarios.

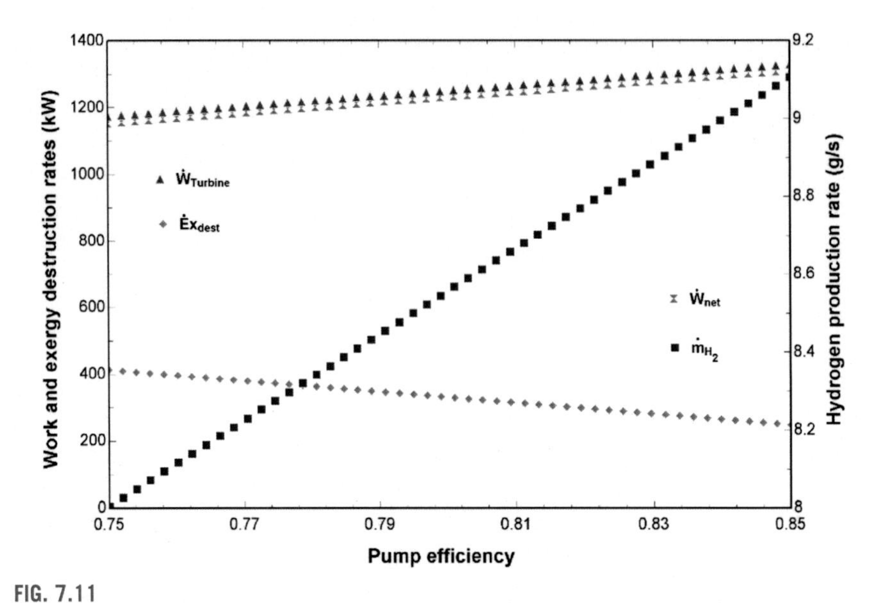

FIG. 7.11

Effect of pump efficiency on the work and exergy destruction rates of turbine, overall work rate, and hydrogen production.

efficiency on the work and exergy destruction rates of the turbine, overall work rate, and hydrogen production. The pump efficiency ranged from 75% to 85% for the parametric study. The rise in pump efficiency causes the turbine work rate to rise from 1173 to 1329 kW, turbine exergy destruction rate to drop from 414.2 to 248.5 kW, overall work rate to rise from 1153 to 1311 kW, and hydrogen production rate to rise from 8.005 to 9.107 g/s.

The effect of ambient temperature on the system performance in terms of energetic and exergetic efficiencies of two different case scenarios is significant to explore the variability in the designed parameters and is imperative to be investigated. Fig. 7.12 exhibits the effect of ambient temperature on the efficiencies of power and hydrogen production case scenarios. The ambient temperature ranged from 0°C to 40°C for the parametric study. The rise in ambient temperature does not affect the energetic efficiencies of the power and hydrogen production systems while the exergetic efficiency of the power production system rises from 6.426%, reaches the maximum of 22.53% at 16.33°C, and drops down to 3.821% and the exergetic efficiency of the hydrogen production system rises from 5.266%, reaches the maximum of 18.46% at 16.33°C, and drops down to 3.131% at the maximum pressure of 40°C.

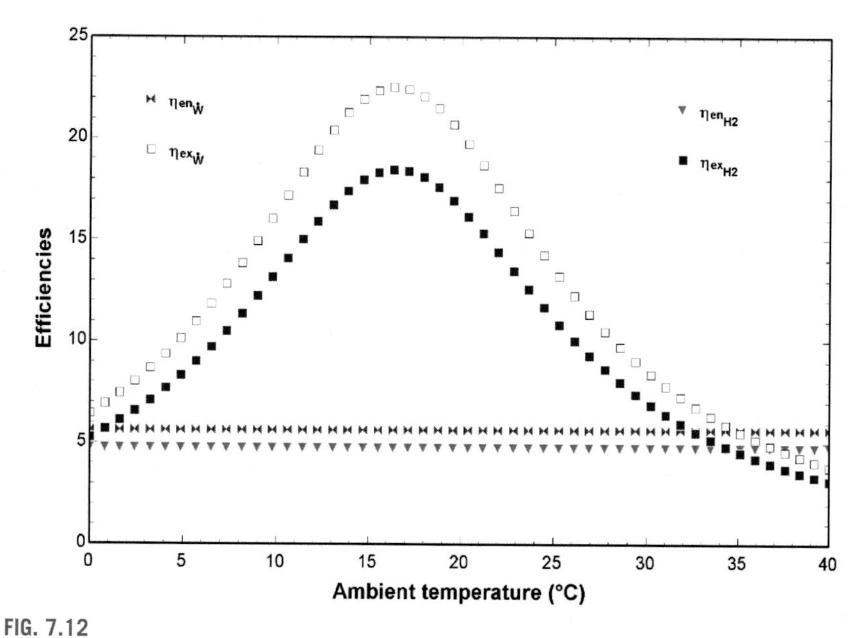

FIG. 7.12

Effect of ambient temperature on the efficiencies of power and hydrogen production case scenarios.

7.7 Closing Remarks

The oceans occupy one of the largest but slightest explored renewable energy sources on earth. Ocean energy source offers the potential to deliver a significant renewable and consistent energy amount around the globe. Ocean power extracts electricity using two types of energy, namely thermal energy through solar heat and mechanical energy through waves and tides. This power production sector is classified as a renewable energy resource as the water cycle is continuously circulated. Ocean power is among the promising global energy resources to produce clean, sustainable, consistent, and environmentally benign hydrogen, and this promising route does not emit GHG emissions to the environment. The ocean power-driven hydrogen production technology can play a vital role in providing the global transition to 100% renewable energy. Closed-cycle systems employ the warm ocean surface water to vaporize the working fluid, and the working fluid offering a low-boiling point is selected, such as ammonia. The vapor expansion turns a turbine that is accompanied by a generator to produce electrical power. The generated electricity can be employed in the water electrolysis process to produce clean hydrogen. Ocean energy comprises all renewable energy forms derived from the ocean. The three significant ocean energy technology types are, namely, tidal energy, wave energy, and ocean thermal energy conversion. All ocean energy forms are still at the early commercialization stage. Tidal energy and ocean thermal energy conversion are found to be cost-competitive as compared with wave-energy technology.

Even though wind and solar renewable energy sources are attracting the significant research scientific community, ocean energy offers consistent and renewable energy sources as well as high energy density as compared with other forms of renewable energy. It also offers more consistent and efficient results as it does not have an intermittent nature in comparison with other renewable energy sources. Tidal turbines offer approximately 80% efficiency that is comparatively much higher than wind and solar energy sources. The ocean thermal energy potential is substantial that offers an environmentally benign power system, even though, it offers very low thermal efficiency followed by the lower temperature difference in surface and deep seawater. Thus, the development of a high-efficient turbine is significant to enhance the thermal efficiency of the OTEC cycle. A case study is designed in this chapter that comprises the ocean thermal energy conversion cycle undergoing two different scenarios of producing power and hydrogen. The OTEC cycle benefits from the temperature between cold deep water and warm surface water and extracts electricity through the organic Rankine cycle. The produced power can be employed in the water electrolyzer process that breaks water into its constituents and produces clean hydrogen. The produced hydrogen can be stored and employed in numerous applications, such as hybrid and fuel cell vehicles, ammonia synthesis, heating, ammonia, and methanol synthesis and to produce power using hydrogen fuel cells. This promising route produces clean, consistent, and environmentally benign hydrogen undergoing zero carbon emissions that can play a vital role to reduce and even replace the conventional power production methods employing fossil fuels and hydrogen production methods undergoing natural gas reforming.

CHAPTER

Biomass Energy-Based Hydrogen Production

8

Biomass is commonly known as bio materials coming from numerous resources, including living species. There are essentially six groups of biomass resources, namely (i) agricultural crops and residues, (ii) forestry crops and residues, (iii) animal residues, (iv) industrial residues, (v) animal residues and (vi) sewage in which the chemical composition may change drastically.

Biomass principally comprises specific energy content that is primarily derived through the sun as solar energy is absorbed by the plants through photosynthesis that is the process of converting water and carbon dioxide into glucose and carbohydrates. Biomass burning is employed to generate heat (thermal energy) that is converted into electrical power or treated into biofuel.

Biomass is well thought out as a renewable source of energy since its intrinsic energy is originated from the sun that can be regrown in a moderately short period of time. Trees absorb the atmospheric carbon dioxide and convert it to biomass, and carbon dioxide is released back to the atmosphere when they die. Biomass is thought out as a renewable source of energy since more crops and trees can always be grown, and waste will exist at all times. Some common examples of bio materials are crops, wood, and manure. The chemical energy associated with biomass is obtained as heat when combusted.

Solid biomass feedstocks like garbage and wood can be combusted directly to generate heat. Bio materials can similarly be employed to produce liquid biofuels namely, biodiesel and ethanol, or into biogas. Animal fats and vegetable oils are employed to produce biodiesel that can be employed as heating oil or in vehicles. Biomass offers a clean and renewable source of energy that carries the potential to improve the economy, environment, and energy security. The biomass energy-based system produces fewer emissions in comparison with fossil fuels and reduces the waste amount directed to the landfills and reduces the foreign oil dependence.

Four common types of biomass that are used today are agricultural and wood products, biogas and landfill gas, solid waste, and alcohol fuels, such as biodiesel and ethanol. The maximum biomass employed today comes from home-grown energy. Wood logs, bark, chips, and sawdust account for approximately 44% of biomass energy.

The amount of biomass is determined by the summation of the dry mass of all entities in the specified land area and reported by identifying areas of concern, for instance, biomass per plot, biome, and ecosystem. To compare the biomass in dissimilar locations, researchers standardize the biomass per unit area. Most bio materials are employed as the primary source of energy for heating.

Renewable Hydrogen Production. https://doi.org/10.1016/B978-0-323-85176-3.00002-0

One noteworthy pollutant formed by biomass burning which appears to be dangerous as well and that is particle pollution known as soot. Biomass burning also emits carbon monoxide that leads to nausea, headaches, and dizziness. Biomass consists of plant waste, organic debris, wood chips, and whole trees while industrial representatives state that burning biomass is carbon neutral. They claim that new growth captures CO_2 and abandons the emissions emitted to the atmosphere from wood burning.

Biomass-based energy is an eco-friendly alternate to generally utilized, climate change-inducing fossil fuels. As a renewable source of energy, bioenergy can condense the carbon footmarks by more than 80%. Biomass is an organic matter that can be employed as an energy carrier in alive or dead form. Biomass examples comprise wood, seaweed, crops, and animal waste. Biomass is a renewable source of energy that extracts energy from the sun.

Bioenergy is a sustainable alternative to fossil fuels because it can be formed from renewable sources like waste and plants, which can be incessantly restocked and condense the gasoline supply that affects national security.

Biomass energy is a comparatively clean source of renewable energy concerning the organic matter utilization that extracted energy from the sun and generated chemical energy once it was alive. It is a source of renewable energy as this stock is continuously increasing and absorbing the energy of the sun, predominantly wherever biomass crops are cultivated. The greatest biomass-based energy is obtained from the plants that gather solar energy through the photosynthesis process. This energy form has been utilized for thousands of years since individuals started burning wood for heat. Technological advancements have permitted biomass-based energy to be employed in the variability of the application together with gases and liquids employed for biofuels to power the transport. Fig. 8.1 exhibits the CO_2 emissions per capita (tonnes per year) in the United Kingdom, United States, Japan, China, India, and Canada from 1900 to 2017.

8.1 Advantages and Disadvantages of Biomass Energy

Biomass-based energy is a rising energy source around the globe. This energy can be generated using many organic products that can be employed to deliver a comparatively clean alternative to conventional transportation fuel sources and electricity. Though, it also accompanies some disadvantages linked with biomass energy.

8.1.1 Advantages

Among the numerous advantages accompanied with biomass energy, a significant advantage is that it emits a reduced amount of harmful greenhouse gas (GHG) emissions in comparison with fossil fuels. As the hike continues in the direction of exploring fossil fuel alternatives, energy derived from biomass is gaining distinction as a prime candidate for possible replacement.

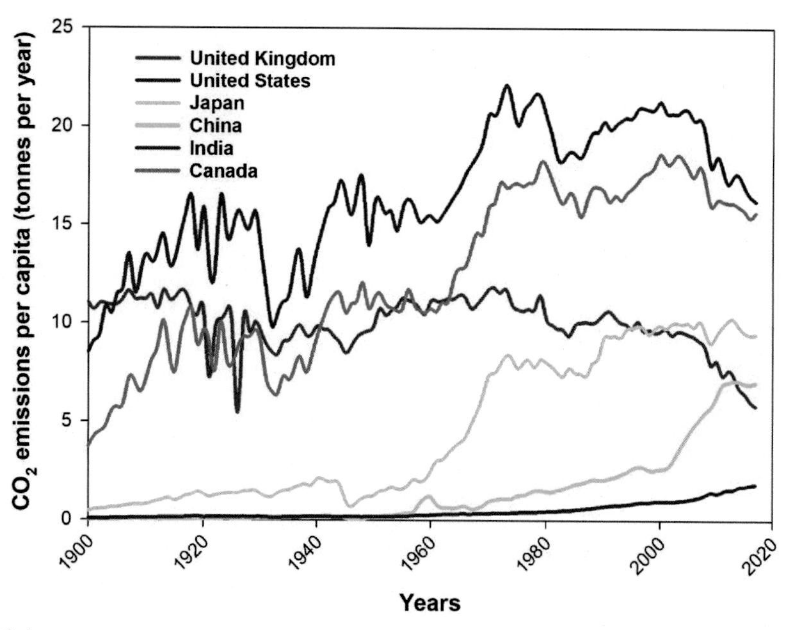

FIG. 8.1

CO_2 emissions per capita (tonnes per year).

Data from [128].

Renewable

Renewable sources of energy are the ones that can be restocked and replenished after usage. Biomass is also categorized as a renewable source as plants and wood can be regrown. Provided that, efforts are implemented to sustain the resources employed to harvest biomass energy by dedicated replenishment and replanting, this fuel source has the probability to offer clean energy solutions and live longer in comparison with conventional fossil fuels.

The significant difference between biomass and other sources of renewable energy like water and sun is the maintenance requirement. Although plant life is plentiful, employing it without implementing efforts to reload stocks can direct to the large quantities of it being wasted, which can be observed from deforestation.

Carbon neutral

The carbon amount that is emitted to the environment is the chief contributor to climate change. Biomass condenses this carbon amount since the fuel is naturally carbon cycle part, unlike coal, oil, and other fossil fuels.

The lone carbon emitted to the environment using bio materials comes from the process of food synthesis of plants throughout their lifecycles.

As these types of plants are restocked, the newly produced plants then capture a similar quantity of carbon that creates neutrality and makes bio materials remarkably clean.

Less fossil fuels dependency

The more we utilize biomass energy, the lesser we will be dependent on fossil fuels that are currently the significant contributors to the change in the environment and other environmental problems.

The accessible biomass material abundance also compensates for that of fossil fuels that make it a readily obtainable fuel source.

Versatile

Biomass energy is among the greatest versatile available alternatives as well. Biomass can be employed to produce different fuel sources that undergo varied applications.

For instance, biomass can be converted to produce biodiesel for vehicles; however, it can also be employed to form a variety of other biofuels and methane gas.

Moreover, wood can also be employed to extract heat, while the produced steam by some biomass forms can also be employed to the turbines to produce power.

Availability

Bio materials are naturally abundant, and this abundance represents that individuals will not face the depletion issues that are currently associated with fossil fuels. These sources can be found almost everywhere on the globe like the water and sun.

Nevertheless, it is critical as well to maintain this abundance. Even though bio materials are always available because these sources are part of the natural lifespan of the planet; however, this must not lead to negligence in their usage.

Low comparative cost than fossil fuels

Biomass energy possesses a low comparative cost than fossil fuels. In comparison with drilling for creating gas pipelines or oil, the cost involved to collect bio materials is tremendously low.

This low cost can be conceded to consumers as well when energy bills will no longer be reliant on the problems, such as company decisions that supply energy and availability.

Such low costs make biomass further attractive to the producers as well, as they will appreciate increased profits for a smaller amount of output.

Waste reduction

Loads of waste that is produced are principally planted matter and biodegradable that possibly will lay further effective usage elsewhere.

Biomass energy can frequently utilize the waste that would regularly be seated and rankle in landfills. This drops the effects of such sites on the natural environment that are chiefly noticeable with respect to the damaging wildlife and polluting local habitats.

This waste reduction opens additional zones for humans to occupy as well, as lesser space is desired to produce landfills.

Domestic production

Biomass may control the produced energy by the bigger companies as well. That means people are no longer required to be obliged to the power companies along with their charges.

The biomass nature represents that anybody can generate power and employ it on a domestic level practically. Even though it takes some work, even burning wood as an alternative to consuming a central heating system may save money and cause an advantageous impact on the environment.

8.1.2 Disadvantages

Even though there are numerous advantages associated with biomass energy, it is also significant to consider the disadvantages as well. This is critical if biomass energy is to be confronted appropriately to allow it to make available for the people.

Some of the disadvantages are directly related to the fuel usage while further are subsidiary or indirect costs of the biomass energy generation or application; however, all of these direct and indirect consequences are significant to be kept in mind to explore the practical usage of biomass energy.

Not entirely clean

Even though biomass is a carbon-neutral source of fuel, it is not entirely clean. The burning of woods and other plants creates some other emissions in the accumulation of carbon that can contaminate the local environment, even though the consequences are not as extreme as those caused by fossil fuels.

The bio materials burning indicated as air polluters by some organizations like Partnership for Policy Integration. In comparison with other renewables like water, this clean emission deficiency is the main disadvantage.

High comparative cost

Even though the extraction cost of bio materials is lesser in comparison with most fossil fuel types, they still surpass many other renewable energy forms. In some scenarios, the projects of extracting biomass energy are not considered to be worth the completion price, specifically when wind, water, and solar substitutes are accessible. Such cost is initiated from the requirement of maintaining biomass resources and for the replantation of extracted biomass. Moreover, the machinery cost that is employed to extract the biomass and biomass transportation are significant factors.

Possible deforestation

Even though biomass energy is extracted from renewable fuels, they still required to be maintained, and widespread deforestation can occur in case of failure. This is a significant environmental subject. It tremendously impacts the habitable zones accessible to scores of wildlife species and leads to extinction. This factor of possible deforestation holds back the large-scale employment of bio materials, as replantation efforts might not be capable of meeting the quantity of fuel needed.

Space

The plantation required a large space to grow the crops or plant life that is utilized in biomass energy. These spaces are not going to be accessible always, especially in built-up areas. This factor also limits the areas wherever biomass energy power plants can be constructed, as the facility needs to be adjacent to the fuel sources to neglect or reduce the transportation costs.

This factor makes biomass energy less promising in comparison with solar power that entails a smaller amount of space and can be conveniently installed in populated areas and cities. Lastly, the land employed might be employed to grow crops as well, which is particularly vital for the large population.

Water requirement

Every so often unnoticed disadvantage associated with biomass energy is the water quantity required for production. All plants require water for living that always indicates sources to be accessible. This factor does not only lead to the improved costs concerning irrigation; nonetheless, it possibly will also result in the sources of water to become less available to the people and wildlife. Moreover, as water is itself an alternative energy form that is cleaner as compared with biomass energy as well, it can raise the question of employing water for the same purpose instead.

Inefficiencies

Even though bio materials are naturally produced, these sources are not comparatively efficient as treated fossil fuels, like gasoline and petroleum. Biofuels and comparable biodiesel are frequently combined with minor quantities of fossil fuels that make them more efficient and effective. This reduces the effectiveness of biofuels as resources to cut down on utilization of fossil fuel resources.

Under development

Further research works and advancements need to be conducted to extract the biomass energy potential. Nevertheless, this source is seized back as an alternative source of fuel due to some of the described disadvantages. In comparison with solar, water, and wind sources, the biomass energy source is inefficient and needs to be further explored and researched. The research scientists are investigating the methods of making it further efficient. Up until that barricade is crossed, it is improbable that biomass energy sources will be implemented on the large scale as a viable alternative source of energy.

Fig. 8.2 displays the production of biofuels in different regions of the world. The regions included in this figure are Africa, Asia Pacific, China, Europe, the Middle East, North America, South and Central America, the United Kingdom, and the United States. From the years 1990 to 2018, the production of biofuels in Africa increases from 0.73 to 50.77 TWh, increases from 0 to 1620.96 TWh in the Asia Pacific, increases from 0 to 360.46 TWh in China, increases from 0.73 to 1854.87 TWh in Europe, increases from 0 to 1.87 TWh in the Middle East, increases from 176.69 to 4598.07 TWh in North America, increases from 727.49 to 2963.06 TWh in South and Central America, increases from 0 to 82.32 TWh in the United Kingdom, and increases from 177 to 4429.6 TWh in the United States.

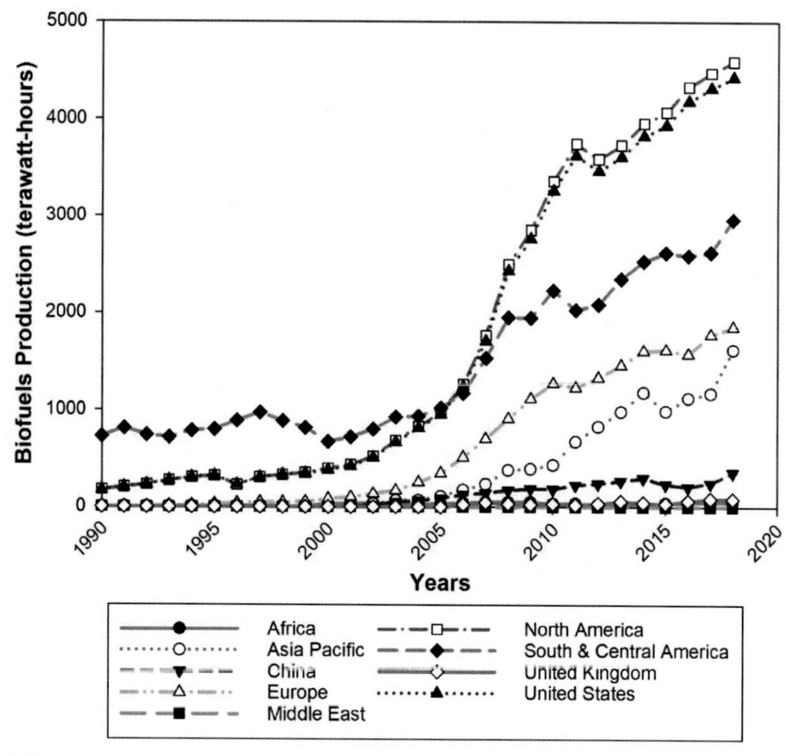

FIG. 8.2

Production of biofuels in different regions of the world.

Data from [129].

The biomass energy is viable and readily available. Biomass is a carbon-neutral source of fuel that desires lower comparative costs than fossil fuels and offers extremely miscellaneous solutions.

Nevertheless, several problems hold it back from large-scale implementation. The efficiency issue of the fuel needs to be investigated specifically to make the process more efficient and problems like cost and space also need to be well thought out.

Up until this is ensured, it is improbable that biomass energy harvesting can be implemented on the conventional scale to replace fossil fuels as a viable alternative; however, it can be employed for other uses. Especially, biomass energy usage on the local and domestic levels can lead to a decrease in energy bills. The following are some of the significant types of biomass products that are employed to harvest biomass energy:

- Wood waste and residues
- Fuelwood
- Pellets and briquettes

- Biogas
- Biogenic municipal solid waste (MSW)
- Waste liquor
- Other liquid biofuels
- Other solid biomass
- Biodiesel

8.2 Biomass as a Renewable Energy Resource

Biomass is a renewable energy source that consists of animal and plant materials like animal and human wastes, organic industrial, wood from forests, and leftover material from forestry and agricultural processes. The confined energy in the biomass principally comes from the sun as a plant utilizes atmospheric carbon dioxide during photosynthesis in the presence of solar light to synthesize carbon-containing molecules, such as starches, sugars, and cellulose, and this carbon dioxide is released back to the environment when biomass is burnt. The stored chemical energy in animals that eat plants or other animals and plants or in their waste is termed as bioenergy or biomass energy.

Biomass is originated from a diversity of sources that include the following:

- Wood from woodlands and natural forests
- Forestry residues
- Forestry plantations
- Agroindustrial wastes, such as rice husk and sugarcane bagasse
- Agricultural residues, such as straw, cane trash, stover, and agricultural wastes
- Animal wastes, such as cow manure and poultry litter
- Sewage
- Industrial wastes like black liquor from the paper manufacturing
- Food processing wastes
- Municipal solid wastes

Fig. 8.3 represents different biomass sources. Biomass energy resources that exist renewably are either employed directly as fuel or transformed into energy products, or another form are generally mentioned as feedstocks.

8.2.1 Biomass Feedstocks

The feedstocks of biomass consist of devoted energy crops, forestry residues, agricultural crop residues, animal waste, algae, MSW, wood processing residues, and wet waste (wood wastes, municipal solid wastes, and sewage and industrial wastes).

Devoted energy crops

Devoted energy crops are also termed as nonfood crops that are capable to be grown on peripheral land that is not appropriate for conventional crops like soybeans and

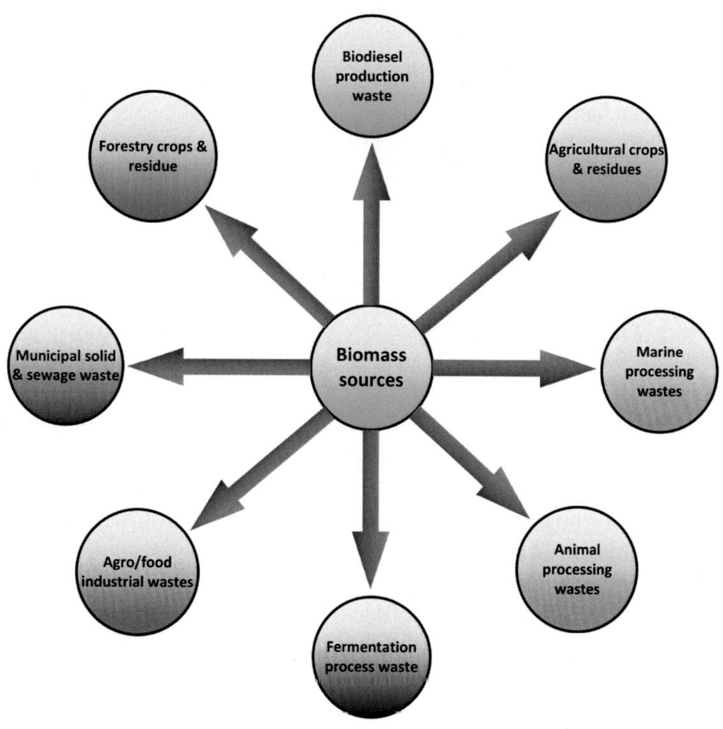

FIG. 8.3

Representation of different biomass sources.

corn, explicitly to deliver biomass. These are classified into two wide-ranging categories of herbaceous and woody. Herbaceous crops of energy are perennial grasses that are cultivated annually after captivating approximately 2 years to achieve full productivity. These comprise switchgrass, bamboo, miscanthus, sweet sorghum, tall fescue, wheatgrass, kochia, and others. Short-rotation crops of wood are quick-growing hardwood trees that are planted within 5–8 years. These comprise hybrid willow, hybrid poplar, silver maple, green ash, eastern cottonwood, black walnut, sycamore, and sweetgum. Several of such species can assist in improving soil and water quality, improving wildlife habitat comparative to the annual crops, diversifying income sources, and improving inclusive farm productivity.

Forestry residues

Forest biomass feedstocks are classified into two main categories of forest residues remaining after timber logging (including limbs, culled trees, tops and components of trees that are unmerchantable) or whole-tree biomass collected specifically for biomass. Dead, poorly formed, diseased, and unmerchantable trees are habitually left in the woods after the timber harvest. This woody wreckage can be employed for utilization in bioenergy, whereas leaving sufficient behind to deliver habitat and uphold hydrologic features and proper nutrients. There are prospects to utilize the excess biomass on masses of acres in the forests as well. Collecting extreme woody biomass can condense the hazard of pests and fire and assists in forest restoration, vitality, productivity, and resilience. Such biomass can be harvested to extract bioenergy without destructively impacting the stability and health of the functions and ecological structures of the forest.

Agricultural residues

Crop residues incorporate all agricultural wastes, namely bagasse, stem, straw, stalk, husk, leaves, shell, pulp, peel, and stubble. Large crop residues measures are formed annually worldwide and remain immensely underutilized. Rice harvests both rice husks and straws at the processing plant that can be conveniently converted into energy without difficulty.

Significant biomass quantities remain in cob form in the fields when maize is gathered that can be employed to produce energy. Harvesting sugar cane leads to yield residues while processing harvests fibrous bagasse, which is a good energy source. The coconuts processing and harvesting generate quantities of fiber and shell that can be employed.

Present farming practices are generally to cultivate these residues again into the soil or burnt, grazed by cattle or left to decompose. Such residues might be treated through thermochemical processes or liquid fuels to generate heat and electricity. Agricultural residues can be categorized based on the characteristics and seasonal availability that are different from other solid fuels namely, wood, char briquette, and charcoal. The key differences are high volatile matter content and lower burning time and density.

Fig. 8.4 displays the classification of the biomass primary energy consumption by different biomass types. Wood waste and residues contain 33%, pellets and briquettes encompass 6%, fuelwood comprises 30%, biogenic MSW involves 4%, biogas takes in 3%, waste liquor occupies 12%, biodiesel contains 5%, other solid biomass materials consume 3%, and other liquid biofuels comprises 4% of biomass primary energy consumption.

Animal waste

A wide range of animal wastes exist that can be employed as biomass energy sources. The greatest communal sources are poultry and animal manure. Such waste was recovered, traded, and sold employing a fertilizer or merely spread over the agricultural land in the past; nevertheless, the tougher environmental control introduction on the water pollution and odor shows that some sort of waste management is essential nowadays that can deliver additional incentives aimed at waste-to-energy conversion.

The greatest attractive technique to convert such organic waste materials into advantageous form is termed as anaerobic digestion that produces biogas that can be employed as a fuel in internal combustion engines (ICEs) to produce electrical power using small gas turbines, can be burnt directly and employed for cooking, or can also be employed for water and space heating.

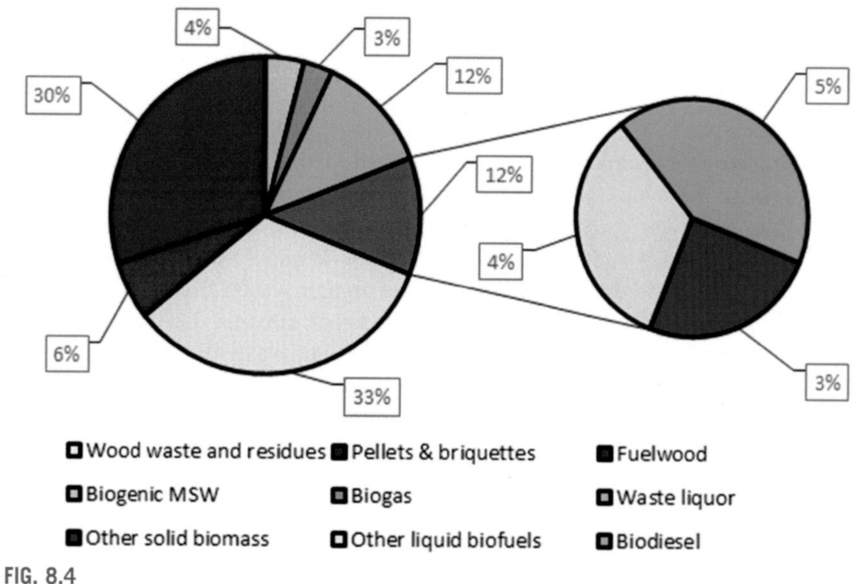

FIG. 8.4

Classification of the biomass primary energy consumption by different biomass types.

Modified from [130].

Algae

Algae as bioenergy feedstock denote a miscellaneous group comprising highly productive organisms including microalgae, cyanobacteria, and macroalgae (seaweed). Numerous utilize nutrients and sunlight to produce biomass that encompasses key components, such as proteins, lipids, and carbohydrates that can be transformed and promoted to a diversity of products and biofuels. Contingent to the strain, algae can be grown by consuming fresh, brackish, or saline water from groundwater, surface water sources, or seawater. Furthermore, algae can also grow in water employing second-use sources like processed industrial wastewater; municipal, aquaculture or agricultural wastewater; or water produced from gas and oil drilling operations.

Sorted municipal waste

Municipal solid waste resources contain mixed residential and commercial garbage like yard trimming, plastics, paper and paperboard, rubber, food wastes, textiles, and leather. Municipal solid waste for bioenergy presents a commercial and residential waste reduction opportunity by distracting substantial volumes from landfills to a refinery.

Wood processing residues

The processing of wood produces waste streams and byproducts that are mutually termed as wood processing residues and offer the significant potential of energy. For instance, the wood processing for pulp or other products produces unemployed sawdust, branches, bark, and leaves. Such residues are then converted into bioproducts or biofuels. As such residues are collected at the processing point, they can offer relatively inexpensive and appropriate biomass energy sources.

Wet waste

Wet waste feedstocks contain commercial, residential, and institutional food wastes (predominantly those that are disposed of in the landfills); manure slurries as of concentrated livestock processes; organic-rich biosolids (like processed sewage sludge by municipal wastewater); biogas (gaseous organic matter decomposition product produced in the absence of oxygen); and organic wastes from industrial processes derived through any of the mentioned feedstock streams. The conversion of such waste streams into energy can support resolving waste-disposal issues and assist in creating supplementary revenue for rural economies.

Wood wastes

Industries undergoing wood processing principally consist of sawmilling, wood panel, plywood, furniture, flooring, building component, particleboard, jointing, molding, and craft industries. Wood wastes usually employed are concentrated at processing factories, for instance, sawmills and plywood mills. The quantities of waste produced by different wood processing industries differ from one industry to another and depend upon the raw material type and finished product.

Largely, the waste generated by the wood industries like veneer, saw plywood and millings, other sawdust, trims, off-cuts, and shavings. Sawdust arises from cutting, resawing, sizing, edging, shaving, and while trims are the penalties of wood smoothing and trimming. All together 1000 kg wood processing in the furniture industry can result in approximately 45% of wood. Likewise, 1000 kg wood processing in sawmill results in approximately 52% of wood.

Municipal solid wastes and sewage

Masses of household waste are composed every year with the immense disposed majority in open fields. Biomass resources in municipal solid waste include putrescible, plastic, and paper and almost 80% of overall municipal solid waste is collected. The MSW can be employed to extract energy through natural anaerobic digestion or direct combustion in an engineered landfill.

At the landfill sites, the produced gas by the natural decomposition of municipal solid waste is termed as landfill gas and contains nearly 50% carbon dioxide and 50% methane, is gathered through the stored material, scrubbed and cleaned before feeding into ICE or gas turbine to produce power and heat. The municipal solid waste organic fraction can be stabilized anaerobically in the high-rate digester to employ biogas for steam generation or electricity production.

Manure is a biomass energy source that is quite comparable to other animal wastes. Anaerobic digestion can be employed to extract energy from sewage consuming the produced biogas. The remaining manure sludge can undergo incineration or pyrolysis to generate further biogas.

Industrial wastes

The food industry produces an abundance of by-products and residues that can be employed as sources of biomass energy. Such waste materials are produced through all food industrial sectors with a whole lot of meat production to confection waste products that can be employed as a source of energy. Solid wastes contain scraps and peelings from vegetables and fruits that do not encounter the quality control standards, fiber and pulp from starch and sugar extraction, filters coffee ground, and sludges. Such wastes are generally relinquished in landfill dumps.

Liquid wastes are produced by cleaning meat, fish, and poultry, blanching vegetables and fruits, precooking meats, and cleaning and processing processes. This wastewater contains starches, sugars, and other solid organic and dissolved matters. These industrial wastes have the potential to be processed through anaerobic digested to generate biogas or fermented to form ethanol and quite a few conventional waste-to-energy conversion examples also exist.

Paper and pulp industries are counted among the extremely polluting industries and consume bulky energy amounts and water in numerous unit operations. The discharged wastewater by these industries is extremely heterogeneous as it encompasses complexes of wood and other raw materials, treated chemicals; in addition to the formed compounds throughout processing. The black liquor can be cautiously employed for biogas production employing anaerobic UASB technology.

Fig. 8.5 displays the installed bioenergy generation and installed capacity trends from 2010 to 2017. Figure exhibits the biogas-based energy generation and installed

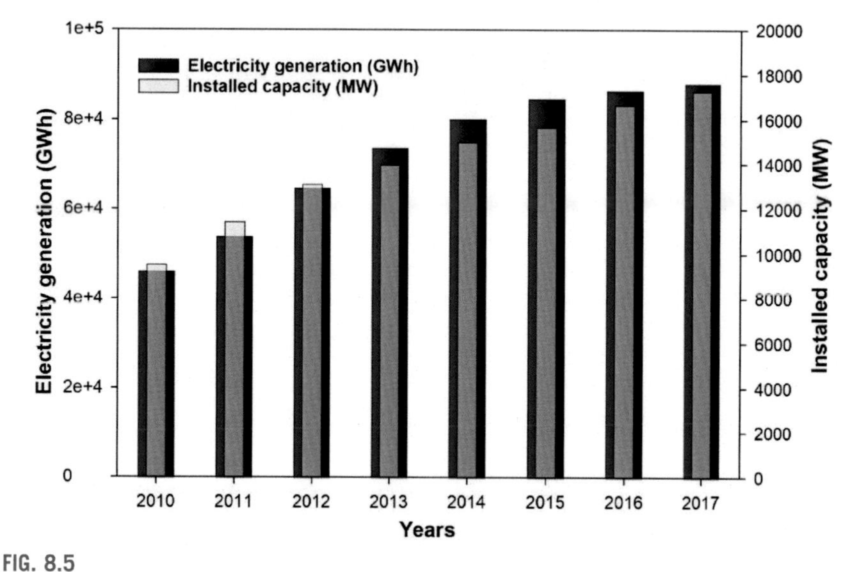

FIG. 8.5

Biogas based electricity generation and installed capacity trends from 2010 to 2017.

Modified from [131].

capacity trends from 2010 to 2017. A steady rise in bioenergy generation and installed capacities using biogas can be depicted from figure. A steady increase in electricity generation can be depicted from 46,129 to 87,935 GWh from 2010 to 2017 and meanwhile, installed capacity increased from 9518 to 17,268 MW.

Fig. 8.6 exhibits the liquid biofuels-based energy generation and installed capacity trends from 2010 to 2017. A steady rise in bioenergy generation and installed capacities using biogas can be depicted from figure. A steady increase in electricity generation can be depicted from 5296 to 6507 GWh from 2010 to 2017 and meanwhile, installed capacity increased from 1857 to 3233 MW. Fig. 8.7 exhibits the solid biofuels-based energy generation and installed capacity trends from 2010 to 2017. A steady rise in bioenergy generation and installed capacities using biogas can be depicted from the figure. A steady increase in electricity generation can be depicted from 225,933 to 343,863 GWh from 2010 to 2017 and meanwhile, installed capacity increased from 47,572 to 78,285 MW.

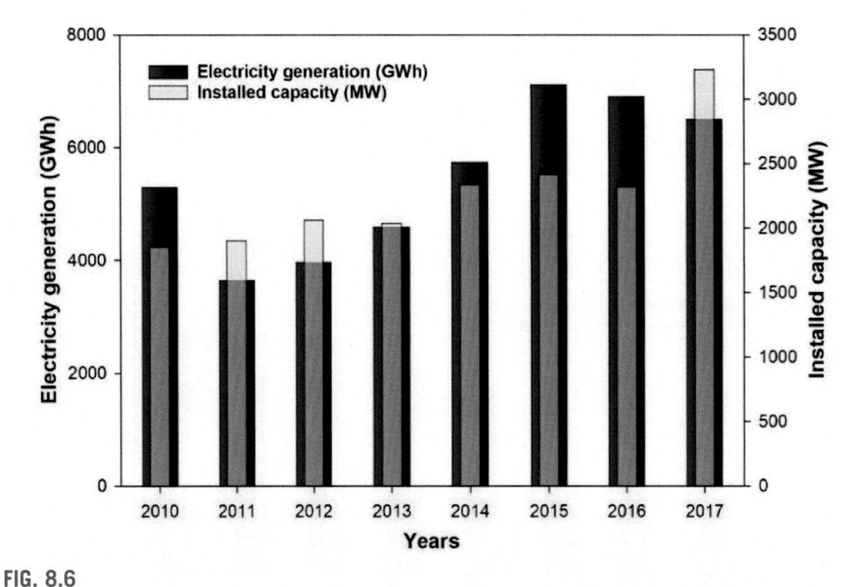

FIG. 8.6

Liquid biofuels-based electricity generation and installed capacity trends from 2010 to 2017.

Modified from [131].

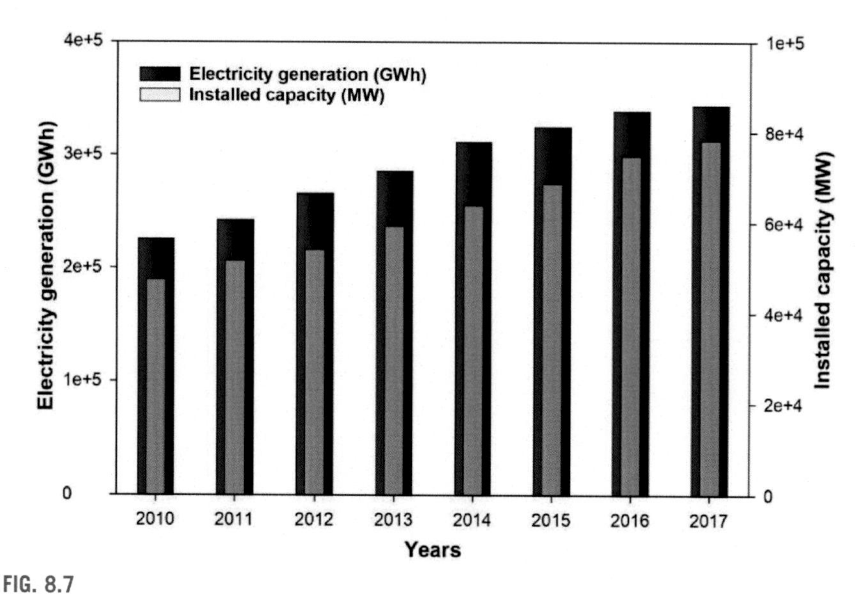

FIG. 8.7

Solid biofuels-based electricity generation and installed capacity trends from 2010 to 2017.

Modified from [131]

8.2.2 Types of Biomass-Based Hydrogen Production Methods

Classification of different methods that are used for biomass-based hydrogen production is displayed in Fig. 8.8 and listed as follows:

- Biological
  - Dark fermentation
  - Photo fermentation
  - Microbial electro-hydro-genesis cell
  - Direct/indirect biophotolysis
- Thermochemical
  - Gasification
  - High-pressure aqueous
  - Pyrolysis

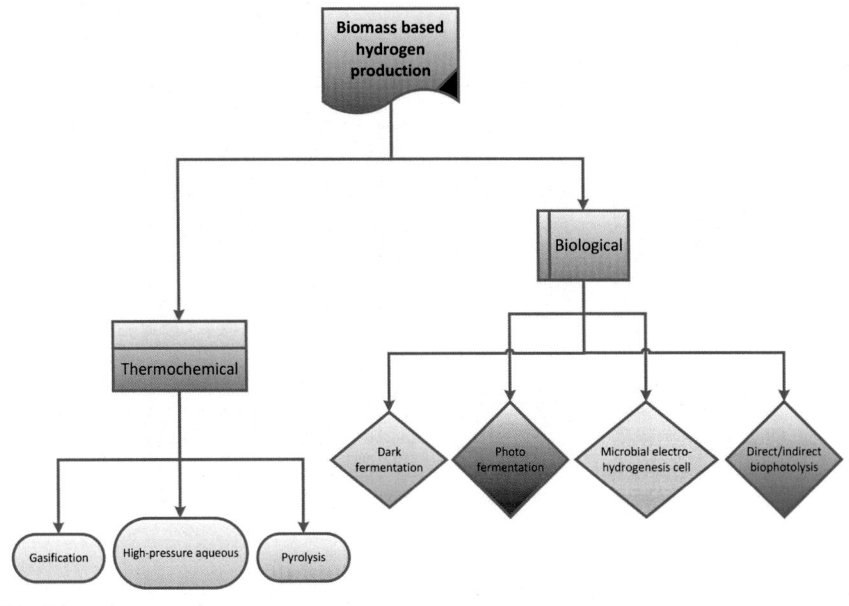

FIG. 8.8

Classification of biomass based hydrogen production methods.

8.3 Pyrolysis

Pyrolysis is well known as a thermochemical treatment that can be employed for any organic product. Pyrolysis can be done using pure materials and mixtures as well. In pyrolysis treatment, the material is exposed to high temperature, and the process undergoes in the absence of oxygen through physical and chemical separation into dissimilar molecules. The high operating temperature also causes a reduction in water volume. By consuming the formed gases as fuel, the external fuel supply can be limited. The process of pyrolysis is complicated and entails high investment and operational costs.

Pyrolysis is a thermal conversion process of solid fuels in the whole absence of oxygen that is used as an oxidizing agent or with limited supply, which indicates that gasification does not happen to any noticeable extent. Conventional applications are either explicitly focused on charcoal production or liquid product production, the biooil. The latter one is practically interesting as a fuel oil alternate and as a feedstock for diesel fuel or synthetic gasoline production. Fig. 8.9 exhibits a basic schematic of biomass-based hydrogen production by fast pyrolysis and in-line steam reforming.

During the pyrolysis process that takes place at high temperatures of 400–800°C, the greatest of the hemicellulose and cellulose and part of lignin disintegrates to yield lighter and smaller molecules that are gases released at pyrolysis temperature. As these gases are slightly cooled, some vapors are condensed to produce a liquid that is biooil. The remaining biomass parts, principally lignin parts are left as charcoal. To some amount, it is probable to affect the mixed product to promote gases, solid charcoal, or condensable vapors. Pyrolysis products are distributed as a function of residence time, heating rate, and extreme reaction temperature.

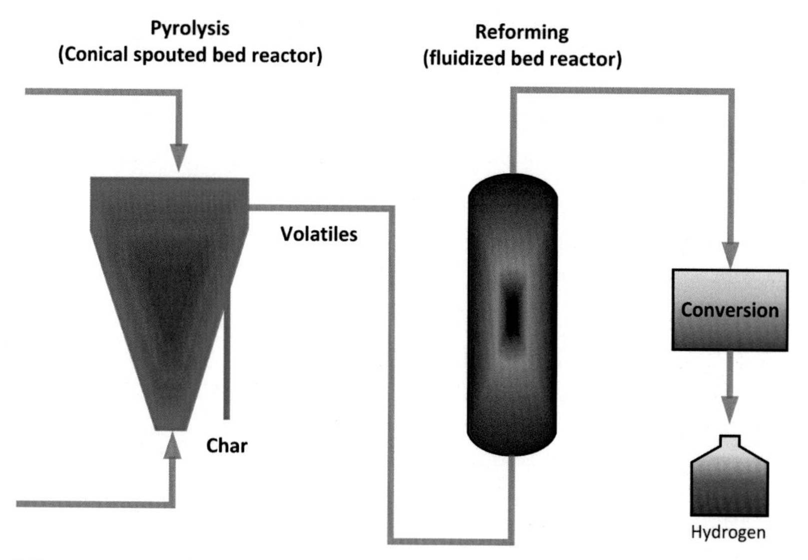

FIG. 8.9

Biomass-based hydrogen production by fast pyrolysis and in-line steam reforming.

Pyrolysis is generally employed to convert carbon-based materials into solid residue encompassing carbon and ash with insignificant amounts of gases and liquid. On the other hand, extreme pyrolysis produces carbon as residue, and the process is termed as called carbonization. Contrasting the other high-temperature operations, such as combustion and hydrolysis, pyrolysis does not include reaction with oxygen, water, or other reagents. Nevertheless, as it is unlikely for the practical applications to establish an oxygen-free atmosphere, however, insignificant oxidation always exists in any pyrolysis unit.

8.3.1 Types of Pyrolysis Reactions

There are three significant types of pyrolysis reactions that are distinguished on the basis of process temperature and processing time.

- Slow pyrolysis
- Flash pyrolysis
- Fast pyrolysis

Slow pyrolysis

Slow pyrolysis is specified by extensive residence times for gas and solids, slow heating rates of biomass and low temperatures. In this approach, heating temperature varies from 0.1 to 2°C per second and the dominant temperature is nearly 500°C. The gas residence time might be approximately 5 s and for biomass, it might vary from minutes to days.

Throughout the slow pyrolytic process, char and tar are released as core products on the slowly devolatilized of biomass. After the primary reaction, repolymerization or recombination reactions occur.

Flash pyrolysis

Flash pyrolysis occurs at moderate temperatures ranging from 400 to 600°C and rapid heating rates. Nevertheless, the process vapor residence time is less than 2 s. Smaller amounts of tar and gas are produced in the flash pyrolysis process in comparison with slow pyrolysis.

Fast pyrolysis

Fast pyrolysis process is principally employed to yield gas and biooil. During the process, biomass is heated rapidly to temperatures of 650–1000°C depending upon the desired quantities of gas or biooil products. Char is collected in large amounts that need to be removed regularly.

This process is shown to get the advantage of microwave heating usage. Biomass characteristically captures microwave radiation splendidly that makes the material heating highly efficient like microwave food heating, it may condense the time required to initiate the pyrolytic reactions and also reduces the process energy requirement significantly. As microwave heating is capable of initiating the pyrolysis process at comparatively lower inclusive temperatures ranging from 200 to 300°C, it is established that the produced biooil encompasses high concentrations of higher-value chemicals and thermally labile, signifying that microwave biooil possibly will be employed as a crude oil feedstock replacement for some chemical processes.

8.3.2 Advantages

The significant advantages of the pyrolysis process include the following:

- Pyrolysis is a simple and inexpensive processing technology for an extensive variability of feedstocks.
- It reduces GHG emissions and wastes going to the landfill.
- It reduces water pollution risk.
- It is probable to reduce the dependence of a country on the imported energy resources by means of producing energy using domestic resources.
- Waste management is inexpensive with the assistance of advanced pyrolysis technology in comparison with landfill disposal.
- The pyrolysis power plant construction is a comparatively quick process.
- It also initiates numerous job opportunities for the local public depending upon the waste generation quantities in the region that also offers public health advantages through waste cleanup.

8.3.3 Applications of Pyrolysis

Some significant applications of the pyrolysis process are as follows:

- Pyrolysis is widely employed in the chemical industry to produce charcoal, activated carbon, methanol, and other materials from the wood.
- The waste conversion process using pyrolysis produces synthetic gas that can be employed in steam or gas turbines for electricity generation.
- A blend of stone, glass, soil, and ceramics extracted from the waste of pyrolytic process can be utilized for numerous applications, such as construction slag, building material, and landfill cover liners filling.
- The pyrolytic process plays a significant part in mass spectrometry and carbon-14 dating.
- This process can also be utilized for numerous cooking procedures, such as caramelizing, frying, grilling and baking.

8.4 Biomass Gasification

Gasification is a process of converting organic carbonaceous materials into carbon dioxide, carbon monoxide, and hydrogen. This is accomplished by high-temperature (more than 700°C) material reaction without combustion employing controlled oxygen and/or steam amount. The pyrolysis process is known as the thermal decomposition of volatile organic components under the high temperature ranging from 200 to 760°C and forming syngas in the absence of oxygen while gasification utilizes only a controlled oxygen amount that is required to burn the material.

Biomass gasification is an established technology that employs a measured process involving steam, heat, and oxygen to produce hydrogen along with other products using biomass without combustion. As increasing biomass eradicates atmospheric carbon dioxide, the overall carbon emissions can be lowered for this technique, particularly if integrated with the carbon capturing, utilization, and storage in the long-standing. Biofuels-based gasification plants are being constructed and functioned and capable of providing the best lessons and practices learned for producing hydrogen. The United States Department of Energy predicts that biomass gasification will possibly be deployed in the near timeframe.

A recent study[132] focused on biomass gasification-based hydrogen production. The objectives of this study were to explore biomass gasification to yield syngas and biohydrogen in the fluidized bed. α-cellulose and further agricultural wastes were gasified at high-temperature ranging from 600 to 1000°C concerning diverse equivalents, and the ratios of steam to biomass were investigated. A kinetic model has been proposed to govern the reaction order and activation energy. The outcomes revealed that at 0.2 equivalent ratios and 1000°C temperature without steam established the maximum yield of 29.5% biohydrogen and 23.6% carbon monoxide, and the concentration of carbon dioxide was found to be 10.9%. Another research study[133] investigated the biomass steam gasification for the production of hydrogen-enriched gas in the presence of CaO. Biomass steam gasification might be a striking opportunity for the sustainable production of hydrogen. Biomass that is regarded as carbon-neutral fuel can be claimed as carbon-negative fuel if produced carbon dioxide is captured and not emitted to the environment. Thus, this study conducted an experimental investigation to explore the hydrogen production potential from biomass steam gasification in presence of CaO sorbent and the effect of dissimilar operating parameters, such as steam/biomass ratio, CaO/biomass ratio, and temperature. Results revealed that the product gas with 54.43% hydrogen concentration was established at CaO/biomass ratio of 2, steam/biomass ratio of 0.83, and temperature of 670°C.

A recent review study[134] was conducted on the thermochemical biomass gasification to produce biopower, biofuels, and chemicals to explore the current technological status. The upstream gasification processes are like other biomass processing approaches. Nevertheless, challenges persist in the downstream processing and gasification for feasible commercial applications. The gasification challenges are to recognize the impact of operating conditions on the gasification reactions to consistently predict and optimize the product compositions and to obtain the maximum efficiencies. Produced gases can be transformed into chemicals and biofuels, for instance, Fischer–Tropsch fuels, hydrogen, green gasoline, dimethyl ether, methanol, ethanol, and higher alcohols. This study also summarized the processes and challenges for these conversions. Technoeconomic investigation of hydrogen and ammonia production through integrated biomass gasification was conducted in a recent study.[135] The core objective of this study was to conduct the technoeconomic assessment of ammonia production through biomass gasification in an existing paper and pulp mill. The results revealed that energy performance and process economics are favorable for the integrated case in comparison with stand-alone production.

Comprehensive performance evaluation of an incessant solar-assisted biomass gasifier was conducted in this study.[136] The experimental performance evaluation of an incessant solar-assisted biomass gasifier utilizing real high-flux concentrated solar irradiance as a heat source was performed. Numerous parametric studies were conducted to explore dissimilar lignocellulosic biomass feedstocks, steam/biomass ratios, biomass feeding rates, carrier gas flow rates, and reaction temperatures to achieve the optimized syngas production capacity. An insignificant amount of excess water relating to the stoichiometry was found to be advantageous for biomass gasification in terms of the increased hydrogen and carbon monoxide and decreased methane and carbon dioxide production. An upsurge in gas residence time improved the syngas quality and yields. Noteworthy improvement in the production rates and syngas yields through the operating temperature rise was emphasized with the activation energy ranging from 24 to 29 kJ/mol. An increase in the biomass feeding rate enhanced the gasification rates and syngas yields, allowing effective solar energy storage into syngas and improving energy upgrade factor more than 1.20, solar-to-fuel energy conversion efficiency more than 29%, and thermochemical reactor efficiency of more than 27%.

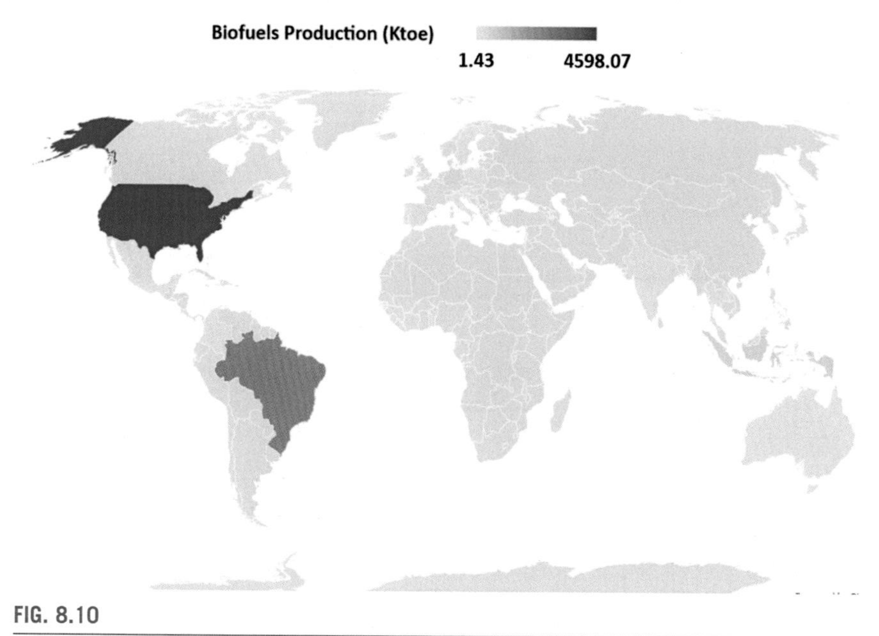

FIG. 8.10

Production of biofuels around the globe in 2019.

Modified from [137].

Fig. 8.10 displays the production of biofuels around the globe in 2019. The biofuels production capacity was found in the significant global countries and regions of Africa, Argentina, Asia Pacific, Australia, Austria, Belgium, Brazil, CIS, Canada, China, Colombia, Europe, Europe (other), Finland, France, Germany, India, Indonesia, Italy, Mexico, Middle East, Netherlands, North America, other Asia and Pacific, Other South and Central America, Poland, Portugal, South and Central America, South Korea, Spain, Sweden, Thailand, the United Kingdom, and the United States.

Biomass gasification is a process of converting organic carbonaceous materials into carbon dioxide, carbon monoxide, and hydrogen. This is accomplished by high-temperature (more than 700°C) material reaction without combustion utilizing controlled oxygen and/or steam amount. The carbon monoxide is then fed to the water–gas shift reactor where it reacts with water to form more hydrogen and carbon dioxide. The produced hydrogen can be separated from the gas stream using different adsorbers, special membranes, or pressure swing adsorption. The simplified example of the reaction can be expressed as follows:

$$C_6H_{12}O_6 + O_2 + H_2O \rightarrow CO + CO_2 + H_2 + \text{other species}$$

The water–gas shift reaction (WGSR) can be written as follows:

$$CO + H_2O \rightarrow CO_2 + H_2 + \text{Heat}$$

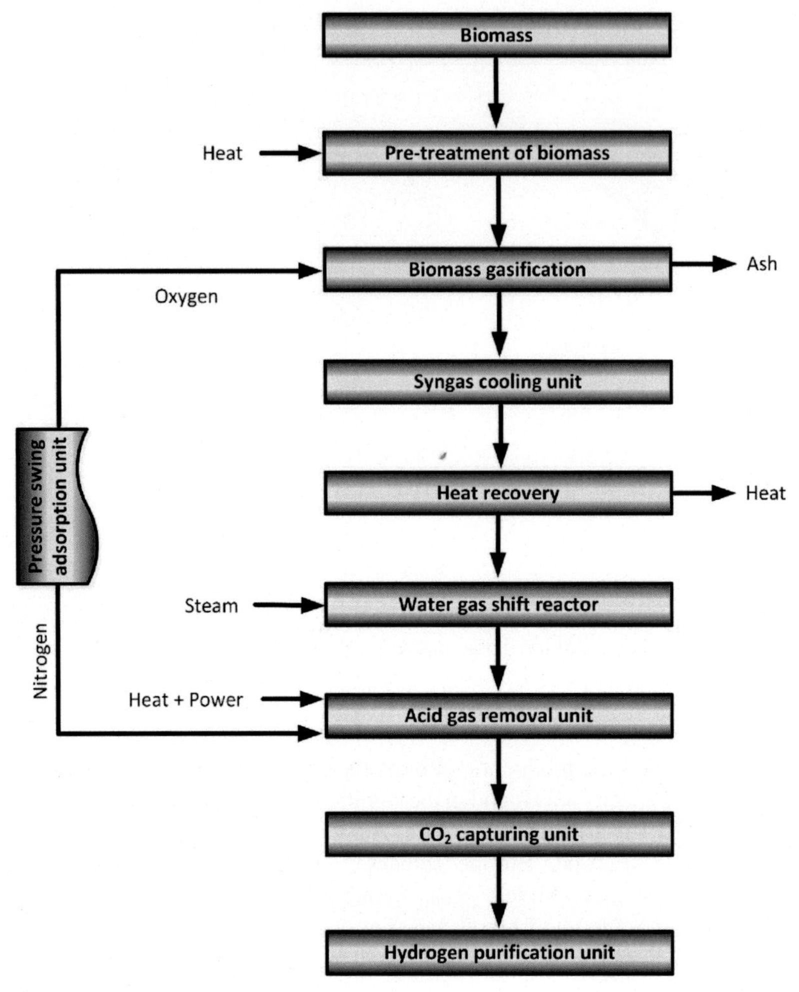

FIG. 8.11

Biomass gasification steps for hydrogen production.

Fig. 8.11 displays the steps followed in biomass gasification-based hydrogen production. Pyrolysis is known as the gasification of biomass in oxygen absence. Biomass does not gasify easily in comparison with coal, and it yields some other hydrocarbon compounds as well in the mixture of gases leaving the gasifier. Therefore, an extra step is taken characteristically to reform such hydrocarbon gases to produce a clean syngas mixture of carbon monoxide, hydrogen, and carbon dioxide using a catalyst. This reaction is followed by a shift reaction known as WGSR that converts carbon monoxide into hydrogen and carbon dioxide using steam.

8.4.1 Biomass Power to Hydrogen

Biomass gasification is known as the process that is employed to convert solid biomass into gaseous combustible gas known as producer gas, over classification of thermochemical reactions. The producer gas is low-heating value fuel having a calorific value ranging from 1000 to 1200 kcal/Nm^3.

Gasification of biomass produces syngas, which is employed to generate power using the different techniques, namely Brayton cycle, heat recovery steam generator followed by the steam Rankine cycle, and steam turbine. Fig. 8.12 displays the schematic illustration of the biomass power-to-hydrogen technique. The power generated by biomass gasification is employed in the electrolyzer for hydrogen production, which is also known as biomass power to hydrogen.

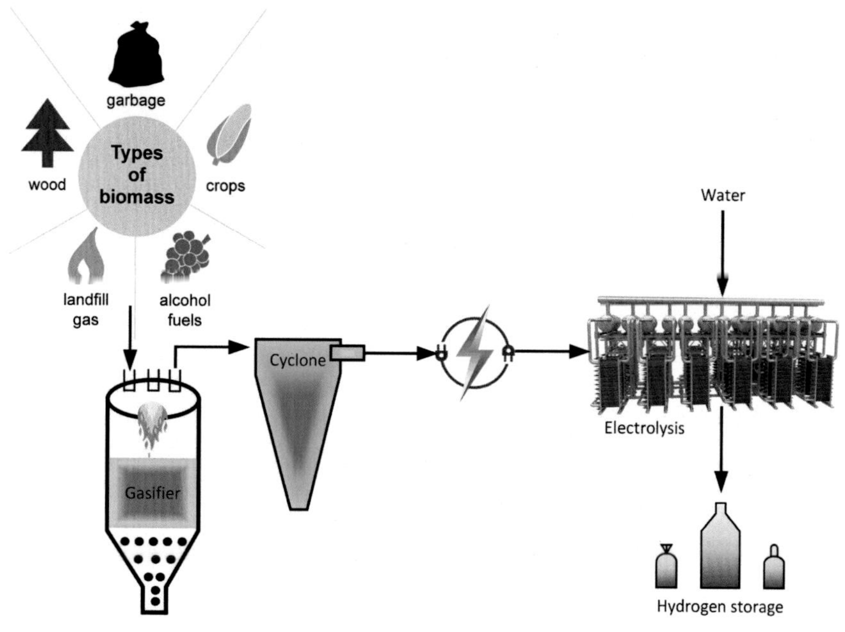

FIG. 8.12

Biomass power to hydrogen.

8.5 Types of Gasifiers

There are different types of gasifiers listed as follows:

- Updraught gasifier
- Downdraught gasifier
- Fluidized bed gasifier
- Cross-draught gasifier

8.5.1 Counter Current or Updraught Gasifier

In the updraught gasifier, the air intake is designed at the bottom of the gasifier while the gas exits from the top. The combustion reactions occur adjacent to the grate at the bottom that is trailed by the reduction reactions to some extent higher up and about in the gasifier. In the upper gasifier part, the heating and pyrolysis of feedstock exist because of forced convective and radiative heat transfer from lower zones. The volatiles and tars formed throughout this process are passed through the gas stream while the ash is separated from the bottom of the gasifier. The most significant pros of this gasifier type are the high charcoal burn-out, simplicity and internal exchange of heat that leads to the high equipment efficiency, low exit gas temperatures, and the operation possibility with numerous types of feedstocks, such as cereal hulls and sawdust.

The significant disadvantages have resulted from the channeling possibility in the equipment that may cause the break-through of oxygen and explosive and dangerous conditions and the requirement to mount them automatically moving grates, in addition to the complications related to the tar-containing condensate disposal that is resulted from gas cleaning processes. To be utilized for the direct heating applications, the gas cleaning process is insignificant where tar is simply burnt.

8.5.2 Cocurrent or Downdraught Gasifiers

The designed downdraught or cocurrent gasifiers offer a resolution to the tar entrainment problems in the gas stream where the oxidation zone of the gasifier is used to introduce the gasification air. The producer gas moves through the gasifier bottom and removed, intending to move both gas and fuel in an identical direction. On the flow down toward the tarry and acid distillation products, the fuel is required to move through the glowing charcoal bed and consequently are transformed into the permanent gases of carbon dioxide, hydrogen, methane, and carbon monoxide. A comparatively complete tars breakdown is accomplished depending upon the residence time and hot zone temperature of the tarry vapors.

The core advantage of downdraught gasifiers is based on the prospect of tar-free gas production appropriate for engine applications. Nevertheless, a tar-free gas is hardly ever accomplished over the complete equipment operating range: the tar-free operational turn-down ratios of three are well thought out as standard while a ratio of 5–6 is considered to be excellent. The downdraught or cocurrent gasifiers suffer through a lesser number of environmental objections as compared with updraught gasifiers, as the lower organic components are employed in the condensate.

The insignificant disadvantages of the cocurrent gasifier are the lower efficiency to some extent after the deficiency of exchange in internal heat in addition to the lower heating value (LHV) of gas. In addition to this, the requirement to sustain the uniform high temperature through the specified cross-sectional area makes the employment of a downdraught gasifier impractical for the power range higher than 350 kW.

A noteworthy downside of the downdraught gasifier lies in the incapability to operate for a variety of untreated fuels. More specifically, low-density materials cause extreme pressure drop and flow problems, and solid fuel needs to be briquetted or pelletized before usage. Downdraught gasifier also suffers from complications related to the high ash content in comparison with updraught gasifiers.

8.5.3 Fluidized Bed Gasifier

The process of both updraught and downdraught gasifiers is affected by the chemical, physical, and morphological fuel properties. Complications that are generally encountered are as follows: deficiency of bunker flow, immense pressure drop through the gasifier, and slagging. A fluidized bed gasifier is a design approach eliminating the above-mentioned problems and complications.

Air is blown through the solid particle bed at an adequate velocity to retain these at the suspension state. The bed is formerly heated externally, and the introduction of the feedstock occurs right after the adequately high temperature is achieved. The particles of fuel that are entered through the reactor bottom are mixed with bed material rapidly and practically heated up to reach the bed temperature. This treatment causes the fuel to be pyrolyzed rapidly that results in the component mix with comparatively large gaseous materials amount. Additional tar-conversion and gasification reactions occur in the gas phase. Greatest systems are employed with an inner cyclone to minimalize the char blow-out. Ash particles are passed through the reactor top as well that need to be removed from the gas stream if this gas is employed in engineering applications.

The substantial advantages of the fluidized bed gasifier are the stem from the flexibility of feedstocks after the easy temperature control that can be reserved under the fusion or melting point of rice husks and their capability to treat the fine-grained and fluffy materials such as sawdust without the requirement of preprocessing. Complications with feeding, bed instability, and sintering of the fly-ash in gas channels may occur with some biofuels.

Other downsides of fluidized bed gasifier are in terms of the high tar content in the product gas, a poor response in the changes in load, and incomplete carbon burnout. As the control equipment is needed particularly to accommodate the latter difficulty, very small-sized fluidized bed gasifiers are not predicted, and the range of application must be timidly set at directly above 500 kW of shaft power.

8.5.4 Cross-Draught Gasifier

Cross-draught gasifiers are adapted for charcoal utilization. Charcoal gasification undergoes a temperature of 1500°C or higher in the oxidation zone that can cause material complications. In the cross-draught gasifiers, charcoal fuel itself provides insulation against such high temperatures.

System advantages lie on a very small scale at which cross-draught gasifiers can be operated. Installations of small-scale units undergoing around 10 kW shaft power can be economically feasible under certain conditions. A simple gas-cleaning train including a hot filter and cyclone is the reason, and cyclone can be used with filters while employing cross-draught gasifiers in combination with the small engines. Another disadvantage of these gasifiers is the minimal capabilities of tar-converting, and the subsequent is essential for high-quality charcoal having low volatile content.

To overcome the charcoal quality uncertainty, many charcoal gasifiers used the downdraught principle to maintain a minimal tar-cracking capability.

8.5.5 Entrained-Flow Gasifier

In the entrained-flow gasifiers, biomass is fined and the oxidant along with steam is entered from the top of the gasifier. The steam and oxidant surround or entrain the fuel particles as biomass flows through the gasifier. Entrained-flow gasifiers operate at very high temperature to melt the biomass ash into the inert slag. The fine biomass feed and high operational temperature permit the gasification reaction to occur at a very high rate and residence time is a few seconds and undergoes high carbon conversion efficiencies of 98%–99.5%. The tar, phenols, oil and other liquids formed from the biomass devolatilization inside the entrained-flow gasifier are decomposed into hydrogen, carbon monoxide, and insignificant quantities of hydrocarbon gases. Entrained-flow gasifiers can practically handle any biomass feedstock and yield a clean and tar-free syngas. The fine biomass feed can be employed to the entrained-flow gasifier in a dry or slurry form. The prior employs a hopper system while the latter counts on the utilization of high-pressure slurry pumps. Feeding the slurry is a simple process; however, the water is introduced into the reactor with a slurry that is desired to be evaporated. The outcome of the additional water content is the syngas product having improved H_2 to CO ratio, however, with a lower thermal efficiency of the gasifier. The preparation system for the feed also requires to be assessed as other process design substitutes, for a specific application. Entrained-flow gasifiers characteristically display numerous characteristics listed as follows:

- Fuel flexibility
- Large oxidant requirements
- Can employ either air or oxygen as an oxidant
- Uniform temperature inside the reactor
- Slagging operation
- Short residence time
- High carbon conversion
- High sensibility of heat in the product gas and heat recovery is mandatory to improve the efficiency

8.6 Case Study 7

A case study is designed to investigate the biomass energy-based hydrogen production system. An entrained flow gasifier is employed to gasify the input biomass in the presence of steam and oxygen. The designed case study is simulated employing the Aspen Plus simulation software. The property method employed in the Aspen Plus software was RK-SOAVE that deals with real fluids and gases.

8.6.1 System Description

Biomass that is regarded as the carbon-neutral pathway, can be claimed as carbon-negative fuel if produced carbon dioxide is captured and not emitted to the environment. Fig. 8.13 displays the biomass gasification-assisted hydrogen production system while Fig. 8.14 exhibits the Aspen Plus simulation flowsheet of biomass

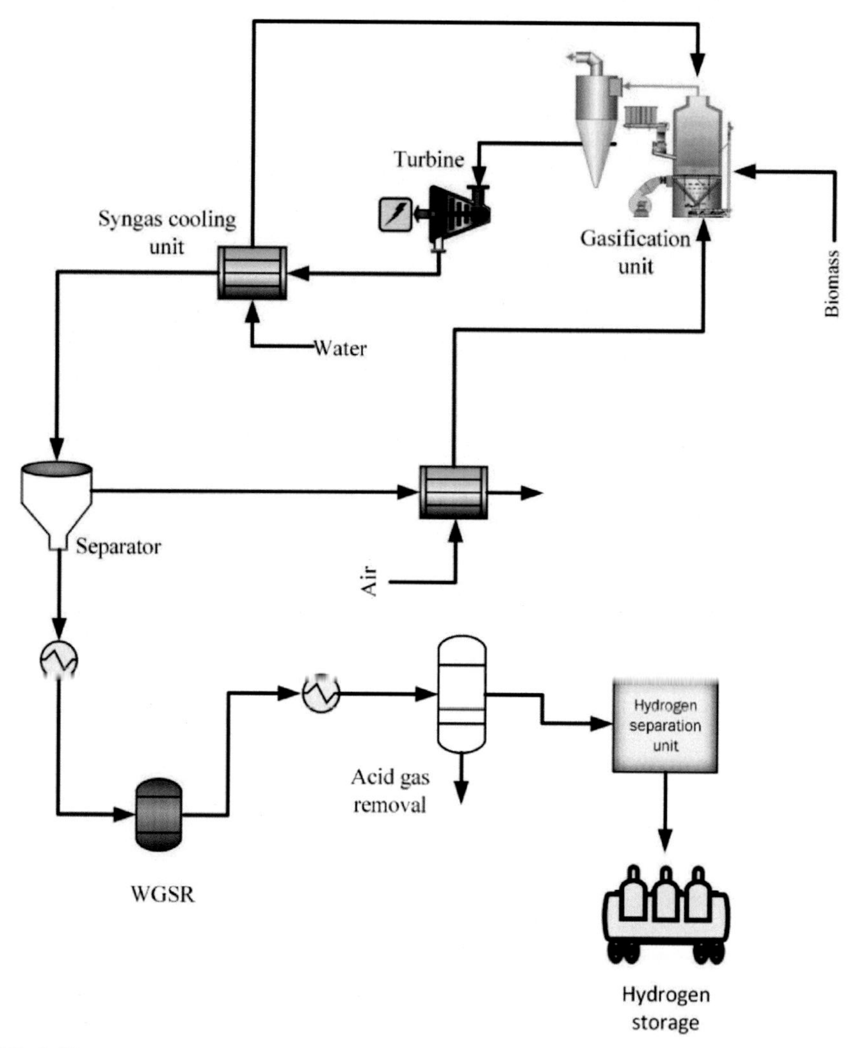

FIG. 8.13

Biomass gasification assisted hydrogen production system.

gasification-assisted hydrogen production system. The entrained flow gasifier is employed for the biomass gasification, and heat is recovered from the high-temperature syngas to convert water into steam and to heat the input air. The designed case study is simulated employing the Aspen Plus simulation software under the property method of RK-SOAVE that deals with real fluids and gases.

The syngas is produced through the biomass gasification unit C2 that passes through a turbine C3 that expands the high-temperature and pressure syngas to generate electrical power. The expanded syngas passes through the heat exchanger

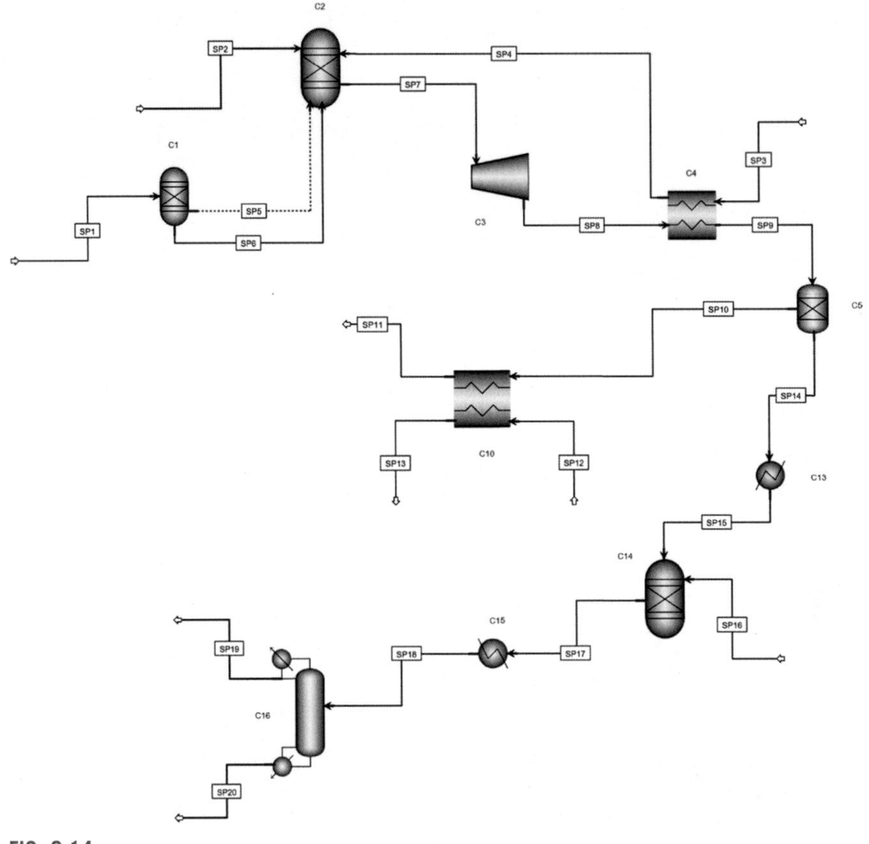

FIG. 8.14

Aspen Plus simulation flowsheet of biomass gasification assisted hydrogen production system.

C4 that recovers the additional heat from the syngas to convert water into steam that is fed to the gasification reactor through stream SP4. The ultimate and proximate analysis constraints of the bamboo wood biomass gasification system are tabulated in Table 8.1. The high-grade syngas is desired to achieve which requires the biomass gasifier to operate at the suitable operating temperature, and chemical composition is completely based on the biomass composition balance. The chemical reactions of the decomposition of char, volatile matter combustion, and chemical species decomposition can be expressed as follows:

$$\text{Biomass} \rightarrow \text{Char} + (C_6H_6 + CO + CO_2 + H_2 + N_2 + H_2O + CH_4 + H_2S)$$

$$\text{char} \rightarrow C + O_2 + N_2 + H_2 + S + \text{Ash}$$

$$H_2 + 0.5\,O_2 \rightarrow H_2O$$

Table 8.1 Ultimate and Proximate Analysis Constraints of the Biomass Gasification System [138].

Composition	Value
Biomass type	Bamboo wood (dry basis by weight%)
Pyrolysis Analysis	
Moisture	0.0
Volatile matter	86.8
Fixed carbon	11.24
Ash	1.95
Ultimate Analysis	
C	48.8
H	6.32
N	0.2
O	42.77
S	0.0
Ash	1.95
Higher heating value	20.55 MJ/kg

$$CO + 0.5\ O_2 \rightarrow CO_2$$

$$CO + H_2O \rightarrow CO_2 + H_2$$

$$CH_4 + 2O_2 \rightarrow 2H_2O + CO_2$$

$$C_6H_6 + 7.5\ O_2 \rightarrow 3H_2O + 6CO_2$$

$$CH_4 + H_2O \rightarrow CO_2 + H_2$$

$$C + O_2 \rightarrow CO_2$$

$$C + H_2O \rightarrow CO + H_2$$

$$C + 0.5\ O_2 \rightarrow CO$$

$$C + 2H_2 \rightarrow CH_4$$

$$C + CO_2 \rightarrow 2CO$$

$$S + H_2 \rightarrow H_2S$$

From the heat exchanger C4, the significant portion of the carbon dioxide, hydrogen, and carbon monoxide is separated using separator C5 and the remaining syngas is employed to the heat exchanger C10 through stream SP10. This heat exchanger is employed to recover the low-grade heat that is employed for space heating. A significant portion of the syngas including carbon dioxide, hydrogen, and carbon monoxide passes through a heater C13 that is employed to maintain the pressure, and the output of the heater is employed to the water–gas shift reactor C14.

The water–gas shift reactor converts carbon monoxide into carbon dioxide by reacting with steam, and additional hydrogen is produced. The steam is fed to the water–gas shift reactor C14 through stream SP16, and the chemical reaction is as follows:

$$CO + H_2O \rightarrow CO_2 + H_2 + \text{Heat}$$

The output of the water–gas shift reactor passes through a separation unit C16 that separates hydrogen from the other gases included in the syngas. The syngas composition is a significant parameter that is used to investigate the fraction of each gas in the syngas.

8.6.2 Analysis and Assessment

The bamboo wood is the biomass type that is employed by the entrained flow gasifier. The ultimate and proximate analysis constraints of the bamboo wood biomass gasification system are tabulated in Table 8.1. The model design equations employed to explore the performance of each component are presented in this section.

Biomass gasification unit

The correlation that is employed to evaluate the biomass chemical exergy is expressed as follows[139]:

$$ex^{f}_{\text{ch}} = \left[(\text{LHV} + \omega h_{\text{fg}}) \times \beta + 9.417S\right] \tag{8.1}$$

The LHV depends on the chemical composition of the biomass and the expression to calculate β is as follows[140]:

$$\beta = 0.1882\frac{H}{C} + 0.061\frac{O}{C} + 0.0404\frac{N}{C} + 1.0437 \tag{8.2}$$

The mass, energy, entropy, and exergy balance equations of each component of the designed system are as described in this section.

Yield reactor C1

The mass, energy, entropy, and exergy balance equations of the yield reactor are as follows:

$$\dot{m}_{\text{SP1}} = \dot{m}_{\text{SP6}} \tag{8.3}$$

$$\dot{m}_{\text{SP1}}\text{LHV}_{\text{SP1}} + \dot{Q}_{\text{Decomp}} = \dot{m}_{\text{SP6}}\text{LHV}_{\text{SP6}} \tag{8.4}$$

$$\dot{m}_{\text{SP1}}s_{\text{SP1}} + \frac{\dot{Q}_{\text{Decomp}}}{T} + \dot{S}_{\text{gen}} = \dot{m}_{\text{SP6}}s_{\text{SP6}} \tag{8.5}$$

$$\dot{m}_{\text{SP1}}ex_{\text{SP1}} + \dot{Q}_{\text{Decomp}}\left(1 - \frac{T_o}{T}\right) = \dot{m}_{\text{SP6}}ex_{\text{SP6}} + \dot{E}x_d \tag{8.6}$$

Gasification reactor C2

The mass, energy, entropy, and exergy balance equations of the gasification reactor are as follows:

$$\dot{m}_{SP2} + \dot{m}_{SP4} + \dot{m}_{SP6} = \dot{m}_{SP7} \quad (8.7)$$

$$\dot{m}_{SP2}h_{SP2} + \dot{m}_{SP4}h_{SP4} + \dot{m}_{SP6}\mathrm{LHV}_{SP6} - \dot{Q}_{Decomp} = \dot{m}_{SP7}h_{SP7} \quad (8.8)$$

$$\dot{m}_{SP2}s_{SP2} + \dot{m}_{SP4}s_{SP4} + \dot{m}_{SP6}s_{SP6} - \frac{\dot{Q}_{Decomp}}{T} + \dot{S}_{gen} = \dot{m}_{SP7}h_{SP7} \quad (8.9)$$

$$\dot{m}_{SP2}h_{SP2} + \dot{m}_{SP4}h_{SP4} + \dot{m}_{SP6}ex_{SP6} - \dot{Q}_{Decomp}\left(1 - \frac{T_o}{T}\right) = \dot{m}_{BG7}ex_{BG7} + \dot{E}x_d \quad (8.10)$$

Turbine C3

The mass, energy, entropy, and exergy balance equations of the turbine are as follows:

$$\dot{m}_{SP7} = \dot{m}_{SP8} \quad (8.11)$$

$$\dot{m}_{SP7}h_{SP7} = \dot{m}_{SP8}h_{SP8} + \dot{W}_{out} \quad (8.12)$$

$$\dot{m}_{SP7}s_{SP7} + \dot{S}_{gen} = \dot{m}_{SP8}s_{SP8} \quad (8.13)$$

$$\dot{m}_{SP7}ex_{SP7} = \dot{m}_{SP8}ex_{SP8} + \dot{W}_{out} + \dot{E}x_d \quad (8.14)$$

Heat exchanger C4

The mass, energy, entropy, and exergy balance equations of the heat exchanger are as follows:

$$\dot{m}_{SP3} = \dot{m}_{SP4} \text{ and } \dot{m}_{SP8} = \dot{m}_{SP9} \quad (8.15)$$

$$\dot{m}_{SP3}h_{SP3} + \dot{m}_{SP8}h_{SP8} = \dot{m}_{SP4}h_{SP4} + \dot{m}_{SP9}h_{SP9} \quad (8.16)$$

$$\dot{m}_{SP3}s_{SP3} + \dot{m}_{SP8}s_{SP8} + \dot{S}_{gen} = \dot{m}_{SP4}s_{SP4} + \dot{m}_{SP9}s_{SP9} \quad (8.17)$$

$$\dot{m}_{SP3}ex_{SP3} + \dot{m}_{SP8}ex_{SP8} = \dot{m}_{SP4}ex_{SP4} + \dot{m}_{SP9}ex_{SP9} + \dot{E}x_d \quad (8.18)$$

Separator C5

The mass, energy, entropy, and exergy balance equations of the separator are as follows:

$$\dot{m}_{SP9} = \dot{m}_{SP10} + \dot{m}_{SP14} \quad (8.19)$$

$$\dot{m}_{SP9}h_{SP9} = \dot{m}_{SP10}h_{SP10} + \dot{m}_{SP14}h_{SP14} \quad (8.20)$$

$$\dot{m}_{SP9}h_{SP9} + \dot{S}_{gen} = \dot{m}_{SP10}h_{SP10} + \dot{m}_{SP14}h_{SP14} \quad (8.21)$$

$$\dot{m}_{SP9}ex_{SP9} = \dot{m}_{SP10}ex_{SP10} + \dot{m}_{SP14}ex_{SP14} + \dot{E}x_d \quad (8.22)$$

Heat exchanger C10

The mass, energy, entropy, and exergy balance equations of the heat exchanger are as follows:

$$\dot{m}_{SP10} = \dot{m}_{SP11} \text{ and } \dot{m}_{SP12} = \dot{m}_{SP13} \tag{8.23}$$

$$\dot{m}_{SP10}h_{SP10} + \dot{m}_{SP12}h_{SP12} = \dot{m}_{SP11}h_{SP11} + \dot{m}_{SP13}h_{SP13} \tag{8.24}$$

$$\dot{m}_{SP10}s_{SP10} + \dot{m}_{SP12}s_{SP12} + \dot{S}_{gen} = \dot{m}_{SP11}s_{SP11} + \dot{m}_{SP13}s_{SP13} \tag{8.25}$$

$$\dot{m}_{SP10}ex_{SP10} + \dot{m}_{SP12}ex_{SP12} = \dot{m}_{SP11}ex_{SP11} + \dot{m}_{SP13}ex_{SP13} + \dot{E}x_d \tag{8.26}$$

Heater C13

The mass, energy, entropy, and exergy balance equations of the heater are as follows:

$$\dot{m}_{SP14} = \dot{m}_{SP15} \tag{8.27}$$

$$\dot{m}_{SP14}h_{SP14} = \dot{m}_{SP15}h_{SP15} + \dot{Q}_{out} \tag{8.28}$$

$$\dot{m}_{SP14}s_{SP14} + \dot{S}_{gen} = \dot{m}_{SP15}s_{SP15} + \frac{\dot{Q}_{out}}{T} \tag{8.29}$$

$$\dot{m}_{SP14}ex_{SP14} = \dot{m}_{SP15}ex_{SP15} + \dot{Q}_{out}\left(1 - \frac{T_o}{T}\right) + \dot{E}x_d \tag{8.30}$$

Water–gas shift reaction C14

The mass, energy, entropy, and exergy balance equations of the WGSR are as follows:

$$\dot{m}_{SP15} + \dot{m}_{SP16} = \dot{m}_{SP17} \tag{8.31}$$

$$\dot{m}_{SP15}h_{SP15} + \dot{m}_{SP16}h_{SP16} = \dot{m}_{SP17}h_{SP17} + \dot{Q}_{out} \tag{8.32}$$

$$\dot{m}_{SP15}s_{SP15} + \dot{m}_{SP16}s_{SP16} + \dot{S}_{gen} = \dot{m}_{SP17}s_{SP17} + \frac{\dot{Q}_{out}}{T} \tag{8.33}$$

$$\dot{m}_{SP15}ex_{SP15} + \dot{m}_{SP16}ex_{SP16} = \dot{m}_{SP17}ex_{SP17} + \dot{Q}_{out}\left(1 - \frac{T_o}{T}\right) + \dot{E}x_d \tag{8.34}$$

Separator C15

The mass, energy, entropy, and exergy balance equations of the separator are as follows:

$$\dot{m}_{SP18} = \dot{m}_{SP19} + \dot{m}_{SP20} \tag{8.35}$$

$$\dot{m}_{SP18}h_{SP18} = \dot{m}_{SP19}h_{SP19} + \dot{m}_{SP20}h_{SP20} \tag{8.36}$$

$$\dot{m}_{SP18}s_{SP18} + \dot{S}_{gen} = \dot{m}_{SP19}s_{SP19} + \dot{m}_{SP20}s_{SP20} \tag{8.37}$$

$$\dot{m}_{SP18}ex_{SP18} = \dot{m}_{SP19}ex_{SP19} + \dot{m}_{SP20}ex_{SP20} + \dot{E}x_d \tag{8.38}$$

Performance indicator

The energetic and exergetic efficiency equations of the designed case study can be written as follows:

$$\eta_{ov} = \frac{\dot{m}_{H_2}\text{LHV}_{H_2} + \dot{m}_{SP12}(h_{SP13} - h_{SP12}) + \dot{W}_{C3}}{\dot{m}_{biomass}\text{LHV}_{biomass}} \tag{8.39}$$

$$\psi_{ov} = \frac{\dot{m}_{H_2}ex_{H_2} + \dot{m}_{SP12}(ex_{SP13} - ex_{SP12}) + \dot{W}_{C3}}{\dot{m}_{biomass}ex_{biomass}} \tag{8.40}$$

8.6.3 Results and Discussion

The designed case study is explored to investigate the performance of the biomass-assisted hydrogen production system. The composition of the syngas is a substantial parameter in biomass gasification that specifies the importance of exploring the fractions of each gas included in the syngas. The composition of the syngas is measured in terms of the flowrates of carbon dioxide, methane, oxygen, carbon monoxide, steam, and hydrogen gases.

The effect of biomass and steam flowrates is remarkable to be investigated on the composition of the syngas that is measured in terms of the flowrates of carbon dioxide, methane, oxygen, carbon monoxide, steam, and hydrogen gases. Fig. 8.15 shows the biomass and oxygen flowrates effect on carbon dioxide flowrate, Fig. 8.16 exhibits the biomass and oxygen flowrates effect on steam flowrate, Fig. 8.17 displays the biomass

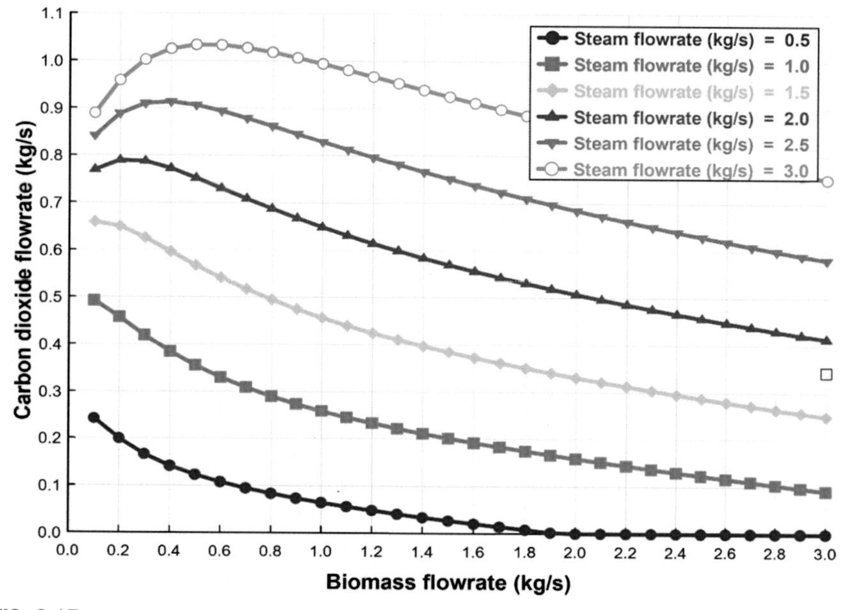

FIG. 8.15

Effect of biomass and steam flowrates on carbon dioxide flowrate.

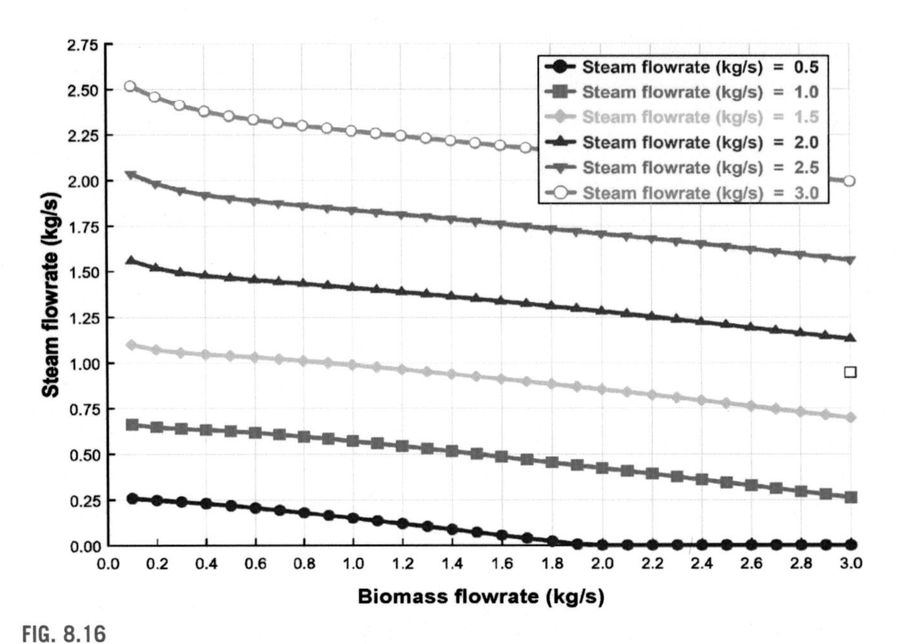

FIG. 8.16

Effect of biomass and steam flowrates on output steam flowrate.

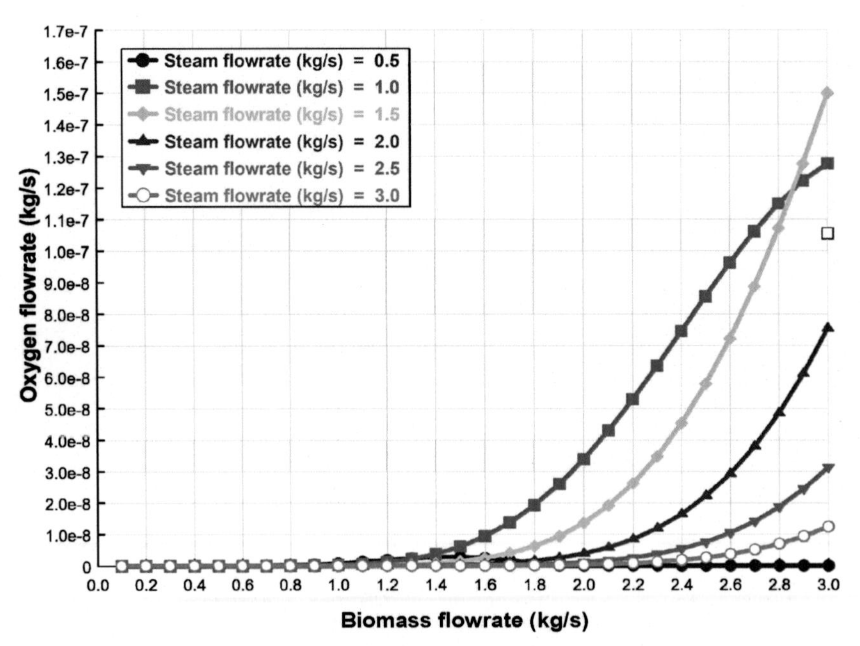

FIG. 8.17

Effect of biomass and steam flowrates on output oxygen flowrate.

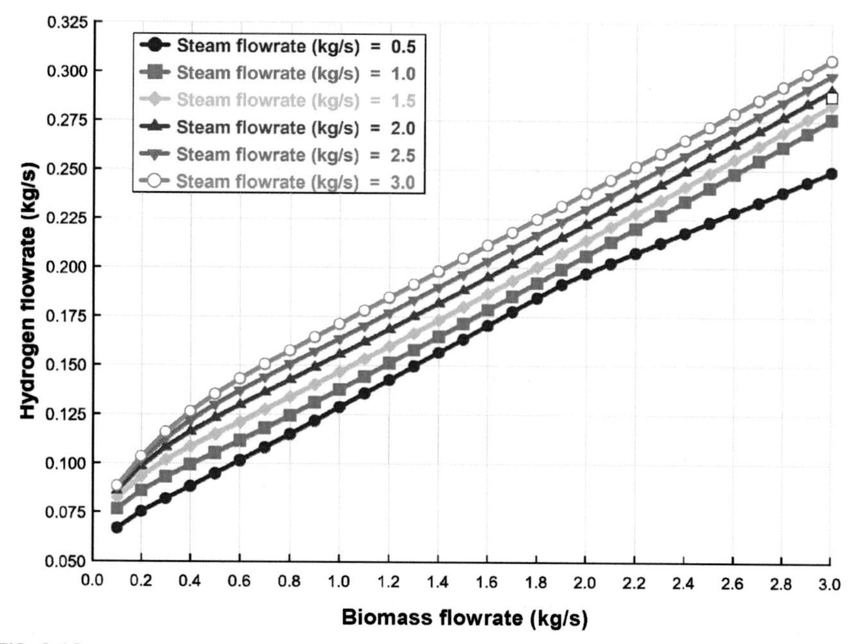

FIG. 8.18

Effect of biomass and steam flowrates on output hydrogen flowrate.

and oxygen flowrates effect on oxygen flowrate, Fig. 8.18 displays the biomass and oxygen flowrates effect on hydrogen flowrate, Fig. 8.19 exhibits the biomass and oxygen flowrates effect on methane flowrate, and Fig. 8.20 represents the biomass and oxygen flowrates effect on carbon monoxide flowrate.

Fig. 8.15 exhibits the effect of biomass and steam flowrate on the flowrate of carbon dioxide. The ranges of the biomass flowrate are specified from 0.1 to 3 kg/s and the steam flowrate ranged from 0.5 to 3.0 kg/s. Each line represents the flow rate of carbon dioxide at the biomass flow rate according to the x-axis and specific steam flow rate according to the legends. One can depict that the carbon dioxide flow rate increases in the early stage of biomass gasification while it decreases gradually with the continuous rise in the biomass and steam input flowrates.

Fig. 8.16 shows the effect of biomass and steam flowrate on the flowrate of output steam. It is displayed that the steam flowrate decreases gradually with the continuous rise in the biomass and steam input flowrates as steam is consumed in the biomass gasification reaction.

The effect of biomass and steam flowrate on the flowrate of output oxygen is displayed in Fig. 8.17. It is shown that all the oxygen supplied to the biomass gasification unit is consumed in the early stage and it starts increasing gradually with the continuous rise in the biomass and steam input flowrates as the reaction will need further biomass and steam.

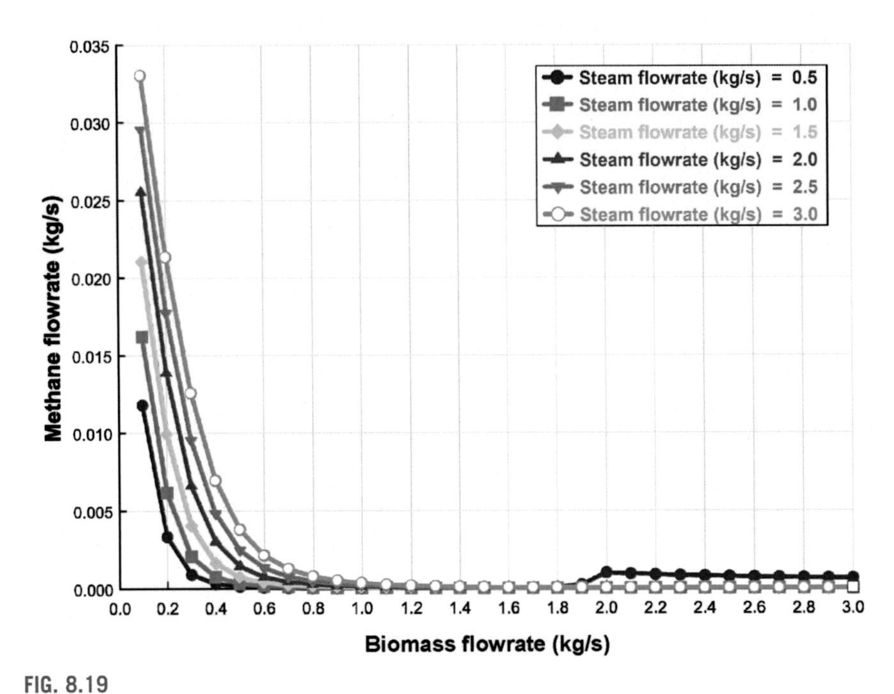

FIG. 8.19

Effect of biomass and steam flowrates on methane flowrate.

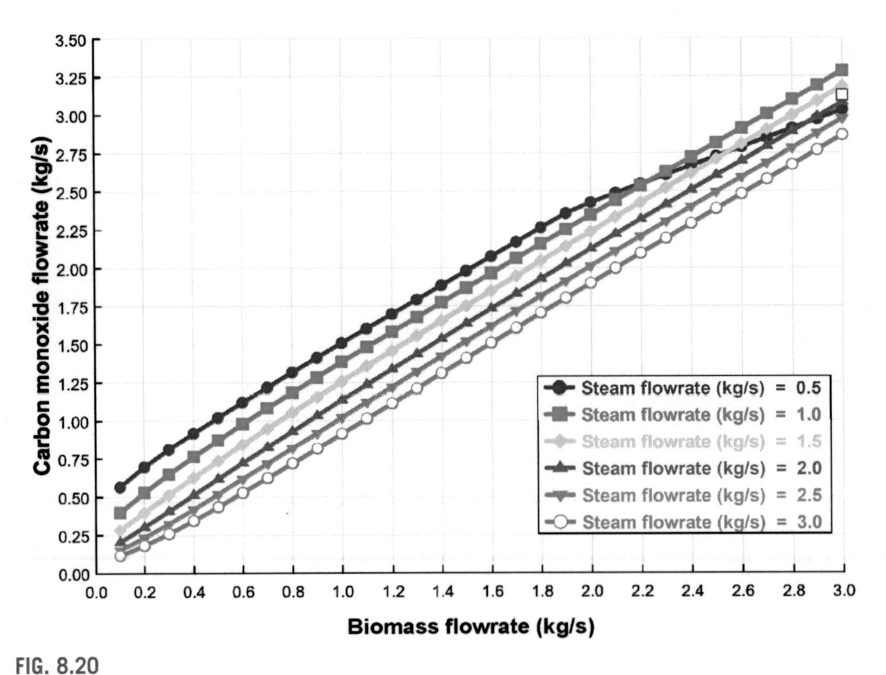

FIG. 8.20

Effect of biomass and steam flowrates on carbon monoxide flowrate.

Table 8.2 Significant Results of Designed Biomass Assisted System.

Component	Result
Power generation via C3 turbine	6.289 MW
Hydrogen production before WGSR	129.5 mol/s
Hydrogen production after WGSR	169.7 mol/s
Overall energy efficiency	48.55%
Overall exergy efficiency	45.08%

Table 8.2 displays the significant results of the designed biomass assisted system. This table comprises noteworthy results, such as power generation, hydrogen production rates before and after the water–gas shift reactor, and energetic and exergetic efficiencies. The designed case study produces the electrical power of 6.28 MW, and the hydrogen production rates before and after the water–gas shift reactor are found to be 129.5 mol/s and 169.7 mol/s while the energetic and exergetic efficiencies are found to be 48.55% and 45.08%, respectively.

Fig. 8.18 exhibits the effect of biomass and steam flowrate on the hydrogen flowrate. The hydrogen flowrate increases gradually with the continuous rise in the biomass and steam input flowrates as hydrogen production is desired from the biomass gasification reaction and additional hydrogen is produced in the water gas shift reactor.

The effect of biomass and steam flowrate on methane gas flowrate is shown in Fig. 8.19. It is displayed that all the methane gas is produced in the early stage of the biomass reaction that is cracked down with a continuous rise in the biomass and steam input flowrates and methane fractions become to zero in the syngas.

The effect of biomass and steam flowrate on the flowrate of carbon monoxide is shown in Fig. 8.20. It is evident that the carbon monoxide flowrate increases gradually with the continuous rise in the biomass and steam input flowrates as carbon monoxide is generated in the biomass gasification reaction, which is further converted to carbon dioxide and additional hydrogen is produced in the water gas shift reactor.

Table 8.3 displays the hydrogen production rates before and after the water–gas shift reactor. The water–gas shift reactor converts the carbon monoxide into carbon dioxide by chemically reacting with steam and produces further oxygen, which improves the hydrogen production flowrates after the water–gas shift reactor. Table displays that the hydrogen production rate increased from 37.73 to 128.8 mol/s before WGSR while it increases from 40.83 to 169.7 mol/s after WGSR.

Table 8.3 Hydrogen Production Rates Before and After the Water–Gas Shift Reactor.

Biomass Input Flowrate (kg/s)	Hydrogen Production Rate before WGSR in Stream SP15 (mol/s)	Hydrogen Production Rate before WGSR in Stream SP17 (mol/s)
0.1	37.73	40.83
0.2	43.08	47.57
0.3	47.23	53.15
0.4	50.56	57.93
0.5	53.53	62.32
0.6	56.38	66.58
0.7	59.21	70.79
0.8	62.04	75.00
0.9	64.90	79.21
1	67.78	83.44
1.1	70.68	87.68
1.2	73.61	91.93
1.3	76.56	96.19
1.4	79.52	100.47
1.5	82.51	104.75
1.6	85.50	109.04
1.7	88.51	113.34
1.8	91.54	117.65
1.9	94.57	121.96
2	97.61	126.28
2.1	100.66	130.60
2.2	103.72	134.93
2.3	106.79	139.26
2.4	109.86	143.60
2.5	112.94	147.94
2.6	116.02	152.28
2.7	119.11	156.62
2.8	122.20	160.97
2.9	125.30	165.32
3	128.40	169.67

8.7 Closing Remarks

Biomass is known as the animal or plant material that is employed to produce energy or used as a raw substance in numerous industrial processes to form a variety of products. Biomass can be found in numerous forms, such as grown energy crops, forest or wood residues, human waste of sewage plants or animal farming, horticulture, food crops waste, and food processing. Biomass chiefly contains the energy that is primarily solar derived and absorbed by the plants through photosynthesis that is the process of converting water and carbon dioxide into glucose and carbohydrates, and this process occurs in the presence of sunlight. The burning of biomass is employed to generate thermal energy that is converted into electrical power or treated into biofuel. Biomass is well thought out as a renewable source of energy as its intrinsic energy is originated from the sun that can be regrown in a moderately short period. Trees absorb the atmospheric carbon dioxide and convert it to biomass, and carbon dioxide is released back to the atmosphere when they die. Solid biomass feedstock, such as garbage and wood can be combusted directly to generate heat. Biomass can similarly be employed to produce liquid biofuels, namely biodiesel and ethanol or biogas. Animal fats and vegetable oils are employed to produce biodiesel that can be employed as heating oil or in vehicles. Biomass offers a clean, consistent, and renewable source of energy that has the potential to improve the economy, environment, and energy security.

Gasification is a process known as the conversion of organic carbonaceous materials into carbon dioxide, carbon monoxide, and hydrogen. This is accomplished by high-temperature (more than 700°C) material reaction without combustion employing controlled oxygen and/or steam amount. The pyrolysis process is recognized as the thermal decomposition of the volatile organic components under the high temperature and forming syngas in the absence of oxygen. A case study is designed to investigate the biomass gasification-assisted hydrogen production system. An entrained flow gasifier is installed to gasify the biomass input in the presence of oxygen and steam. The designed case study is simulated employing the Aspen Plus simulation software under the property method of RK-SOAVE, which deals with real fluids and gases. The composition of the syngas is a noteworthy constraint that is explored and determined in terms of the flowrates of each gas included in the syngas. The designed case study offered a hydrogen production rate of 169.7 mol/s along with a power generation of 6.28 MW. The design case study offered energetic and exergetic efficiencies of 48.55% and 45.08%, respectively. Biomass that is regarded as carbon-neutral fuel can be claimed as carbon-negative fuel if produced carbon dioxide is captured and not emitted to the environment. The biomass energy-based hydrogen production can offer promising results on a commercial scale, and it can also place a leading role in the global transition from conventional to renewable energy.

CHAPTER

Integrated Systems for Hydrogen Production

9

Integration of energy systems is intended to combine the energy carriers, for instance, thermal pathways, electricity, and fuels with infrastructures, such as communications, transportation, water, hydrogen, and combined heating and cooling to maximize the efficiency and effective utilization of renewable energy sources. The compatible integration of the energy components and subsystems is a significant prospect to pursue the optimal utilization of the newly established technologies and influence the global transition.

In engineering, system integration is well defined as the process of linking the components and subsystems into a single system (a combination of cooperating subsystems so that the system can deliver predominant functionality) and confirming the functionality of the subsystems together as a system. Renewable energy integration requires a sturdy transmission grid in an energy system that can deal with variations from wind and solar energy. Energy systems integration across heating and cooling, electricity, hydrogen, water, and transportation are known as efficient approaches. This integration boosts the overall energy system flexibility and assists in balancing the renewable energy source variations in an economically viable way.

Energy system integration combines distributed generation technologies or onsite power with thermally stimulated technologies to deliver energy storage, cooling, humidity control, heating, and/or additional process functions employing thermal energy that is generally unexploited in electricity/power production. Energy system integration produces electricity and on-site thermal energy byproduct with 80% or more conversion potential of fuel into operational energy. Energy system integration has the potential to propose the advantages of consumer choice, extraordinary energy efficiency advances, and energy security. It can also reduce the commercial and industrial building sector carbon emissions drastically and upsurge the source energy efficiency.

A comprehensive review[38] was published on renewables and their role in achieving more sustainable communities with the reduced environmental impact. Their study then focused on the clean-energy solutions to establish better sustainability, and hence discussed the prospects and challenges considering various dimensions, such as social, energetic, economic, and environmental aspects and also evaluated the existing potential statuses and applications of probable clean-energy systems. It is further discussed that renewable energy-assisted systems offered multiple advantages in terms of energy utilization, system cost, and environmental emissions. The residential sector covers approximately one-third of global energy consumption.

Renewable Hydrogen Production. https://doi.org/10.1016/B978-0-323-85176-3.00013-5

Distributed energy applications and combined heating and power in institutional and commercial buildings have, nevertheless, factually been limited because of the inadequate practice of byproducts, such as thermal energy, predominantly during summer when minimum heating is required. In recent years, custom manufactured systems have advanced integrating potentially high-value facilities from thermally stimulated technologies, such as humidity control and cooling. Such thermally stimulated technology equipment can be integrated into a combined heating and power system to exploit the heat output byproduct efficiently to deliver desiccant humidity control or absorption cooling for the buildings. Energy system integration can thus enlarge the thermal energy services potential and, by this means, encompass the conventional combined heating and power market into the applications of the building sector that combined heating and power systems cannot serve solely. Today, such combined heating, cooling, and humidity control systems can decrease carbon emissions more than ever, whereas improving the efficiency of source energy in the buildings sector.

Fig. 9.1 exhibits the global fuel shares of (a) electricity generation, (b) CO_2 emissions, and (C) total primary energy supply in 2011. Fig. 9.1a displays the global share of fuels in electricity generation. The total share involves 41% petroleum/coal, 22% natural gas, 12% nuclear, 16% hydro, 5% oil, and 4% other including wind, solar, geothermal, heat, and waste. Fig. 9.1b shows the global share of fuels in CO_2 emissions. The total share involves 44% petroleum/coal, 35% natural gas, 20% oil, and 1% other. Fig. 9.1c exhibits the global share of fuels in the primary energy supply. The total share involves 32% oil, 29% petroleum/coal, 21% natural gas, 10% biofuels/waste, 5% nuclear, 2% hydro, and 1% other sources.

Even with such developments over conventional combined heating and power systems, the integrated energy system faces substantial technological and economic challenges. Of critical importance to the accomplishment of integrated energy systems is the aptitude to treat the hydrogen production, heating, air conditioning, ventilation, lighting, water heating, and power as subsystems of an integrated system, accommodating these loads directly or indirectly from the combined heating and power system output. The combined heating and power technology roadmaps are focusing the research on the comprehensive integration approach: equipment integration, component integration, modular and packaged system development, grid integration, and system integration with process loads and building. This noticeable change in research technology and development has directed to the formation of a new IES acronym to well reflect the development of nature in this significant energy efficiency area as integrated energy systems.

9.1 Status of Integrated Energy Systems

While integrated energy systems practice in US buildings is at the early stages, combined heating and power systems have been in confined usage for decades in the buildings sector. Numerous data sources contradict the number of buildings that

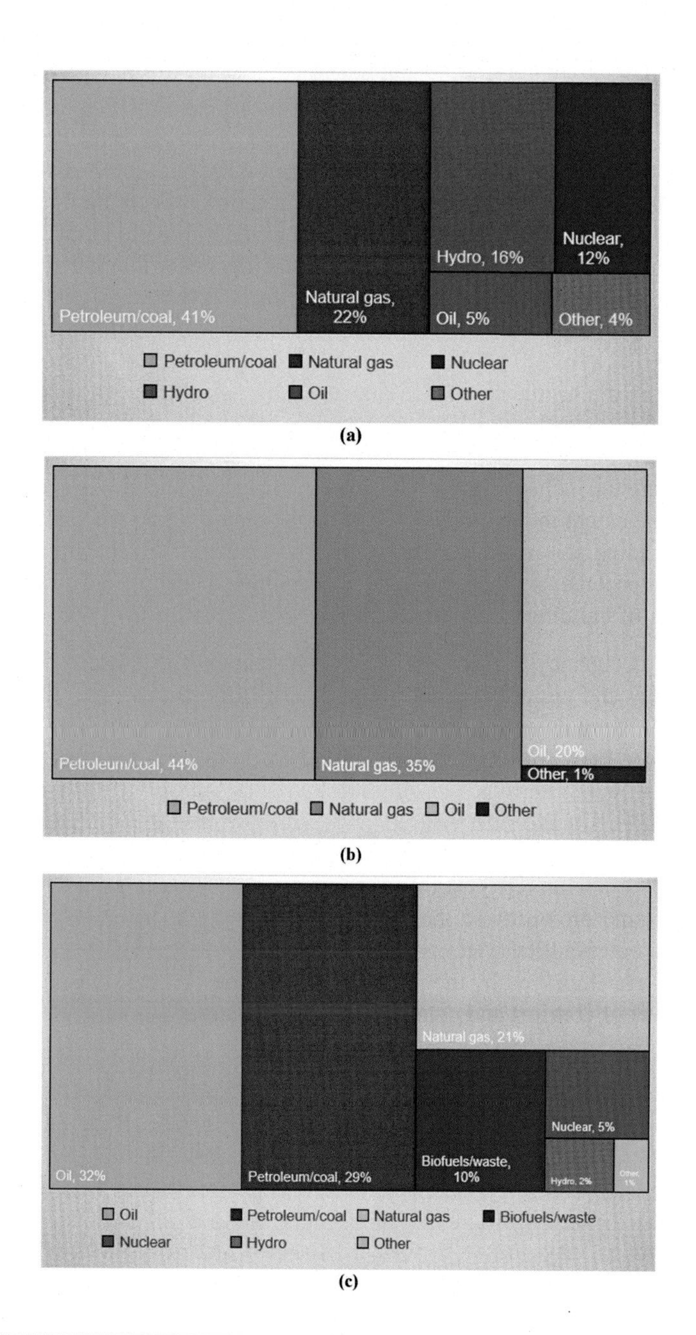

FIG. 9.1

Global fuel shares of (a) electricity generation, (b) CO_2 emissions, and (c) total primary energy supply.

Modified from [141].

currently employ combined heating and power systems, as utility data institute statistics display around 2600 MW and energy information administration post total of approximately 1900 MW. Independent organizations of energy, namely district energy library, hosted by the University of Rochester, displays combined heating and power systems installations in educational institutes that exceed the figures quoted by the above-mentioned sources. Mark Spurr from the international district energy association prepared a report on the district energy systems integrated with combined heating and power, and Spurr examined these data sources and concluded that total combined heating and power serving buildings over district energy systems were estimated to be 3500 MW. Relating these figures with the combined heating and power systems in the industries and buildings, the figure of approximately 46,000 MW in 1998 out of which 5%—10% of the installed base was in the buildings sector.

Even though the inclusive size of the combined heating and power systems is uncertain to some extent in the buildings sector, the greatest agree that educational facilities are leading this market followed by the health care sector. Some of the aspects that impact the favorable economics of combined heating and power systems in these types of buildings are as follows:

- Levels of occupancy are normally high with patients or students lodging the services day and night, generating high-load factors that assist to pay back the investment in combined heating and power systems.
- The stability between electric and thermal loads in such buildings is high in comparison with other types of buildings.
- Numerous buildings under shared ownership, so that heating, cooling, and electricity loads can be accumulated and assisted by the central system that is larger and comparatively cost-effective over more than a few smaller systems.
- Close building proximity so that linking the building distribution piping to the hot steam/water/chilled water is not cost extensive.
- Buildings are employed by the property owners and not leased; consequently, a higher level of comfort and control is usually required.

Fig. 9.2 exhibits the CO_2 emissions (a) by sector (b) by energy source from 1990 to 2015. The CO_2 emissions by different energy sources from 1990 to 2015 displayed in Fig. 9.2a involved coal energy source rise from 8296 to 14,635 Mt, oil energy source rise from 8505 to 11,150 Mt, natural gas energy source rise from 3677 to 6456 Mt, and other energy sources rise from 44 to 189 Mt.

The CO_2 emissions by different sectors from 1990 to 2015 displayed in Fig. 9.2b involved the electricity and heat production sector rise from 7625 to 13,405 Mt, other energy industries sector rise from 977 to 1653 Mt, the transportation sector rise from 4595 to 7702 Mt, industrial sector rise from 3959 to 6361 Mt, residential sector rise from 1832 to 1850 Mt, commercial sector rise from 774 to 832 Mt, and agricultural sector rise from 398 to 413 Mt.

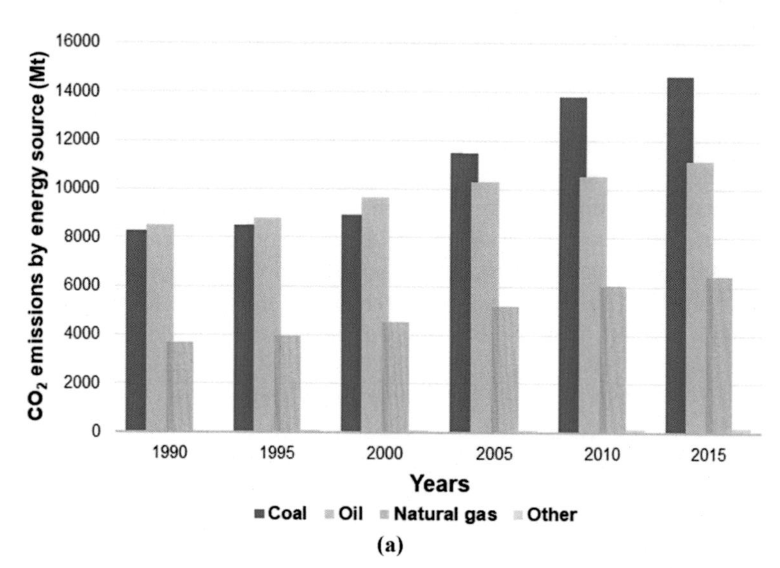

(a)

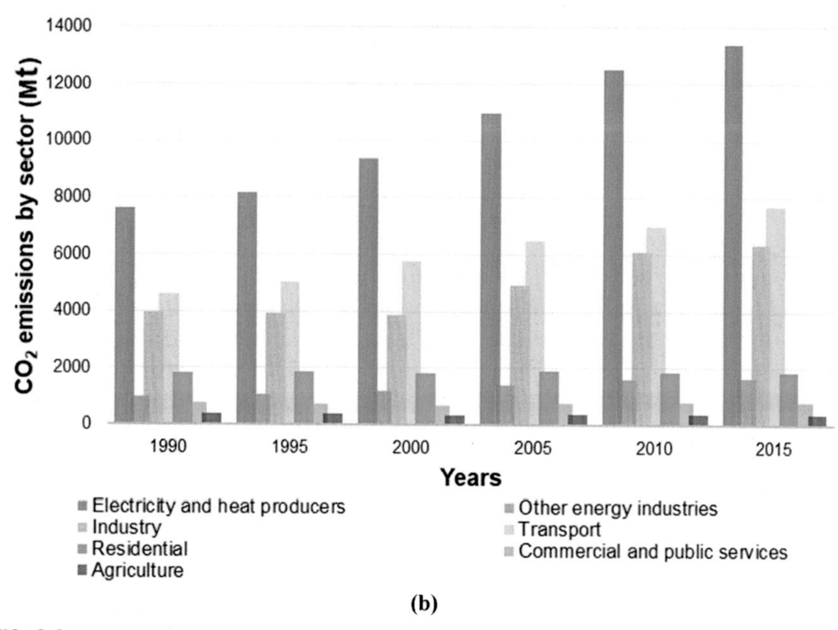

(b)

FIG. 9.2

Global CO_2 emissions from 1990 to 2015 (a) by energy source (b) by sector.

Data from [3].

9.1.1 Integrated Energy Systems for Buildings

It is extensively agreed that an integrated energy system is the underutilized technological option in the building sector. Even though some of this is because of the inadequate economic returns associated with the seasonal cooling and heating loads, the reasons for not employing an integrated energy system widely in buildings are established. Numerous building owners decide on the cost basis and an integrated energy system offers higher costs as compared with conventional alternatives. Moreover, the community building design inclines to be risk opposing, supporting the conventional and established alternatives and not endorsing the possibilities that have not been quantified before. Consequently, most of the vast building does not employ integrated energy systems.

However, numerous trends are generating an integrated energy system-favorable environment for the buildings sector. Restructuring of the electricity industry, whereas auspicious lower rates for more users, has numerous building owners disturbed over increasing prices and reducing grid reliability. Moreover, innovative standards of inside air quality cause the augmented ventilating rates and assisted renew concentration in desiccant dehumidification that varies the humidity control economics in buildings and launches additional applications for combined heating and power waste heat. Lastly, self-governing third parties like ESCOs and utilities are capitalizing on district combined heating and power systems, offering new building savings opportunities without large investments.

Other aspects that are assisting to create a further integrated energy system-favorable environment are the global warming problems that have been raised after the Kyoto protocol. Increased energy efficiency that can be achieved from the extensive usage of combined heating and power and integrated energy system is being considered in numerous policy scenarios that resulted from the Kyoto compliance strategies. Even though industrial combined heating and power is predicted to be critical for such scenarios, an integrated energy system for buildings is being counted upon as well.

9.1.2 Integrated Energy Systems for Hydrogen

Even though hydrogen is being recognized as a potential output candidate from the integrated energy system but is still an underutilized technological option. Renewable energy sources, such as wind, solar, hydro, geothermal, ocean thermal energy conversion, and biomass have shown the potential to be employed for hydrogen production using an integrated energy system. Different renewable energy methods follow different methods and approach for producing hydrogen, and a method classification is shown in Fig. 9.3.

The solar energy source can be employed in the integrated energy system undergoing the photovoltaic, photoelectrochemical, photonic, and solar thermal processes. The solar photovoltaic (PV) panels are employed in the integrated energy

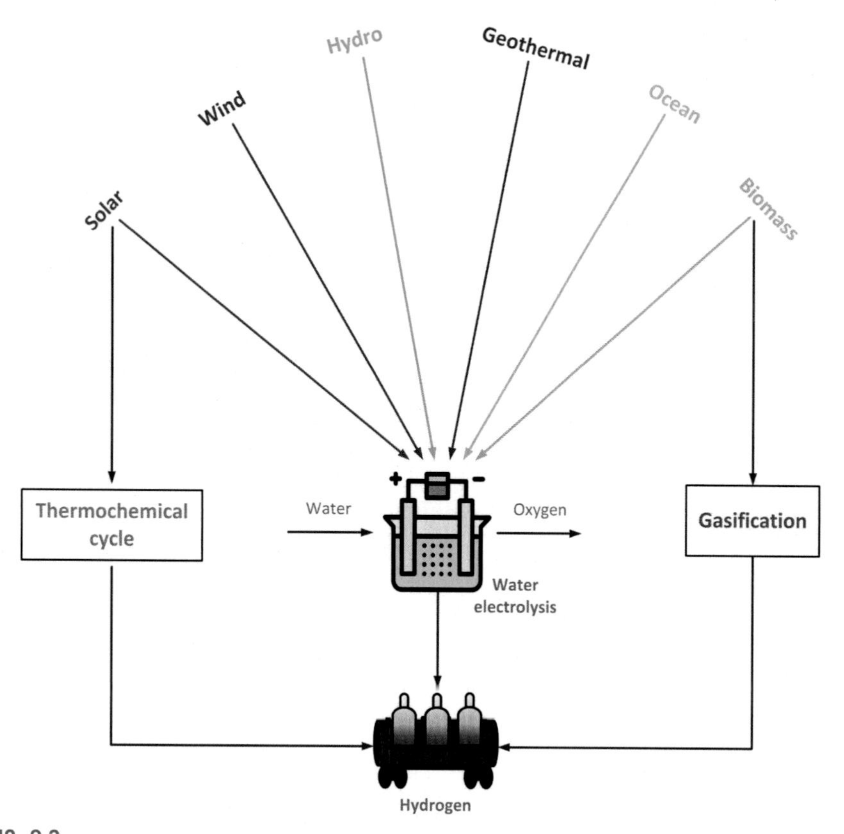

FIG. 9.3

Renewable energy-based hydrogen production routes.

system to generate electricity using a photovoltaic source that is employed in the water electrolysis process for hydrogen production. The photoelectrochemical, photonic sources are integrated with the water electrolysis step that splits water into its components using electrical power. The solar thermal source is further split into two categories of solar thermal collector and solar heliostat that are further integrated with the thermochemical cycle for producing hydrogen using thermal energy and electricity. The other sources, such as wind, hydro, geothermal, and ocean thermal energy conversion (OTEC) generate electrical power that is directly employed in the water electrolysis process for hydrogen production. The biomass route of the hydrogen production undergoes the gasification process that produces syngas and hydrogen is separated from other gases. The produced hydrogen is stored and can be employed for numerous applications, such as power production using fuel cells, as fuel, as an energy carrier, for heating and cooling, and in electric and hybrid electric vehicles.

Hydrogen is a flawless electricity partner, and an integrated energy system is formed together based on the usage and distributed power generation. Electricity and hydrogen are interchangeably employing an electrolyzer that converts electricity into hydrogen and a fuel cell that converts hydrogen into electricity. Electricity and hydrogen both are known as energy carriers, as they require and are produced by employing a primary source of energy dissimilar to the naturally existing hydrocarbon fuels. Fossil fuels and nuclear are the most significant energy sources offering electricity and hydrogen production but renewable energy sources, namely solar, geothermal, wind, hydro, and biomass, are also playing a significant role in electricity and hydrogen production. Concerning future transportation, time plays a significant part to determine the most viable technology and resource for hydrogen production as an automotive fuel. To reduce the environmental problems accompanied by conventional fossil fuel-based hydrogen production methods, renewable energy is the chief focus of the scientific community to replace the traditional energy sources.

9.2 Significance of Integrated Energy Systems

Integrated energy appears to be a significant technology for the global energy transition. It integrates the individual heat, electricity, and mobile energy sectors and consequently certifies efficient renewable energy usage. Integrated energy systems are the individual and potential route to establish a decarbonized global economy. With advanced opportunities offered by digitalization, integrated energy systems technology is leading toward global energy transition. Significant advantages of renewable energy integration are listed as follows:

- Lower operational costs
- Less maintenance requirement
- Less CO_2 emissions
- Longer life span and reliable
- Better renewable heat incentive
- Easier generation of multiple outputs
- Better environmental outcome
- Better economic growth
- Better availability of useful commodities

9.2.1 Efficient Energy Utilization

Efficient energy utilization is one of the most significant reasons that can be employed to trace the importance of integrated energy systems. The power generation proportion characterized by renewable energy sources is continuously increasing. Renewable energy sources covered approximately 20% of the global energy consumption in 2016 according to the REN21 community report[142] on global renewable energy status. Fraunhofer Institute for solar energy systems established that approximately 41% of total electricity production in Germany was found to be renewable in 2018. Power generation using PV and wind energy-based systems thus contribute substantially to the global energy supply. The only disadvantage is that wind and solar are unpredictable as they produce fluctuating energy amounts.

9.2.2 Sustainable Energy Supply

Power-to-energy-carriers technologies and battery storage units convert excess renewable electricity into a different energy form and thus upsurge the percentage of renewables in the transport and heating sectors. Battery storage systems, heat, and gas are the greatest common representatives of such technologies.

Power-to-gas

The standing gas network infrastructures enable the storage and transportation of large energy quantities. Employing power-to-gas technology, this storage can similarly be exploited for renewable energies. For instance, this permits the CO_2 to rise from the biogas production to be employed for the methane production that is then accessible in the natural gas grid as raw material for the chemical industry as the propulsion energy for aircraft and vehicles, or reconversion at gas-fired power plants.

Power-to-heat

Excess electricity from renewable energy sources can similarly be employed to supply heat. This is then accessible for producing hot water or heating energy. In comparison with power-to-gas technologies, the conversion efficiency from electricity to heat is nearly 100%.

The power-to-heat mechanism undergoes hybrid schemes that always employ a heat generator supported by conventional fuel, such as natural gas or wood. In the incident of excess electricity, heat is extracted using electrical energy.

The heat can be supplied to the district or local heating grid or employed to support the local heat supply to the specific industrial companies or buildings. A supplementary buffer tank is employed to store the heat temporarily and retrieved to deliver negative balancing energy.

Battery storage

Battery storage systems are similarly consistent local stores for excess renewable energy. Currently, rechargeable chemical cells absorb energy and supply it on demand. In the home storage area, battery storage systems integrated with wind or photovoltaic systems confirm superior energy efficiency and store the electricity supply on grid failures. In bulky storage power plants in the megawatt range, the batteries offer an operating replacement. Furthermore, currently, battery inverters subsidize the instantaneous reserve and fast frequency stabilization. This plays a significant role when supplementary power is required at short notice since, at all times, a big

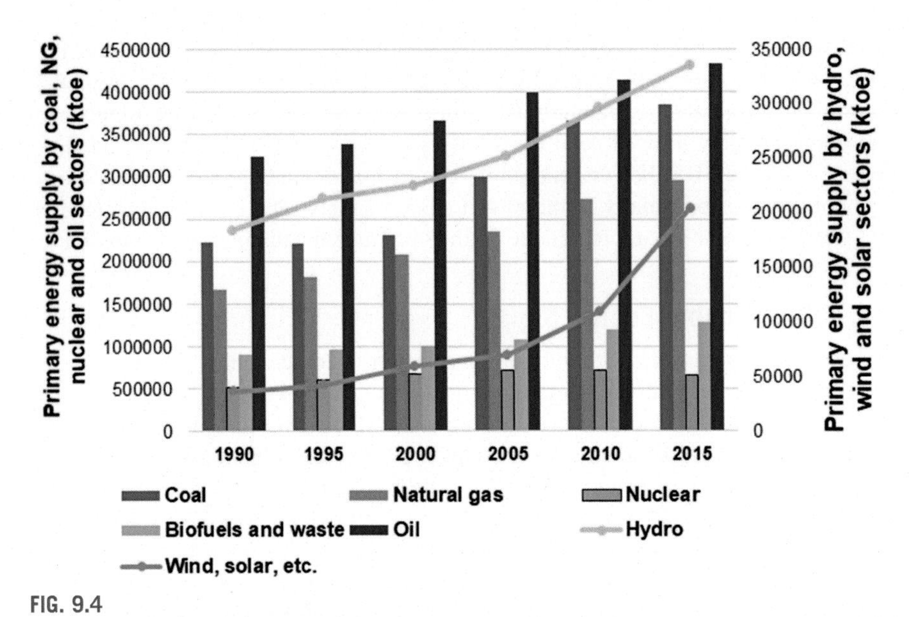

FIG. 9.4

Primary energy supply by conventional and renewable energy sources.

Data from [3].

load always requires loads of power in the short term. If, for instance, a soccer stadium floodlight system switches on, the generator momentum distributes stored energy and consequently, stabilizes the electricity grid. If battery inverters substitute the power plants, this provides another substantial benefit. A conventional power plant can provide a power portion as an operating reserve, and a battery storage system is capable of providing full nominal power.

Fig. 9.4 exhibits the primary energy supply by conventional and renewable energy sources. In the primary energy supply, the oil sector is leading in the conventional sectors followed by the coal and natural gas sectors while the hydro sector is leading in the renewable energy sources followed by the wind and solar energy sources.

9.2.3 Energy Independence

For the consumers, integrated energy grasps the prospect to actively support the energy transition. Countless small wind turbines and solar systems are installed in private properties or public parks to display sustainable energy supply awareness. This is not solely for financial incentives that make renewable energy so attractive, there is similarly improved interest in the innovative technologies that allow enterprises and individual households to develop the source of independent power supply.

Through intelligent energy management systems, the operators of the PV system can synchronize the power generation and consumption to use the largest probable share of self-generated renewable power.

9.2.4 Grid Quality

Unquestionably, as renewable energy generation varies, energy demand is similarly not always constant. Thus, energy balancing markets compensate for the variations between power consumption and generation in the utility grid. To ensure the grid-maintenance quality, power plants must ensure the energy balancing obtainable wherever needed.

If power demand surpasses the power production capacity, electricity is required to be quickly employed to the grid for positive balancing energy. If the electricity supply is more than the demand, electricity is required to be taken out for negative energy balancing. The operators of the power plant that participate in the energy balancing market get a feed-in tariff to maintain the energy balancing.

9.2.5 Global Climate Support

The predominant goal of integrating the distinct energy and consumption divisions is to decarbonize the global economy comprehensively.

Integrated energy systems offer the lone route to prevent CO_2 emissions and thus support the climate. The energy transition can only be established in an all-inclusive, renewable energy system. We must now exploit the flexible systems offering the advancement of capable decentralized energy generation technologies and digitalization that offers inspiring opportunities. The significant contributions of the integrated energy are as follows:

- Global energy transition through sustainable energy production and reduced climate-damaging CO_2
- Fewer deaths/illnesses are caused by lowering the quantity of hazardous substances in the air. The burning of fossil fuels generates not only CO_2 but sulfur, nitrogen, and carbon oxides as well along with particulate matter, lead, mercury, nickel, arsenic, and copper
- Reduced operational and waste nuclear risks as a result of replacing nuclear power production
- Reduced water consumption
- Improved supply dependency: finite resources replacement with infinitely available energy production raw materials
- Low electricity prices: wind and solar-assisted power is becoming further cost-effective.
- More flexibility
- Development assistance for the off-grid regions: renewable energy technologies ensure economic development and sustainable power supply

The residential sector covers approximately one-third of global energy consumption. Bocci et al.[143] published a research article targeting renewable energy integration for households. Renewable energy sources can play a significant role in terms of energy efficiency and hydrogen to condense the consumptions and emissions and also improve energy security. This study analyzed a real 100 m^2 residential house to analyze and experimentally explore the renewable energy source technologies and energy efficiency, sizing the hydrogen and storage for power backup. The hydrogen backup was found expensive in comparison with traditional gasoline generators and batteries but satisfied all-electric requirements, allowed net metering, and increased security. Furthermore, the low-pressure hydrogen storage unit through metal hydrides assured system safety. Fig. 9.5 displays the low-carbon energy footprints in different regions of the world.

National renewable energy laboratory presented a report[144] on renewable power generation and storage technologies. Energy storage is employed in electric grids to meet the high-demand periods in the United States and around the globe. Pumped-storage hydropower plays a dominant role in energy storage globally. In the United States, pumped-storage hydropower was constructed in the 1970s chiefly in response to the market conditions counting the high natural gas and oil and prices, monitoring restrictions on the plants combusting gas and oil, and low-efficiency steam plants dependence for high-demand periods. In addition to pumped-storage hydropower, a 110 MW compressed air energy storage facility was constructed in the United States back in 2003. Storage deployment in the United States has been restricted by low prices of natural gas, accessibility to the high-efficient and flexible gas turbines, and restricted cost reductions in storage technologies over the past few decades. Moreover, the regulatory storage treatment, costly licensing and approval, storage valuation challenges, and utility risk aversion have also restricted storage development.

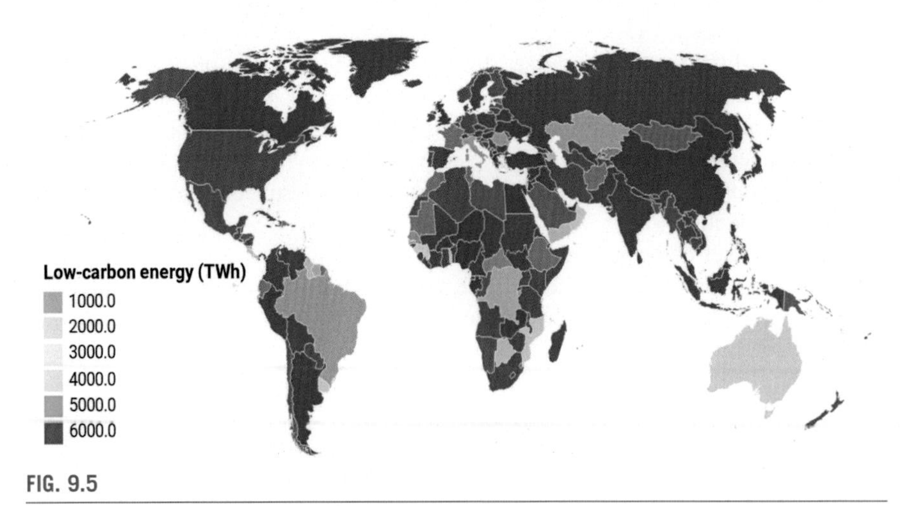

FIG. 9.5

Low-carbon energy footprints in different regions of the world.

Data from [128].

Renewable and appropriate energy laboratory at the University of California, Berkeley is actively working on renewable energy technologies, decentralization, clean-energy solutions, and innovations in energy storage.[145] Simultaneously, policies are being designed to build market growth, and a revolution in battery storage may accomplish cost reductions across clean-energy technologies. Further integration of research and development and new storage technological deployment surfaces a strong path toward low-carbon and cost-effective electricity. They analyzed innovation and deployment employing a two-factor model that integrates the materials innovation investment value and technological deployment by time using an empirical dataset involving battery storage technology. Further advancements in battery storage are significant to decarbonization combined with advances in renewable electricity sources.

Energy storage systems and batteries can deliver the research leading to the new revolutionary technologies. Even though evolving the existing understanding and basic science about energy storage, the title role of the innovative energy storage hub is to develop completely innovative scientific methods together with the exploration of new devices, materials, novel approaches, and systems for the transportation sector and utility-scale storage. The energy storage hub should substitute new designs and develop functional and scalable prototypes that demonstrate completely new approaches and methods for electrochemical storage, to overcome the existing manufacturing restrictions through novelty to decrease the cost and complexity. The eventual target is to exceed the existing technical restrictions for energy storage and condense the risk level to further develop and scale up the novel energy storage techniques.

Fig. 9.6 displays the heat and electricity generation: (a) heat generation from renewables; (b) electricity generation from renewables. Fig. 9.6a exhibits the heat generation from renewables. Primary solid biofuels are leading in renewable heat generation followed by geothermal and liquid biofuel sources. Fig. 9.6b shows the power generation from renewables. The hydro energy source is leading in the renewable electricity generation followed by the wind and solar photovoltaic energy sources.

A research team from the TU Munich in Germany and the University of California found that research and development investments for the energy storage projects effected remarkably in dropping down the cost of the lithium-ion battery from \$10,000/kWh that was found in the early 1990s to a trajectory that might establish \$100/kWh by the next year. The innovation pace is staggering.

Tesla is moving to install the major lithium-ion storage Gigafactory in Nevada that will be the largest facility in the world, new energy storage combinations in terms of scale, size, and chemistry are evolving faster than ever.

Tesla energy storage Gigafactory projects are not the lone example. Cities such as Berlin have already contained grid-scale storage. Berlin city plans to mount an underground battery with a capacity of 120 MW to support solar and wind efforts at the lower prices of 15 cents/kWh. California is the first energy storage home directed to the grid, demanding utilities obtain 1.325 GW storage by 2020. Such

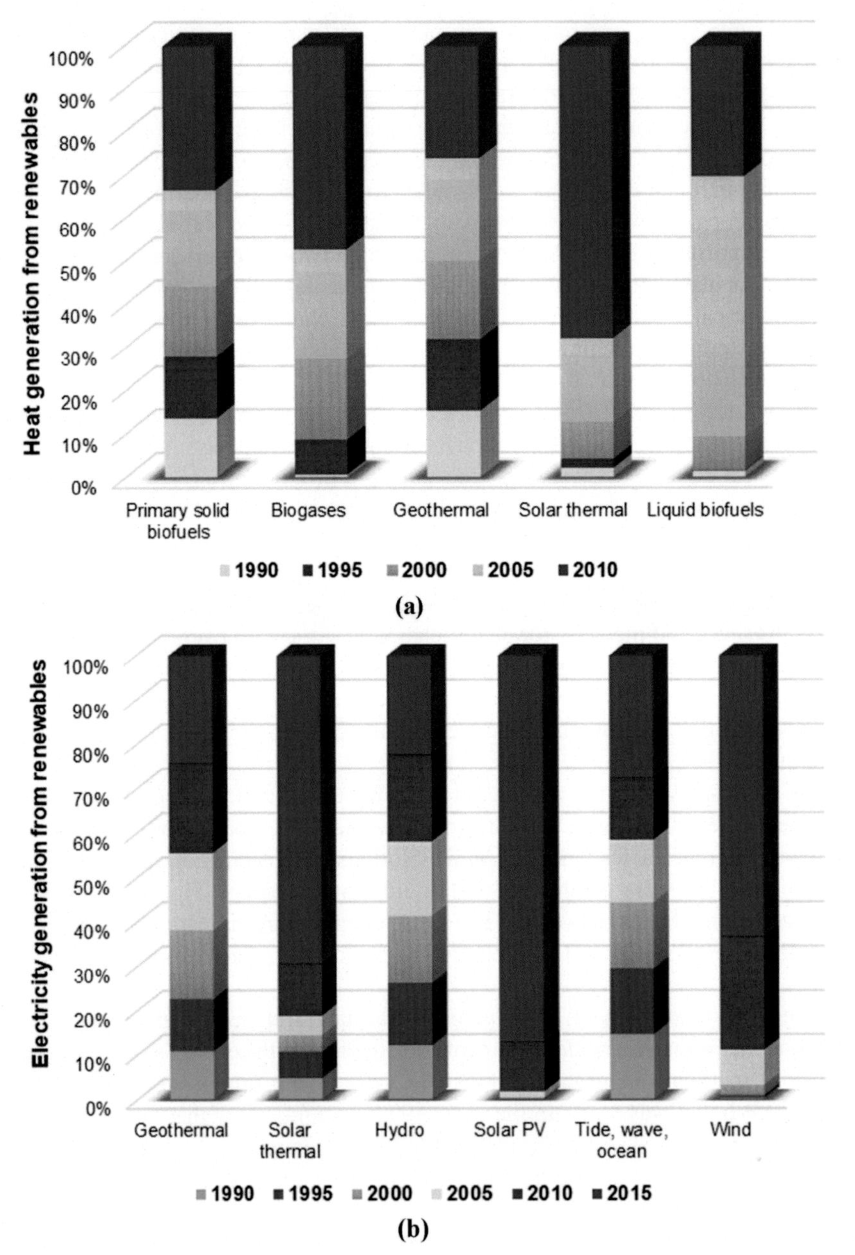

FIG. 9.6

Heat and electricity generation: (a) heat generation from renewables; (b) electricity generation from renewables.

Data from [3].

ground-breaking policies display the storage options range that may assist clean energy, from lower household scale to the city-scale storage offering the back-up to the utility-scale solar and wind farms.

9.3 Case Study 8

This case study is designed to investigate hydrogen production using solar and ocean thermal energy conversion systems. Solar heliostat field is employed to produce hydrogen through the thermochemical Cu—Cl cycle and to generate the electricity using Rankine cycle and produced electricity is employed to the water electrolysis process while the ocean thermal energy conversion system is also employed to generate electrical power. The low-grade heat from the Cu—Cl cycle is recovered to generate cooling through an absorption chiller. The power produced by both solar and ocean thermal energy conversion sources is employed in the electrolyzer for hydrogen production that is compressed at high pressure to store hydrogen.

9.3.1 System Description

Fig. 9.7 displays the hydrogen production system using solar and ocean thermal energy conversion sources while Fig. 9.8 shows the Aspen Plus simulation of the solar-assisted thermochemical Cu—Cl cycle. The significant subsystems of the solar heliostat field, water electrolysis, and ocean thermal energy conversion system are modeled using Engineering Equation solver (EES) while the solar energy-assisted thermochemical Cu—Cl cycle is simulated using Aspen Plus. The solar heliostat field is employed to extract the thermal energy that is partially used to convert water into steam for the Cu—Cl cycle while the remaining portion is employed to generate power using the Rankine cycle. The simulation flowsheet of the solar-assisted Rankine cycle and the Cu—Cl is shown in Fig. 9.8. Heat exchanger C2 is employed for the thermal management that transfers the heat from molten salt to the Cu—Cl cycle and Rankine cycle. The power is generated using turbine C3 that expands the superheated steam. The water converted to the steam reaches the Cu—Cl cycle through stream SP8. The thermochemical Cu—Cl cycle consists of four significant steps that can be expressed as follows:

$$\text{Electrolysis(C14)}: 2\text{CuCl}(aq) + 2\text{HCl}(aq) \xrightarrow{80^\circ\text{C}} \text{H}_2(g) + 2\text{CuCl}_2(aq)$$

$$\text{Drying(C16\&C17)}: \text{CuCl}_2(aq) \xrightarrow{100^\circ\text{C}} \text{CuCl}_2(s)$$

$$\text{Hydrolysis(C7)}: 2\text{CuCl}_2(s) + \text{H}_2\text{O}(g) \xrightarrow{\text{endothermic}\ \ (400^\circ\text{C})} \text{Cu}_2\text{OCl}_2(s) + 2\text{HCl}(g)$$

$$\text{Thermolysis(C10)}: \text{Cu}_2\text{OCl}_2(s) \xrightarrow{\text{endothermic}\ \ (500^\circ\text{C})} 0.5\,\text{O}_2(g) + 2\text{CuCl}(l)$$

Steam is the only input that is added to the Cu—Cl cycle continuously that splits into its constituent while all other constituents are recycled throughout the cycle.

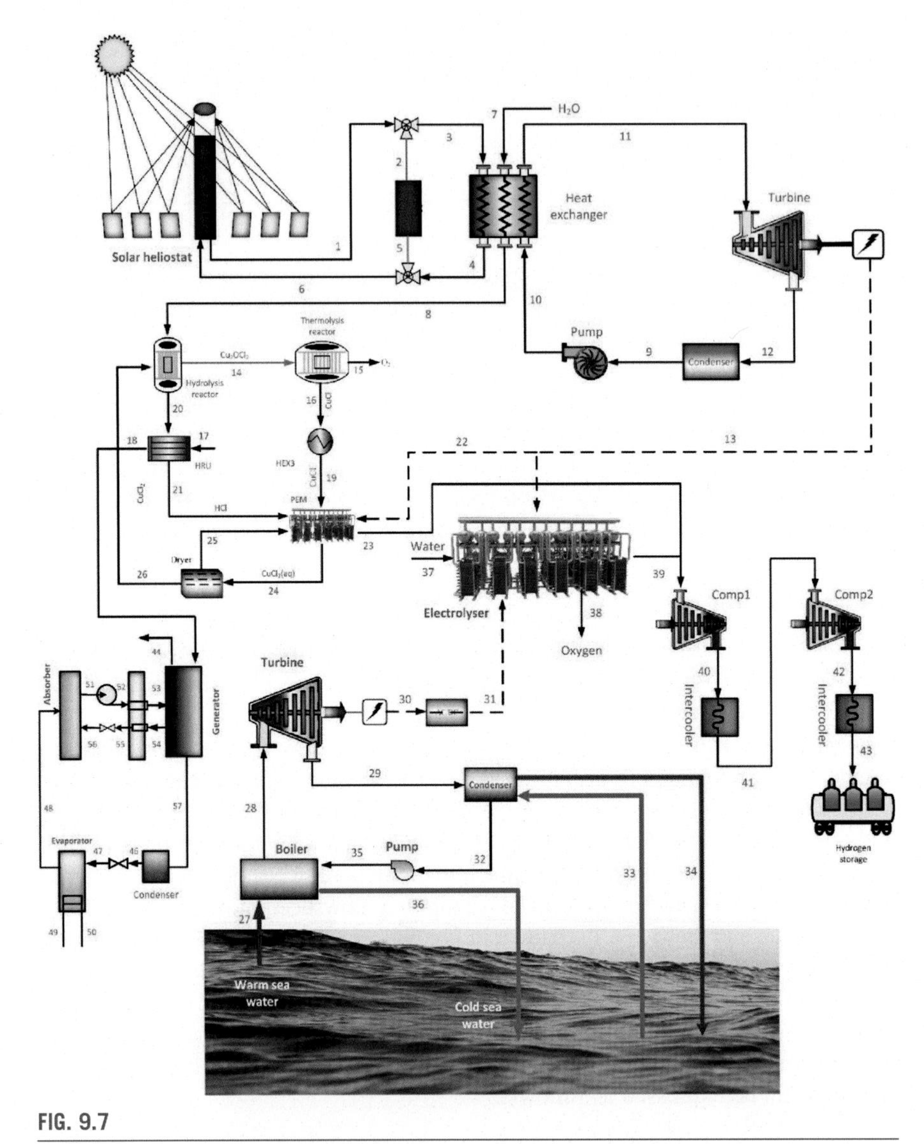

FIG. 9.7

Hydrogen production system using solar and ocean thermal energy conversion sources.

This thermochemical Cu—Cl cycle uses a significant portion of heat and generated electricity is employed to produce clean hydrogen. In the ocean thermal energy conversion system, the temperature difference between warm sea surface water and cold sea deep water is employed to generate power. The warm sea surface water and the working fluid (ammonia) pass through the boiler that works as a heat exchanger. The hot water from the surface is fed to the boiler adjacent to ammonia and warm

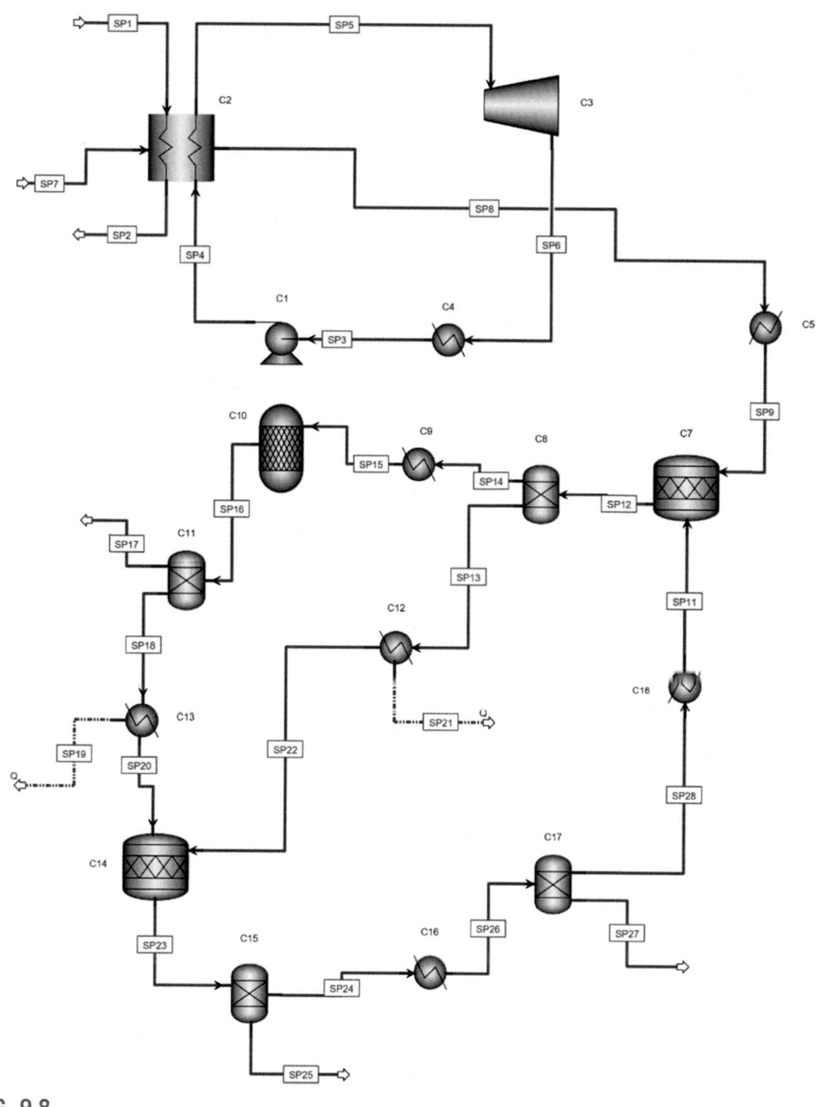

FIG. 9.8

Aspen Plus simulation of solar-assisted thermochemical Cu—Cl cycle.

seawater boils the ammonia and produces vapors by extracting heat from the hot seawater. The warm seawater transfers the heat to the working fluid and is fed back to the sea.

A pump is installed before the boiler which raises the working fluid pressure before reaching the boiler. Thus, the pressurized vapors are then employed to drive the turbine that is further attached to the generator to produce electrical power. Once the ammonia vapors leave the turbine, they are brought down and fed to the condenser. The cold deep seawater is employed in the condenser that operates as the heat exchanger, and heat is transferred from ammonia to the cold deep seawater that gets heated up. The vapors of ammonia are cooled and converted to the liquid again in the condenser and employed in the pump that increases the pressure and recycles back to the boiler to continue the cycle.

The low-grade heat from the Cu–Cl cycle is recovered to generate cooling through an absorption cooling system. The hydrogen produced through the thermochemical cycle and OTEC system is employed in the multistage hydrogen compression unit. This compression unit employs two compressors and intercoolers that are used to store hydrogen at high pressure in the storage unit.

9.3.2 Analysis

Solar thermal energy is extracted through the solar heliostat field to produce power using the Rankine cycle and produces clean hydrogen through the thermochemical Cu–Cl cycle. Table 9.1 displays the operating constraints of the designed case study 8. In this system, molten salt is employed as a working fluid to extract heat from the solar heliostat field and for the heat transfer.

Solar Heliostat Field

The following correlation is employed to evaluate the solar heat input depending upon the heliostat efficiency, direct normal irradiance, area of heliostat, and number of heliostats.

$$\dot{Q}_{solar} = \eta_{he} \dot{I}_b A_{he} N_{he} \tag{9.1}$$

The design model equations for the ocean thermal energy conversion system, water electrolysis unit, and multistage hydrogen compression unit are already described in the previous chapters; thus, this section displays the design model equations of the solar-assisted Rankine cycle and thermochemical Cu–Cl cycle.

Solar-Assisted Rankine Cycle

Pump C1

The energy and exergy balance equations of the pump C1 are as follows:

$$\dot{m}_{SP3} h_{SP3} + \dot{W}_{in} = \dot{m}_{SP4} h_{SP4} \tag{9.2}$$

$$\dot{m}_{SP3} ex_{SP3} + \dot{W}_{in} = \dot{m}_{SP4} ex_{SP4} + \dot{Ex}_d \tag{9.3}$$

Heat exchanger C2

The energy and exergy balance equations of the heat exchanger C2 are as follows:

$$\dot{m}_{SP1} h_{SP1} + \dot{m}_{SP4} h_{SP4} + \dot{m}_{SP7} h_{SP7} = \dot{m}_{SP2} h_{SP2} + \dot{m}_{SP5} h_{SP5} + \dot{m}_{SP8} h_{SP8} \tag{9.4}$$

$$\dot{m}_{SP1} ex_{SP1} + \dot{m}_{SP4} ex_{SP4} + \dot{m}_{SP7} ex_{SP7} = \dot{m}_{SP2} ex_{SP2} + \dot{m}_{SP5} ex_{SP5} + \dot{m}_{SP8} ex_{SP8} + \dot{Ex}_d \tag{9.5}$$

Table 9.1 Operating Constraints of the Designed Case Study 8.

Design Parameters	Value
Solar Heliostat	
Irradiance $\dot{I}_b$	0.8 kW/m^2
Number of heliostats N_{he}	100
Working fluid	Molten salt
Dimensions of heliostat mirror area	121 m^2
Heliostat efficiency	75%
Thermochemical Cu–Cl Cycle	
Cycle operating pressure	1 bar
Hydrolysis temperature	400°C
Thermolysis temperature	500°C
Electrolysis temperature	80°C
Drying temperature	100°C
Water Electrolysis	
Faraday number	96486 C/mol
Cathodic preexponential $\left(J_c^{ref}\right)$	46 × 10 A/m^2
Anodic preexponential $\left(J_a^{ref}\right)$	17 × 10^4 A/m^2
Cell operating pressure	100 kPa
Membrane thickness	80 μm
Cell operating temperature	80°C
OTEC Cycle	
Coldwater temperature	5°C
Warm water temperature	28°C
Multistage Compression Unit	
Pressure ratio of compressors	5
Hydrogen storage pressure	25 bars

Steam turbine C3

The energy and exergy balance equations of the steam turbine C3 are as follows:

$$\dot{m}_{SP5}h_{SP5} = \dot{m}_{SP6}h_{SP6} + \dot{W}_{out} \tag{9.6}$$

$$\dot{m}_{SP5}ex_{SP5} = \dot{m}_{SP6}ex_{SP6} + \dot{W}_{out} + \dot{Ex}_d \tag{9.7}$$

Thermochemical Cu–Cl Cycle

Hydrolysis reactor C7

The energy and exergy balance equations of the hydrolysis reactor C7 are as follows:

$$\dot{m}_{SP9}h_{SP9} + \dot{m}_{SP11}h_{SP11} + \dot{Q}_{in} = \dot{m}_{SP12}h_{SP12} \tag{9.8}$$

$$\dot{m}_{SP9}ex_{SP9} + \dot{m}_{SP11}ex_{SP11} + \dot{Q}_{in}\left(1 - \frac{T_o}{T}\right) = \dot{m}_{SP12}ex_{SP12} + \dot{E}x_d \tag{9.9}$$

Thermolysis reactor C10

The energy and exergy balance equations of the thermolysis reactor C10 are as follows:

$$\dot{m}_{SP15}h_{SP15} + \dot{Q}_{in} = \dot{m}_{SP17}h_{SP17} + \dot{m}_{SP18}h_{SP18} \tag{9.10}$$

$$\dot{m}_{SP15}h_{SP15} + \dot{Q}_{in}\left(1 - \frac{T_o}{T}\right) = +\dot{m}_{SP17}ex_{SP17} + \dot{m}_{SP18}ex_{SP18} + \dot{E}x_d \tag{9.11}$$

Electrolysis reactor C14

The energy and exergy balance equations of the electrolysis reactor C14 are as follows:

$$\dot{m}_{SP20}h_{SP20} + \dot{m}_{SP22}h_{SP22} + \dot{W}_{in} = \dot{m}_{SP23}h_{SP23} \tag{9.12}$$

$$\dot{m}_{SP20}ex_{SP20} + \dot{m}_{SP22}ex_{SP22} + \dot{W}_{in} = \dot{m}_{SP23}ex_{SP23} + \dot{E}x_d \tag{9.13}$$

Separator C15

The energy and exergy balance equations of the separator C15 are as follows:

$$\dot{m}_{SP23}h_{SP23} = \dot{m}_{SP24}h_{SP24} + \dot{m}_{SP25}h_{SP25} \tag{9.14}$$

$$\dot{m}_{SP23}ex_{SP23} = \dot{m}_{SP24}ex_{SP24} + \dot{m}_{SP25}ex_{SP25} + \dot{E}x_d \tag{9.15}$$

Heater C16

The energy and exergy balance equations of the heater C16 are as follows:

$$\dot{m}_{SP24}h_{SP24} + \dot{Q}_{in} = \dot{m}_{SP26}h_{SP26} \tag{9.16}$$

$$\dot{m}_{SP24}ex_{SP24} + \dot{Q}_{in}\left(1 - \frac{T_o}{T}\right) = \dot{m}_{SP26}ex_{SP26} + \dot{E}x_d \tag{9.17}$$

Dryer C17

The energy and exergy balance equations of the dryer C17 are as follows:

$$\dot{m}_{SP26}h_{SP26} + \dot{m}_{SP27}h_{SP27} = \dot{m}_{SP28}h_{SP28} \tag{9.18}$$

$$\dot{m}_{SP26}ex_{SP26} + \dot{m}_{SP27}ex_{SP27} = \dot{m}_{SP28}ex_{SP28} + \dot{E}x_d \tag{9.19}$$

Absorption Cooling System

Generator

The energy and exergy balance equations of the generator are as follows:

$$\dot{m}_{53}h_{53} + \dot{Q}_{ABS} = \dot{m}_{54}h_{54} + \dot{m}_{57}h_{57} \tag{9.20}$$

$$\dot{m}_{53}ex_{53} + \dot{Q}_{ABS}\left(1 - \frac{T_0}{T_{gen}}\right) = \dot{m}_{54}ex_{54} + \dot{m}_{57}ex_{57} + \dot{E}x_d \tag{9.21}$$

Condenser

The energy and exergy balance equations for the condenser can be expressed by

$$\dot{m}_{57}h_{57} = \dot{Q}_{con} + \dot{m}_{46}h_{46} \quad (9.22)$$

$$\dot{m}_{57}ex_{57} = \dot{Q}_{con}\left(1 - \frac{T_0}{T_{con}}\right) + \dot{m}_{46}ex_{46} + \dot{E}x_d \quad (9.23)$$

Throttling valve

The energy and exergy balance equations of the throttle valve are as follows:

$$\dot{m}_{46}h_{46} = \dot{m}_{47}h_{47} \quad (9.24)$$

$$\dot{m}_{46}ex_{46} = \dot{m}_{47}ex_{47} + \dot{E}x_d \quad (9.25)$$

Evaporator

The energy and exergy balance equations of the evaporator are as follows:

$$\dot{m}_{47}h_{47} + \dot{Q}_{evap} = \dot{m}_{48}h_{48} \quad (9.26)$$

$$\dot{m}_{47}ex_{47} + \dot{Q}_{evap}\left(\frac{T_0}{T_{evap}} - 1\right) = \dot{m}_{48}ex_{48} + \dot{E}x_d \quad (9.27)$$

Absorber

The energy and exergy balance equations of the absorber are as follows:

$$\dot{m}_{48}h_{48} + \dot{m}_{56}h_{56} = \dot{m}_{51}h_{51} + \dot{Q}_{abs} \quad (9.28)$$

$$\dot{m}_{48}ex_{48} + \dot{m}_{56}ex_{56} = \dot{m}_{51}ex_{51} + \dot{Q}_{abs}\left(1 - \frac{T_0}{T_{abs}}\right) + Ex_d \quad (9.29)$$

Pump

The energy and exergy balance equations of the pump are as follows:

$$\dot{m}_{51}h_{51} + \dot{W}_{pump} = \dot{m}_{52}h_{52} \quad (9.30)$$

$$\dot{m}_{51}ex_{51} + \dot{W}_{pump} = \dot{m}_{52}ex_{52} + \dot{E}x_d \quad (9.31)$$

Heat exchanger

The energy and exergy balance equations of the heat exchanger are as follows:

$$\dot{m}_{52}h_{52} + \dot{m}_{54}h_{54} = \dot{m}_{53}h_{53} + \dot{m}_{55}h_{55} \quad (9.32)$$

$$\dot{m}_{52}ex_{52} + \dot{m}_{54}ex_{54} = \dot{m}_{53}ex_{53} + \dot{m}_{55}ex_{55} + \dot{E}x_{d,ABS,HX} \quad (9.33)$$

Performance Assessment

The coefficient of performance (COP) of the absorption cooling system are determined using the following expression:

$$\mathrm{COP}_{en,\ \mathrm{ABS}} = \frac{\dot{Q}_{\mathrm{evap}}}{\dot{Q}_{\mathrm{gen}}} \tag{9.34}$$

$$\mathrm{COP}_{ex,\ \mathrm{ABS}} = \frac{\dot{Q}_{\mathrm{evap}}\left(\frac{T_0}{T_{\mathrm{evap}}} - 1\right)}{\dot{Q}_{\mathrm{gen}}\left(1 - \frac{T_0}{T_{\mathrm{gen}}}\right)} \tag{9.35}$$

The energetic and exergetic efficiency equations of the designed case study 8 are as follows:

$$\eta_{\mathrm{ov}} = \frac{\dot{m}_{H_2}\mathrm{LHV}_{H_2} + \dot{Q}_{\mathrm{evap}}}{\dot{Q}_{\mathrm{solar}} + \left(\dot{m}_{27}h_{27} - \dot{m}_{36}h_{36}\right)} \tag{9.36}$$

$$\psi_{\mathrm{ov}} = \frac{\dot{m}_{H_2}ex_{H_2} + \dot{Q}_{\mathrm{evap}}\left(\frac{T_0}{T_{\mathrm{EV}}} - 1\right)}{\dot{Q}_{\mathrm{solar}} + \left(\dot{m}_{27}ex_{27} - \dot{m}_{36}ex_{36}\right)} \tag{9.37}$$

9.3.3 Results and Discussion

The comprehensive analysis of several significant subsystems, such as solar heliostat field, solar-assisted Rankine cycle, ocean thermal energy conversion cycle, water electrolysis system, and multistage compression unit is already conducted and presented in the previous Chapters 3, 4, and 5. This section only displays the significant results of the remaining subsystems of the designed case study 8.

It is significant to investigate the effect of input steam flowrate in the Cu–Cl cycle as steam is the only input that is added to the Cu–Cl cycle continuously while all other constituents are recycled throughout the cycle. The effect of steam input flowrate is investigated against the significant input parameters of heat duties of heater C16, hydrolysis reactor C17 and thermolysis reactor C10 and output flowrates of hydrogen chloride, cupric chloride, cuprous chloride, hydrogen, copper oxychloride, and oxygen in the Cu–Cl cycle.

Fig. 9.9 displays the effect of steam input flowrate on the heat duties of heater C16, hydrolysis reactor C17, and thermolysis reactor C10. The hydrolysis and thermolysis reactions are endothermic by nature that absorb heat and heater C16 converts water into steam to separate it from the cupric chloride that is circulated to the hydrolysis reactor. It is evident that the heat absorbed by both hydrolysis and thermolysis reactors increase with the rise in input flowrate of steam with the rise in the input flowrate, the heater needs to separate more water from the cupric

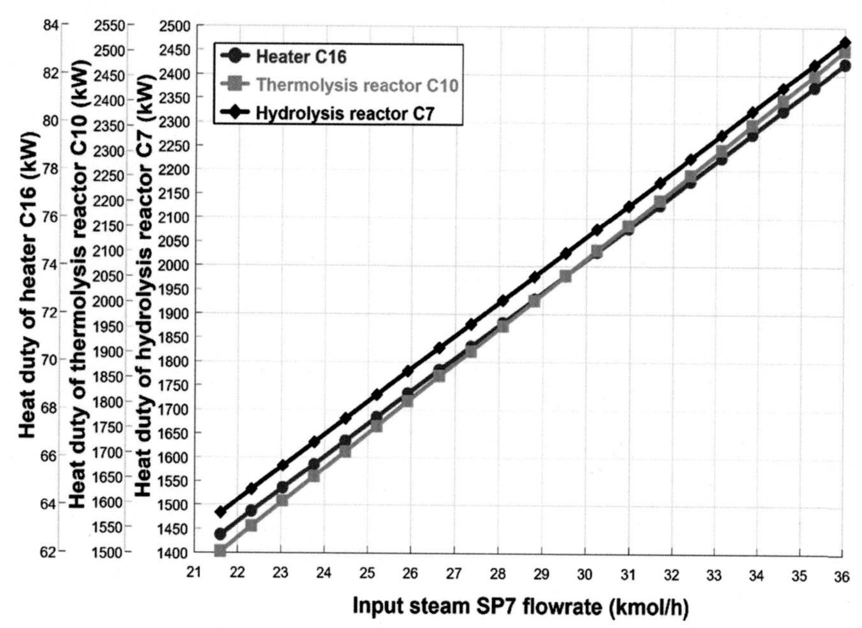

FIG. 9.9

Effect of steam input flowrate on the heat duties of heater C16, hydrolysis reactor C17 and thermolysis reactor C10.

chloride, thus, the heat duties of heater C16, hydrolysis reactor C17 and thermolysis reactor C10 increase with the rise in input steam flowrate.

Fig. 9.10 displays the effect of steam input flowrate on the output flowrates of hydrogen chloride, cupric chloride, and cuprous chloride in the Cu–Cl cycle. The hydrogen chloride gas is formed in the hydrolysis reactor, the cuprous chloride is formed during the thermolysis reaction when copper oxychloride split into its constituents, and cupric chloride is formed in the electrolysis step. The input steam flowrate ranged from 21.5 to 36 kmol/h. The flowrates of these constituents should increase with the rise in input steam flowrate as steam is the only input that is added to the Cu–Cl cycle continuously that splits into its constituent while all other constituents are recycled throughout the cycle. It is evident that the output flowrates of hydrogen chloride, cupric chloride, and cuprous chloride increase with the rise in input steam flowrate.

Fig. 9.11 displays the effect of steam input flowrate on the output flowrates of hydrogen, copper oxychloride, and oxygen in the Cu–Cl cycle. The hydrogen gas is formed in the electrolysis reactor, the oxygen gas formed during the thermolysis reaction when copper oxychloride split into its constituents, and copper oxychloride is formed in the hydrolysis step. The steam is the only input that is added to the Cu–Cl cycle continuously that splits into its constituent while all other constituents are recycled throughout the cycle. It is evident that the output flowrates of hydrogen, copper oxychloride, and oxygen increase with the rise in input steam flowrate.

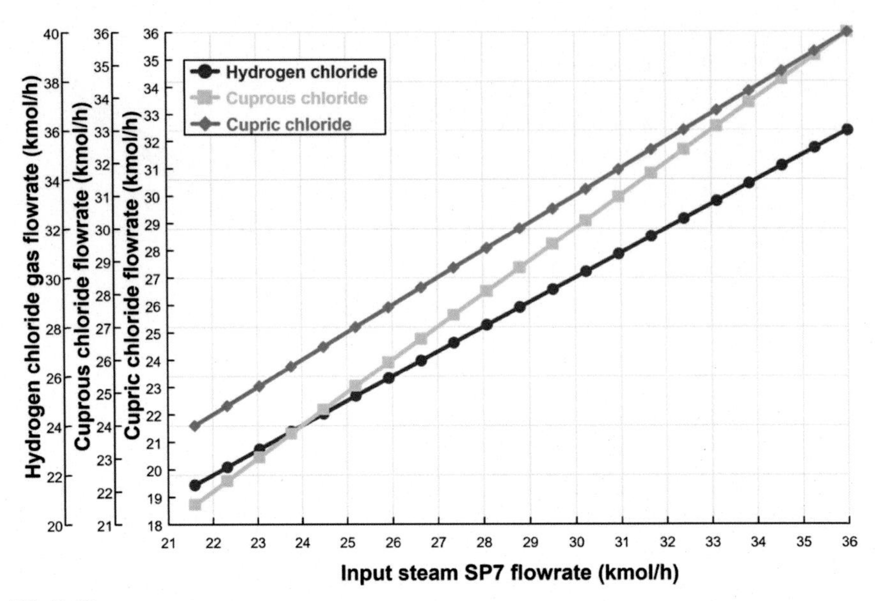

FIG. 9.10

Effect of steam input flowrate on the output flowrates of hydrogen chloride, cupric chloride, and cuprous chloride in the Cu—Cl cycle.

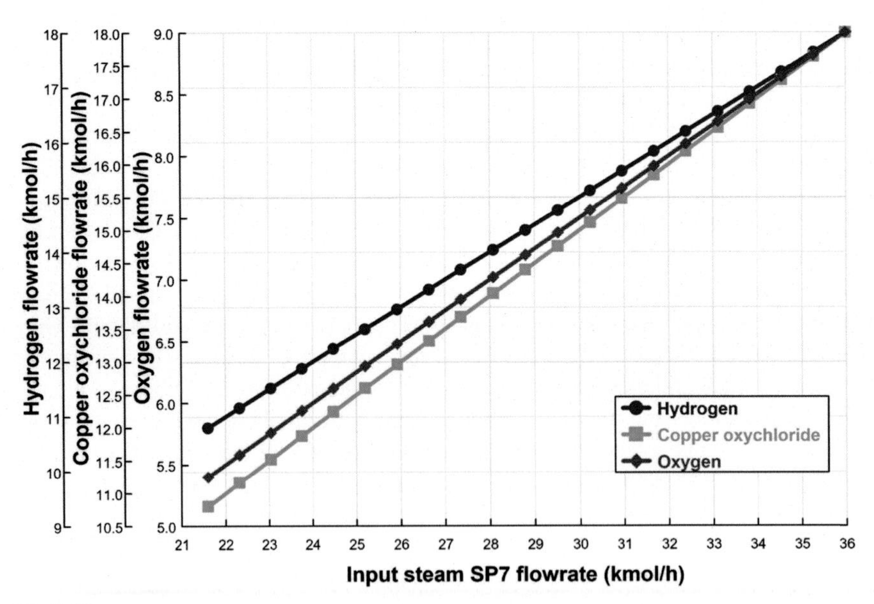

FIG. 9.11

Effect of steam input flowrate on the output flowrates of hydrogen, copper oxychloride, and oxygen in the Cu—Cl cycle.

The seawater flow rate effect on the significant parameters of the designed case study is also imperative to be investigated. Fig. 9.12 displays the effect of seawater flow rate on the work and exergy destruction rates of the turbine, overall work rate, and hydrogen production. The seawater flow rate ranged from 15 to 35 kg/s for the parametric study. The rise in seawater flow rate causes the turbine exergy destruction rate to rise from 198.8 to 463.9 kW, the turbine work rate to rise from 750.4 to 1751 kW, the overall work rate to rise from 739.2 to 1725 kW, and the hydrogen production rate to rise from 5.134 to 11.98 g/s.

To investigate the system performance in terms of energetic and exergetic efficiencies of two different case scenarios is significant to explore the variability in the designed parameters. Fig. 9.13 exhibits the effect of pump efficiency on the efficiencies of power and hydrogen production case scenarios. The rise in pump efficiency causes the energetic efficiencies of the power and hydrogen production systems to rise from 4.684% to 5.533% and 4.121% to 4.687%, respectively, and the exergetic efficiencies of the power and hydrogen production systems to rise from 11.78% to 13.4% and 9.655% to 10.98%, respectively.

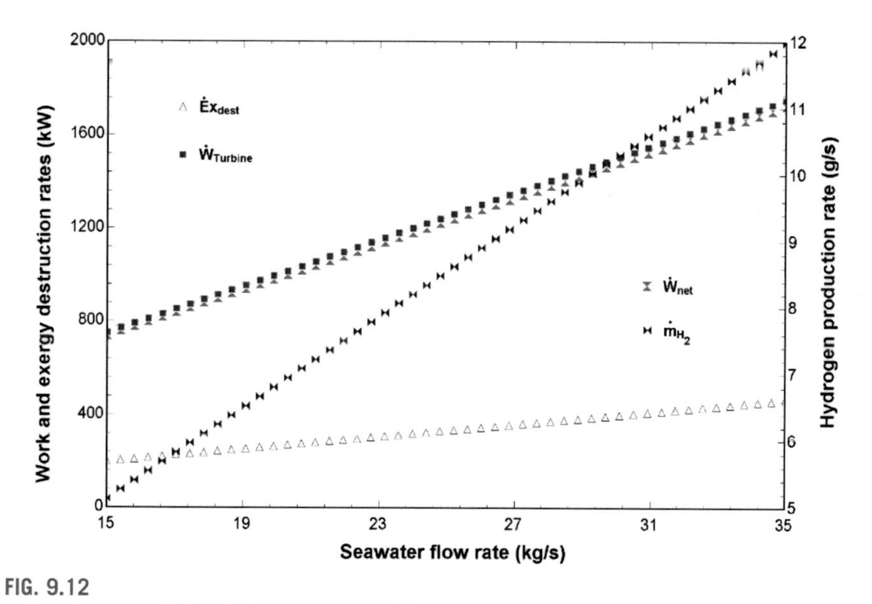

FIG. 9.12

Effect of seawater flow rate on the work and exergy destruction rates of turbine, overall work rate, and hydrogen production.

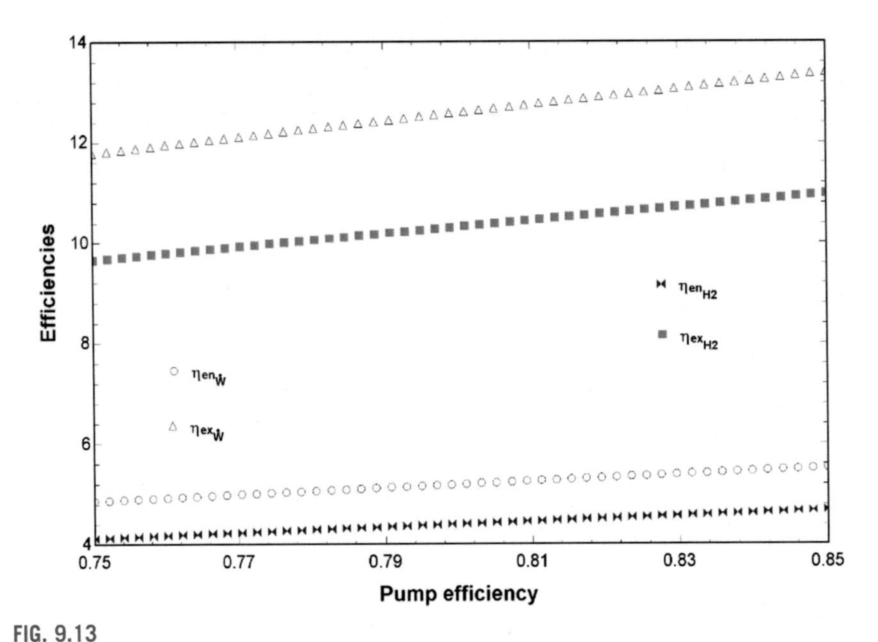

FIG. 9.13

Effect of pump efficiency on the efficiencies of power and hydrogen production case scenarios.

9.4 Case Study 9

This case study is designed to investigate hydrogen production using a solar concentrating collector, wind, and geothermal energy-driven system. Solar concentrating collector is employed to generate electricity using a double-stage Rankine cycle, and the produced electricity is employed in the water electrolysis process while wind and geothermal energy-driven systems are also employed to generate electrical power. The power produced by solar, wind, and geothermal energy sources is employed in the electrolyzer for hydrogen production that is compressed at high pressure to store hydrogen.

9.4.1 System Description

Fig. 9.14 displays the system schematic of the solar concentrating collector, wind, and geothermal energy-driven hydrogen production system while Fig. 9.15 shows the Aspen Plus simulation of solar concentrating collector-assisted double-stage Rankine cycle. The significant subsystems of the solar concentrating collector, wind, and geothermal energy are modeled using EES while the solar energy-assisted double-stage Rankine cycle is simulated using Aspen Plus. The solar concentrating collector is employed to extract the thermal energy that is employed to generate power using a double-stage Rankine cycle. The simulation flowsheet of the solar-assisted double-stage Rankine cycle is shown in Fig. 9.15. The solar thermal energy is extracted using the solar thermal collector and employed in the double-stage Rankine cycle to generate power. This generated electricity is employed in the water electrolysis process for hydrogen production.

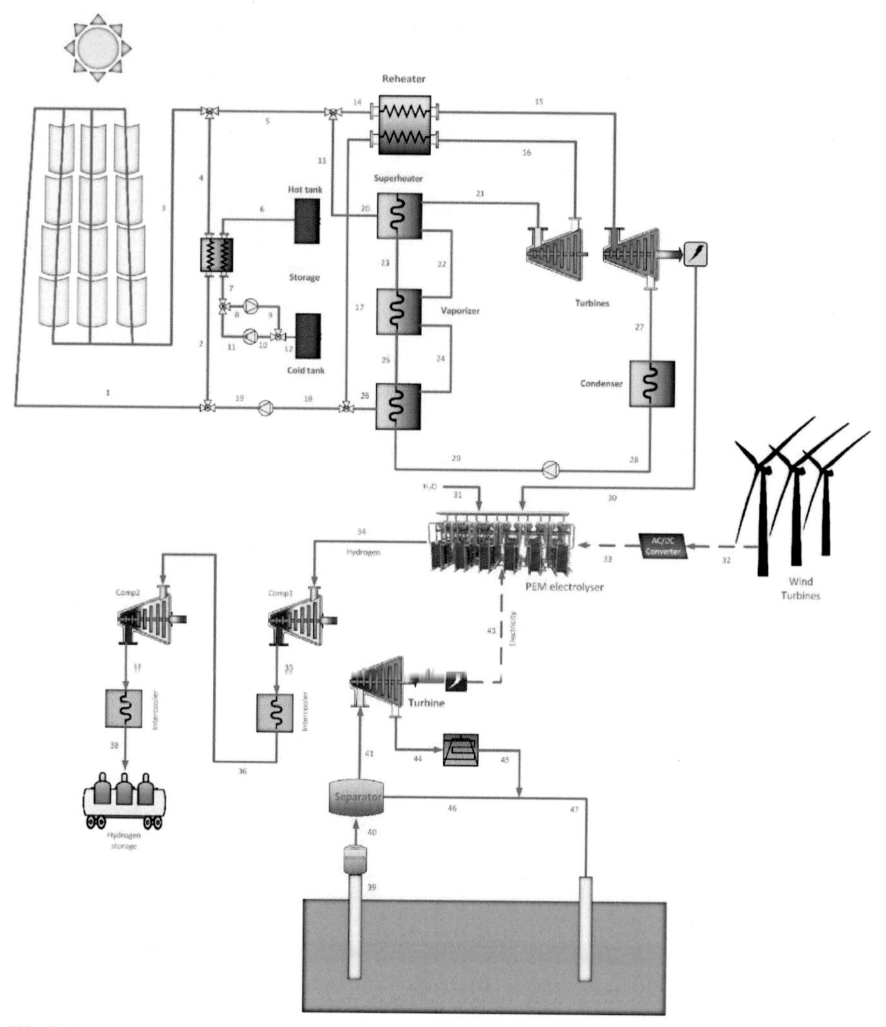

FIG. 9.14

System schematic of solar concentrating collector, wind, and geothermal energy-driven hydrogen production system.

The hot stream SP1 coming from the solar heliostat passes through heat exchanger C1 where water coming from the high-pressure pump extracts some heat and is converted to the superheated steam that expands through high-pressure turbine C2 and generates electrical power. The turbine C2 exit passes through another heat exchanger C3 where hot stream SP1 reaches after exiting the heat exchanger C1. The water stream extracts the remaining heat from the hot SP1 stream and again converted to the superheated steam that is employed in the low-pressure turbine C4 to generate further electricity. The turbine exit then passes through the

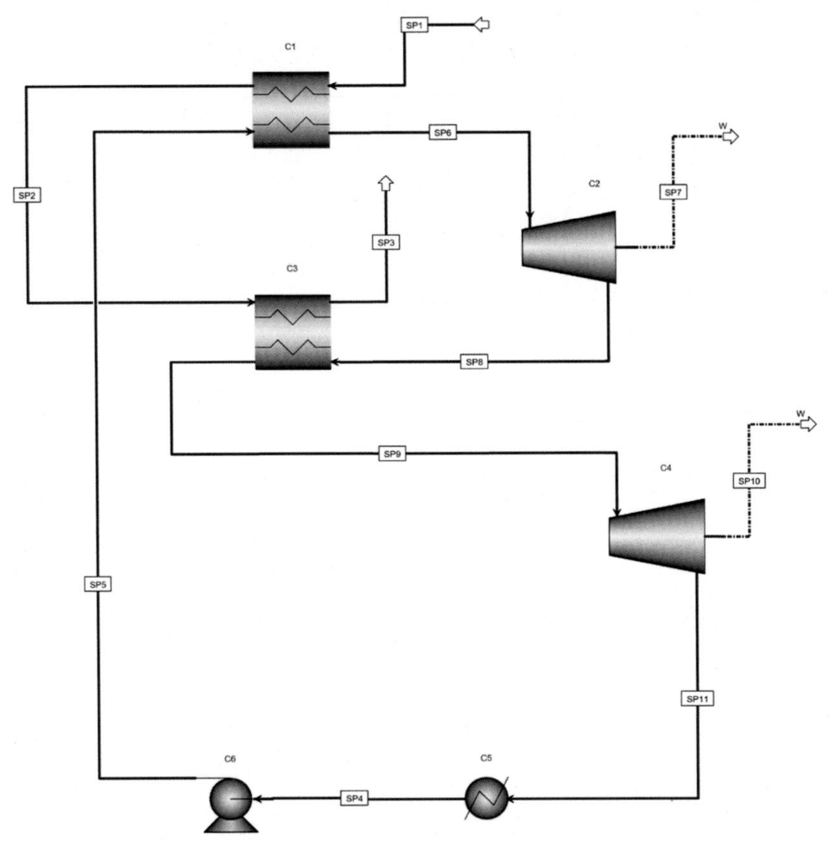

FIG. 9.15

Aspen Plus simulation of solar concentrating collector-assisted double-stage Rankine cycle.

condenser, and the condenser output is recycled to the heat exchanger using a pump and the process continues.

The wind turbine form is employed to generate electrical power using wind velocity. Wind turbine rotates using wind energy and produces mechanical power which is converted to electrical power using a generator accompanied by a turbine. The converter is employed to convert AC into DC and is employed in the water electrolysis process.

The geothermal fluid arrives at the system from the production well in the saturated liquid state and flashed to a comparatively low pressure through a flashing chamber. The flashing reduces the water pressure and thus converts water to the saturated vapor–liquid mixture. This mixture is fed to the separator to separate vapors from the liquid state. The separated liquid from the separator is reinjected to the reinjection well while the saturated vapors leaving the separator are employed in the steam turbine to generate electrical power.

The electrical power produced using the solar concentrating collector, wind, and geothermal energy-driven systems are employed in the water electrolysis process for hydrogen production. The produced hydrogen passes through a double-stage hydrogen compression unit that compresses the hydrogen at high pressure and assists in storing hydrogen at high pressure.

9.4.2 Analysis

Solar thermal energy is extracted through a solar concentrating collector to produce power using a double-stage Rankine cycle that is further employed in the water electrolysis process to produce hydrogen. Table 9.2 displays the operating constraints of the designed case study 9. In this system, all the significant subsystems, namely solar concentrating collector, wind turbine farm, geothermal system, water electrolysis,

Table 9.2 Operating Constraints of the Designed Case Study 8.

Design Parameters	Value
Solar-Assisted Double-Stage Rankine Cycle	
Turbine C2 discharge pressure	8 bars
Turbine C4 discharge pressure	1 bar
Wind Turbine Farm	
Specific heat capacity	0.49
Efficiency of wind turbine	45%
Area	900 m^2
Number of turbines	5
Geothermal Energy Source	
Geothermal fluid inlet condition	Saturated liquid
Geothermal well temperature	220°C
Geothermal fluid mass flow rate	150 kg/s
Geothermal fluid inlet pressure	3347 kPa
Geothermal fluid flashing pressure	550 kPa
Isentropic efficiency of turbine and generator	80%
Water Electrolysis	
Cell operating temperature	80°C
Membrane thickness	100 μm
Cell operating pressure	1 bar
Multistage Compression Unit	
Pressure ratio of compressors	5
Hydrogen storage pressure	25 bars

and multistage hydrogen compression unit are modeled using EES excluding the solar-assisted double-stage Rankine cycle that is simulated using Aspen Plus. The design model equations of the solar-assisted double-stage Rankine cycle, wind turbine farm, and geothermal energy system are described in this section.

Heat exchanger C1

The energy and exergy balance equations of the heat exchanger C1 are as follows:

$$\dot{m}_{SP1}h_{SP1} + \dot{m}_{SP5}h_{SP5} = \dot{m}_{SP2}h_{SP2} + \dot{m}_{SP6}h_{SP6} \tag{9.38}$$

$$\dot{m}_{SP1}ex_{SP1} + \dot{m}_{SP5}ex_{SP5} = \dot{m}_{SP2}ex_{SP2} + \dot{m}_{SP6}ex_{SP6} + \dot{E}x_d \tag{9.39}$$

High-pressure turbine C2

The energy and exergy balance equations of the high-pressure turbine C2 are as follows:

$$\dot{m}_{SP6}h_{SP6} = \dot{m}_{SP8}h_{SP8} + \dot{W}_{out} \tag{9.40}$$

$$\dot{m}_{SP6}ex_{SP6} = \dot{m}_{SP8}ex_{SP8} + \dot{W}_{out} + \dot{E}x_d \tag{9.41}$$

Heat exchanger C3

The energy and exergy balance equations of the heat exchanger C3 are as follows:

$$\dot{m}_{SP2}h_{SP2} + \dot{m}_{SP8}h_{SP8} = \dot{m}_{SP3}h_{SP3} + \dot{m}_{SP9}h_{SP9} \tag{9.42}$$

$$\dot{m}_{SP2}ex_{SP2} + \dot{m}_{SP8}ex_{SP8} = \dot{m}_{SP3}ex_{SP3} + \dot{m}_{SP9}ex_{SP9} + \dot{E}x_d \tag{9.43}$$

Low-pressure turbine C4

The energy and exergy balance equations of the low-pressure turbine C4 are as follows:

$$\dot{m}_{SP9}h_{SP9} = \dot{m}_{SP11}h_{SP11} + \dot{W}_{out} \tag{9.44}$$

$$\dot{m}_{SP9}ex_{SP9} = \dot{m}_{SP11}ex_{SP11} + \dot{W}_{out} + \dot{E}x_d \tag{9.45}$$

Heater C16

The energy and exergy balance equations of the heater C16 are as follows:

$$\dot{m}_{SP11}h_{SP11} = \dot{m}_{SP4}h_{SP4} + \dot{Q}_{in} \tag{9.46}$$

$$\dot{m}_{SP11}ex_{SP11} = \dot{m}_{SP4}ex_{SP4} + \dot{Q}_{in}\left(1 - \frac{T_o}{T}\right) + \dot{E}x_d \tag{9.47}$$

Pump C6

The energy and exergy balance equations of the pump C6 are as follows:

$$\dot{m}_{SP4}h_{SP4} + \dot{W}_{in} = \dot{m}_{SP5}h_{SP5} \tag{9.48}$$

$$\dot{m}_{SP4}ex_{SP4} + \dot{W}_{in} = \dot{m}_{SP5}ex_{SP5} + \dot{E}x_d \tag{9.49}$$

Flash chamber

The energy and exergy balance equations of the flash chamber are as follows:

$$\dot{m}_{39}h_{39} = \dot{m}_{40}h_{40} \tag{9.50}$$

$$\dot{m}_{39}ex_{39} = \dot{m}_{40}ex_{40} + \dot{E}x_{dest} \tag{9.51}$$

Separator

The energy and exergy balance equations of the separator are as follows:

$$\dot{m}_{40}h_{40} = \dot{m}_{41}h_{41} + \dot{m}_{46}h_{46} \tag{9.52}$$

$$\dot{m}_{40}ex_{40} = \dot{m}_{41}ex_{41} + \dot{m}_{46}ex_{46} + \dot{E}x_{dest} \tag{9.53}$$

Turbine

The energy and exergy balance equations of the turbine are as follows:

$$\dot{m}_{41}h_{41} = \dot{m}_{44}h_{44} + \dot{W}_{Turbine} \tag{9.54}$$

$$\dot{m}_{41}ex_{41} = \dot{m}_{44}ex_{44} + \dot{W}_{Turbine} + \dot{E}x_{dest} \tag{9.55}$$

Generator

The equation employed to calculate the work generated is as follows:

$$\dot{W}_{el} - \eta_{generator}\dot{W}_{Turbine} \tag{9.56}$$

Condenser

The energy and exergy balance equations of the condenser are as follows:

$$\dot{m}_{44}h_{44} = \dot{m}_{45}h_{45} + \dot{Q}_{out} \tag{9.57}$$

$$\dot{m}_{44}ex_{44} = \dot{m}_{45}ex_{45} + \dot{Q}_{out}\left(1 - \frac{T_o}{T}\right) + \dot{E}x_{dest} \tag{9.58}$$

Performance assessment

The energetic and exergetic efficiency equations of the designed case study 9 are as follows:

$$\eta_{ov} = \left(\frac{\dot{m}_{H_2}\mathrm{LHV}_{H_2}}{\dot{Q}_{solar} + \dot{P}_{wt} + \dot{m}_{39}h_{39} - \left(\dot{m}_{45}h_{45} + \dot{m}_{46}h_{46}\right)}\right) \tag{9.59}$$

$$\psi_{ov} = \left(\frac{\dot{m}_{H_2}ex_{H_2}}{\dot{Q}_{solar} + \dot{P}_{wt} + \dot{m}_{39}ex_{39} - \left(\dot{m}_{45}ex_{45} + \dot{m}_{46}ex_{46}\right)}\right) \tag{9.60}$$

9.4.3 Results and Discussion

The comprehensive analysis of several significant subsystems, such as solar-assisted double-stage Rankine cycle and geothermal energy-based system is presented and discussed in this section while the remaining subsystems, namely solar concentrating collector, wind farm, water electrolysis system, and multistage compression unit are already analyzed and presented in Chapters 3 and 4. This section only displays the significant results of the remaining subsystems of the designed case study 9.

The effect of some key parameters, such as molten salt flowrate, molten salt temperature, and water input flowrate to the double-stage Rankine cycle is significant to be investigated on the key constraints, such as turbine input temperature and power generated by high- and low-pressure turbines. Fig. 9.16 displays the effect of molten salt flowrate on the turbine input temperature and power generated by high- and low-pressure turbines C2 and C4. Solar thermal energy is extracted through a solar concentrating collector to produce power using a double-stage Rankine cycle that is further employed in the water electrolysis process to produce hydrogen. The molten salt flowrate ranged from 5 to 10 kg/s. It is evident that with the rise in molten salt flowrate from 5 to 10 kg/s, the input temperature of the turbines C2 and C4 rises from 211.9 to 1379.4°C, the work rate of turbine C2 increases from 213.8 to 825.9 kW, and the work rate of turbine C4 increases from 630.6 to 2384.7 kW. Fig. 9.17 displays the effect of water flowrate that is employed as a working fluid

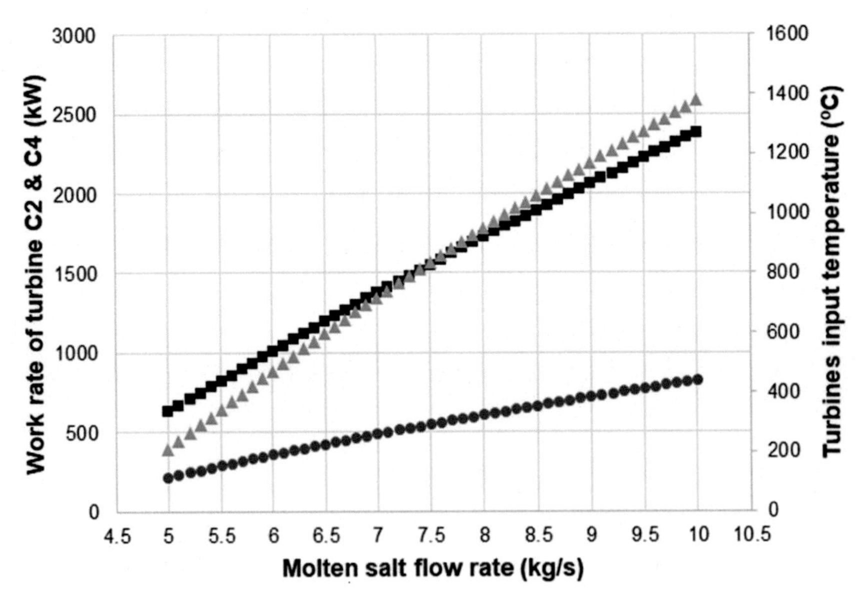

FIG. 9.16

Effect of steam input flowrate on the heat duties of heater C16, hydrolysis reactor C17, and thermolysis reactor C10.

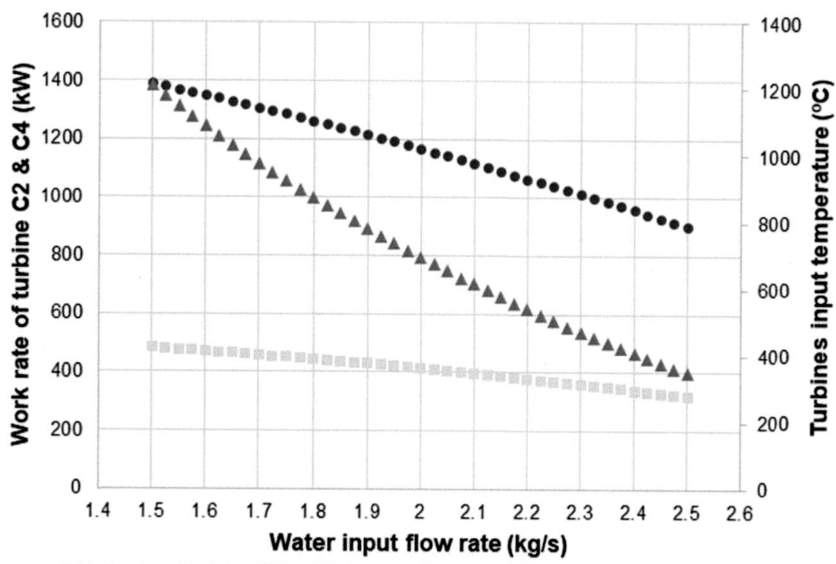

FIG. 9.17

Effect of steam input flowrate on the output flowrates of hydrogen chloride, cupric chloride, and cuprous chloride in the Cu–Cl cycle.

in the double-stage Rankine cycle, on the turbine input temperature and power generated by high- and low-pressure turbines. Solar concentrating collector-assisted thermal management system is employed that converts water into steam and employed to produce power using a double-stage Rankine cycle that is further employed in the water electrolysis process to produce hydrogen. The water input flowrate ranged from 1.5 to 2.5 kg/s. It is evident that with the rise in water input flowrate from 1.5 to 2.5 kg/s, the input temperature of the turbines decreases from 1212.3 to 347.6°C, the work rate of turbine C2 drops down from 482.9 to 318.4 kW, and the work rate of turbine C4 decreases from 1387.7 to 902.1 kW.

The effect of molten salt temperature on the turbine input temperature and power generated by high- and low-pressure turbines C2 and C4 is displayed in Fig. 9.18. Molten salt is employed to extract the solar thermal energy through a solar concentrating collector to generate power that is further employed in the water electrolysis process to produce hydrogen. The molten salt temperature ranged from 5 to 10 kg/s. It can be depicted that with the rise in molten salt flowrate from 5 to 10 kg/s, the input temperature of the turbines C2 and C4 rises from 211.9 to 1379.4°C, the work rate of turbine C2 increases from 213.8 to 825.9 kW, and the work rate of turbine C4 increases from 630.6 to 2384.7 kW.

Fig. 9.18 displays the effect of molten salt flowrate on the turbine input temperature and power generated by high- and low-pressure turbines C2 and C4. Solar thermal energy is extracted through a solar concentrating collector to produce power

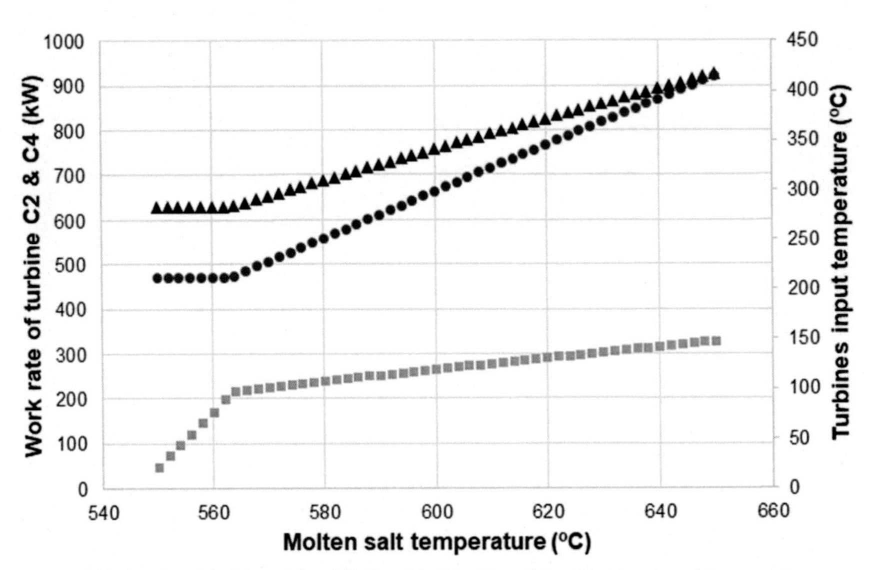

FIG. 9.18

Effect of steam input flowrate on the output flowrates of hydrogen, copper oxychloride, and oxygen in the Cu—Cl cycle.

using a double-stage Rankine cycle that is further employed in the water electrolysis process to produce hydrogen. The molten salt flowrate ranged from 550 to 650°C. The figure shows that with the rise in molten salt temperature from 550 to 650°C, the input temperature of the high- and low-pressure turbines rise from 211.9 to 471.6°C, the work rate of turbine C2 increases from 48.4 to 358.8.9 kW, and the work rate of turbine C4 increases from 630.6 to 1011.7 kW.

The flashing pressure effect on vapor quality and specific enthalpy of steam is shown in Fig. 9.19. The range of flashing pressure is taken from 400 to 800 kPa to investigate the system operation under varying flashing pressures, and the increase in flashing pressure causes the vapor quality to drop from 0.1807 to 00.1315 to and steam specific enthalpy to rise from 2738 to 2768 kJ/kg. The exergetic and energetic efficiencies of the geothermal power-assisted hydrogen production case study are found to be 11.7% and 11.3%, respectively.

Fig. 9.20 displays the flashing pressure effect on work and exergy destruction rates of the turbine and flashing chamber. The system is designed to operate at 650 kPa flashing pressure, and the considered flashing pressure range for sensitivity analysis is 400—800 kPa to explore the system functionality under dissimilar flashing pressures. The increase in flashing pressure causes the exergy destruction rate of the flash chamber to drop from 2724 to 1274 kW and work and exergy destruction rates of the turbine to rise from 2470 to 3062 kW and 478.6 to 593.3 kW, respectively.

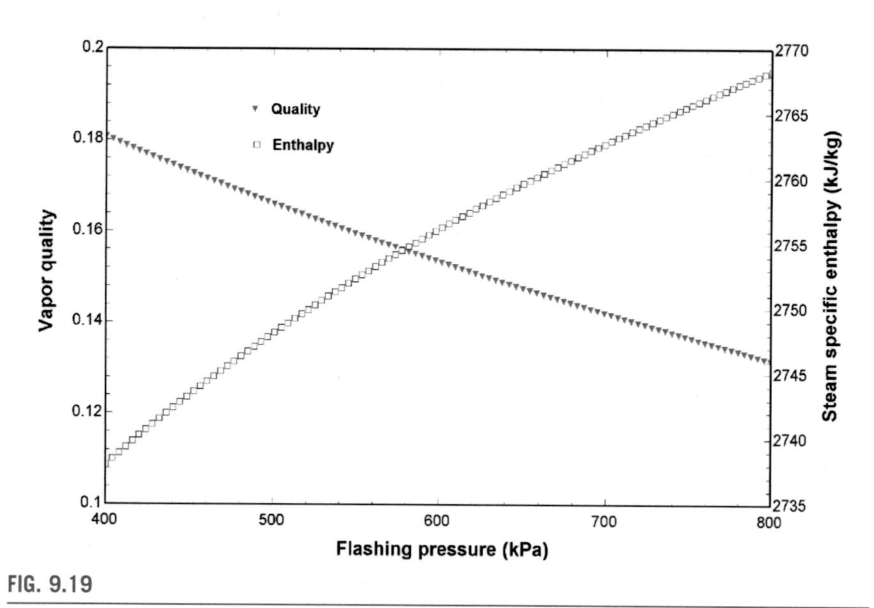

FIG. 9.19

Flashing pressure effect on vapor quality and specific enthalpy of steam.

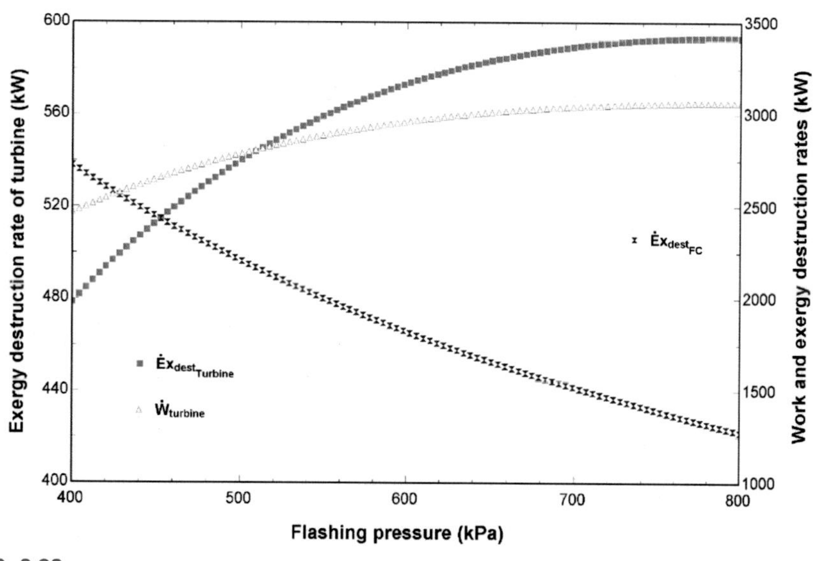

FIG. 9.20

Flashing pressure effect on work and exergy destruction rates of the turbine and flashing chamber.

9.5 Case Study 10

Solar photovoltaic and biomass gasification systems are employed in this case study for producing hydrogen. Solar PV panels are employed to generate electricity using the direct normal irradiance and produced electricity is employed to the water electrolysis process for hydrogen production. A biomass gasification system is also integrated with the designed case study that uses rice husk to produce hydrogen using an entrained flow gasifier. The syngas further passes through a water–gas shift reactor (WGSR) that produces additional hydrogen. Power and hydrogen are the major amenities of the designed integrated system.

9.5.1 System Description

Fig. 9.21 exhibits a hydrogen production system using solar PV and biomass gasification system while Fig. 9.22 displays the Aspen Plus simulation of biomass gasification-assisted hydrogen production. The significant subsystems of solar PV panels, water electrolysis, and hydrogen compression systems are modeled using EES while the biomass gasification-assisted hydrogen production system is simulated using Aspen Plus. The solar PV panels are employed to absorb the direct normal irradiance that produces photons using sunlight to produce electrical power. The biomass gasification system in the designed case study 10 is simulated employing the Aspen Plus software under the RK-SOAVE property method.

The entrained flow gasifier is employed in case study 10 that gasifies the rice husk, and syngas is produced through the biomass gasification unit C2 and C3. A reverse osmosis water desalination unit is employed to supply heat recovery steam generators with fresh water for biomass gasification. This produced syngas passes through a turbine C4 that expands the high-pressure and temperature syngas to produce electrical power. The expanded syngas reaches the heat exchanger C1 that transfers some heat from the hot syngas to the input water and converts it into steam that is required for the biomass gasification reactor.

The ultimate and proximate analysis constraints of the rice husk are tabulated in Table 9.3. The high-grade syngas is desired to achieve which requires the biomass gasifier to operate at the suitable operating temperature and chemical composition is completely based on the biomass composition balance. The syngas composition is significant to be investigated in terms of the fractions of hydrogen, carbon monoxide, hydrogen sulfide, carbon dioxide, steam, oxygen, and methane. The separator C5 separates hydrogen, carbon monoxide, and carbon dioxide from the remaining syngas composition. The mixture of hydrogen, carbon monoxide, and carbon dioxide reaches the WGSR that converts carbon monoxide to produce supplementary hydrogen. The input steam reacts with carbon monoxide and produces hydrogen and carbon dioxide. The produced hydrogen is separated using the separator C10 and fed to the compressor along with hydrogen produced through solar PV panel-assisted hydrogen, and stored hydrogen can be employed for numerous applications, such as fuel cells.

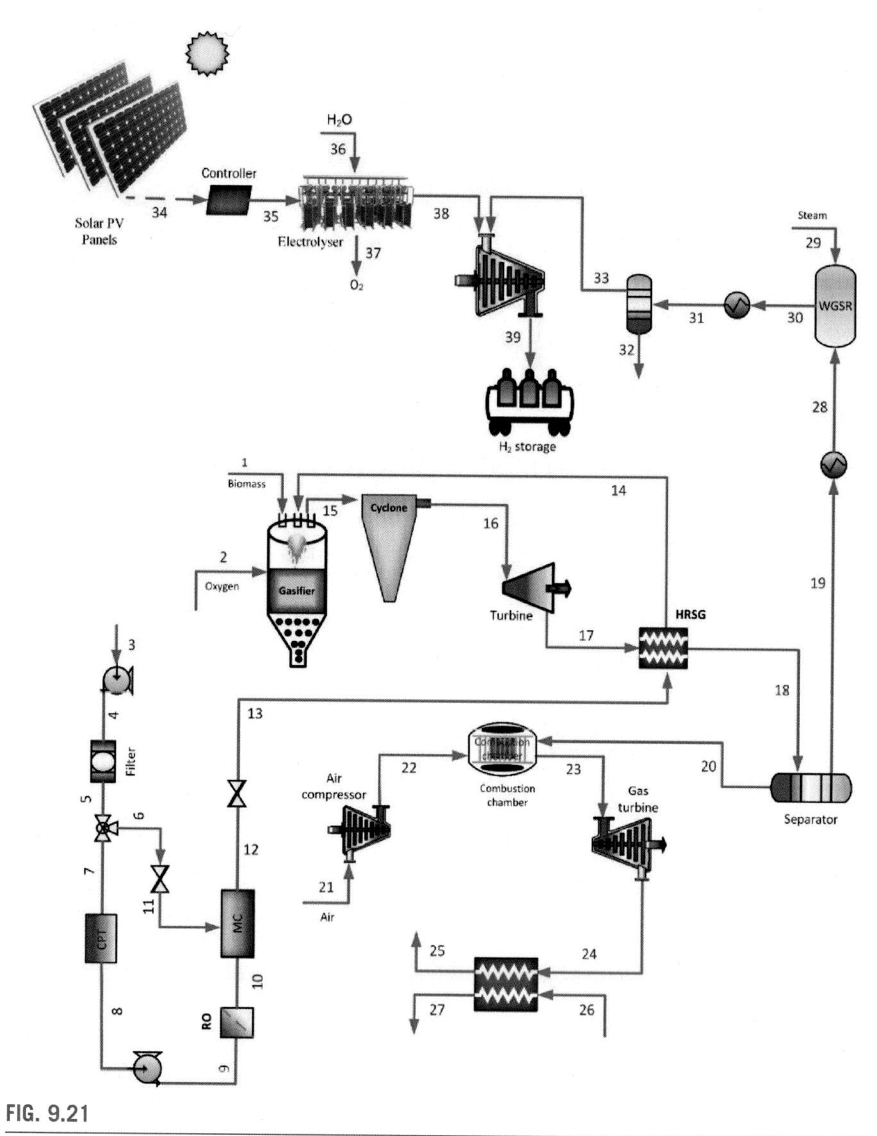

FIG. 9.21

Hydrogen production system using solar PV and biomass gasification system.

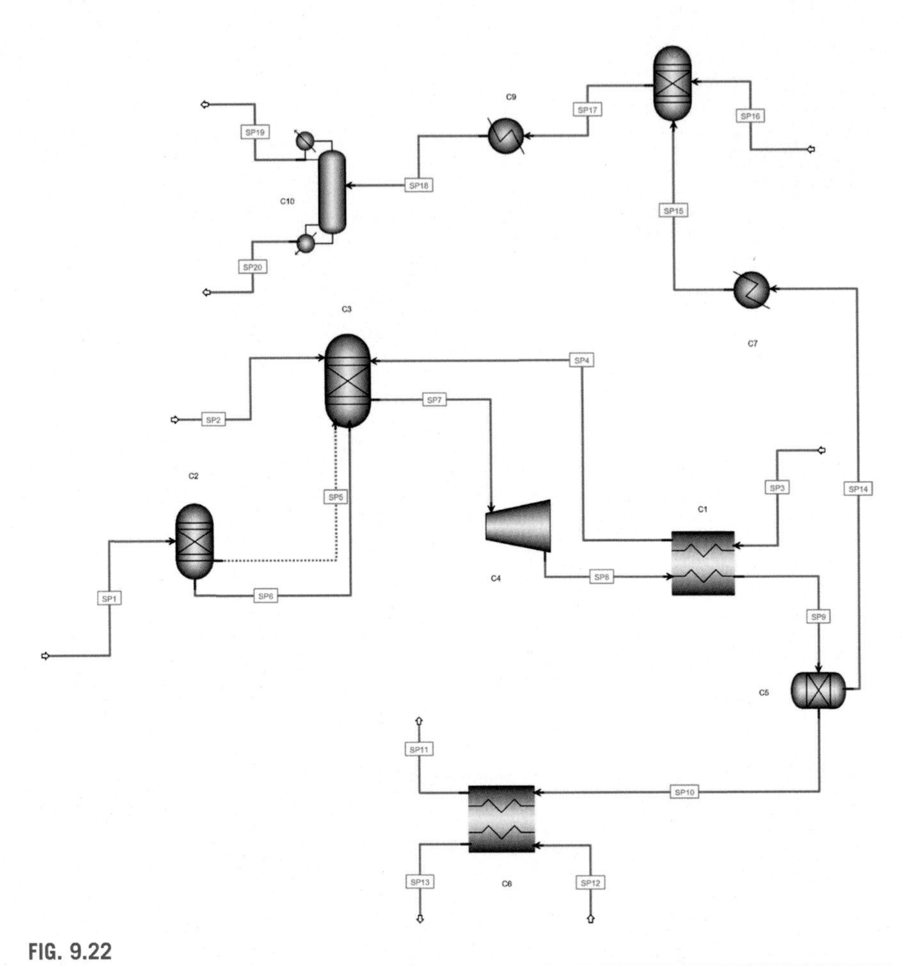

FIG. 9.22

Aspen Plus simulation of biomass gasification-assisted hydrogen production system.

9.5.2 Analysis

Biomass gasification and solar photovoltaic systems are employed in this case study for producing clean hydrogen. The solar PV panels are employed to absorb the direct normal irradiance that produces photons using sunlight to produce electrical power and produced electricity is employed in the water electrolysis process for hydrogen production. Table 9.4 displays the operating constraints of the designed case study 10. The solar PV power is determined using current (I_{PV}) and voltage (V_{PV}) of corresponding PV cells:

$$\dot{P}_{PV} = V_{PV} I_{PV} \tag{9.61}$$

Table 9.3 Ultimate and Proximate Analysis Constraints of the Rice Husk *[138]*.

	Composition	Rice Husk (% Dry Mass Basis)
Proximate analysis	Moisture	0
	Volatile matter	61.81
	Ash	21.24
	Fixed carbon	11.24
Ultimate analysis	C	38.5
	H	5.20
	N	0.45
	O	34.6
	S	0.0
	Ash	21.24
	Higher heating value	14.69 MJ/kg

Table 9.4 Operating Constraints of the Designed Case Study 10.

Design Parameters	Value
Solar PV Panels	
PV cell type	Poly-Si_CSX-310
Efficiency	15%
Number of cells	100
Solar radiation intensity	800 W/m^2
Reverse Osmosis	
Freshwater salinity	450 ppm
Seawater salinity	35,000 ppm
Recovery ratio of membrane	60%
Pump efficiency	85%
Biomass Gasification Unit	
Biomass flow rate	3 kg/s
Biomass stream	NC solids
Entrained flow gasified pressure[146]	25 bars
Gasifier operating temperature	850°C
Water–Gas Shift Reactor	
Operating temperature of WGSR[147]	445°C
Conversion rate	98.2%
Operating pressure of WGSR[147]	14 bars

Continued

Table 9.4 Operating Constraints of the Designed Case Study 10.—*cont'd*

Design Parameters	Value
Water Electrolysis	
Faraday number	96486 C/mol
Cathodic activation energy E_{act_c}	18000 J/mol
Anodic activation energy E_{act_a}	76000 J/mol
Cell operating temperature	80°C
Cell operating pressure	100 kPa
Membrane thickness	120 μm
Water Heating Unit	
Air input temperature	25°C
Air output temperature	65°C

Biomass gasification unit

The correlation that is employed to evaluate the biomass chemical exergy is expressed as follows[139]:

$$ex_{ch}^{f} = \left[\left(\text{LHV} + \omega h_{fg}\right) \times \beta + 9.417S\right] \tag{9.62}$$

The lower heating value (LHV) depends on the chemical composition of the biomass and the expression to calculate β is as follows[140]:

$$\beta = 0.1882\frac{H}{C} + 0.061\frac{O}{C} + 0.0404\frac{N}{C} + 1.0437 \tag{9.63}$$

The energetic and exergetic balance equations of each component of the designed system are as described in this section. Table 9.5 displays the energetic and exergetic balance of each component in the biomass gasification unit.

Performance indicator

The correlations for energetic and exergetic efficiency of solar PV cells can be written as follows:

$$\eta_{PV} = \frac{\dot{P}_{PV}}{\dot{q}_{in,sol} A_{PV}} \tag{9.64}$$

$$\eta_{ex_{PV}} = \frac{\dot{P}_{PV}}{\dot{q}_{in,sol}\left(1 - \frac{T_0}{T_S}\right) A_{PV}} \tag{9.65}$$

where $\dot{q}_{in,sol}$ represent solar radiation intensity, A_{PV} denotes area, T_o signifies ambient temperature, T_s indicates sun temperature, and $\dot{P}_{PV}$ represents PV power.

Table 9.5 Energetic and Exergetic Balance of Each Component in the Biomass Gasification Unit.

Component	Energy Balance	Exergy Balance
Heat exchanger C1	$\dot{m}_{SP3}h_{SP3} + \dot{m}_{SP8}h_{SP8} = \dot{m}_{SP4}h_{SP4} + \dot{m}_{SP9}h_{SP9}$	$\dot{m}_{SP3}ex_{SP3} + \dot{m}_{SP8}ex_{SP8} = \dot{m}_{SP4}ex_{SP4} + \dot{m}_{SP9}ex_{SP9} + \dot{E}x_d$
Yield reactor C2	$\dot{m}_{SP1}LHV_{SP1} + \dot{Q}_{Decomp} = \dot{m}_{SP6}LHV_{SP6}$	$\dot{m}_{SP1}ex_{SP1} + \dot{Q}_{Decomp}\left(1 - \frac{T_o}{T}\right) = \dot{m}_{SP6}ex_{SP6} + \dot{E}x_d$
Gasification reactor C3	$\dot{m}_{SP2}h_{SP2} + \dot{m}_{SP4}h_{SP4} + \dot{m}_{SP6}LHV_{SP6} - \dot{Q}_{Decomp} = \dot{m}_{SP7}h_{SP7}$	$\dot{m}_{SP2}h_{SP2} + \dot{m}_{SP4}h_{SP4} + \dot{m}_{SP6}ex_{SP6} - \dot{Q}_{Decomp}\left(1 - \frac{T_o}{T}\right) = \dot{m}_{BG7}ex_{BG7} + \dot{E}x_d$
Turbine C4	$\dot{m}_{SP7}h_{SP7} = \dot{m}_{SP8}h_{SP8} + \dot{W}_{out}$	$\dot{m}_{SP7}ex_{SP7} = \dot{m}_{SP8}ex_{SP8} + \dot{W}_{out} + \dot{E}x_d$
Separator C5	$\dot{m}_{SP9}h_{SP9} = \dot{m}_{SP10}h_{SP10} + \dot{m}_{SP14}h_{SP14}$	$\dot{m}_{SP9}ex_{SP9} = \dot{m}_{SP10}ex_{SP10} + \dot{m}_{SP14}ex_{SP14} + \dot{E}x_d$
Heat exchanger C6	$\dot{m}_{SP10}h_{SP10} + \dot{m}_{SP12}h_{SP12} = \dot{m}_{SP11}h_{SP11} + \dot{m}_{SP13}h_{SP13}$	$\dot{m}_{SP10}ex_{SP10} + \dot{m}_{SP12}ex_{SP12} = \dot{m}_{SP11}ex_{SP11} + \dot{m}_{SP13}ex_{SP13} + \dot{E}x_d$
Heater C7	$\dot{m}_{SP14}h_{SP14} = \dot{m}_{SP15}h_{SP15} + \dot{Q}_{out}$	$\dot{m}_{SP14}ex_{SP14} = \dot{m}_{SP15}ex_{SP15} + \dot{Q}_{out}\left(1 - \frac{T_o}{T}\right) + \dot{E}x_d$
Water gas shift reaction C8	$\dot{m}_{SP15}h_{SP15} + \dot{m}_{SP16}h_{SP16} = \dot{m}_{SP17}h_{SP17} + \dot{Q}_{out}$	$\dot{m}_{SP15}ex_{SP15} + \dot{m}_{SP16}ex_{SP16} = \dot{m}_{SP17}ex_{SP17} + \dot{Q}_{out}\left(1 - \frac{T_o}{T}\right) + \dot{E}x_d$
Separator C10	$\dot{m}_{SP18}h_{SP18} = \dot{m}_{SP19}h_{SP19} + \dot{m}_{SP20}h_{SP20}$	$\dot{m}_{SP18}ex_{SP18} = \dot{m}_{SP19}ex_{SP19} + \dot{m}_{SP20}ex_{SP20} + \dot{E}x_d$
Pump 1	$\dot{m}_3h_3 + \dot{W}_{in} = \dot{m}_4h_4$	$\dot{m}_3ex_3 + \dot{W}_{in} = \dot{m}_4ex_4 + \dot{E}x_d$
Filter	$\dot{m}_4h_4 = \dot{m}_5h_5$	$\dot{m}_4ex_4 = \dot{m}_5ex_5 + \dot{E}x_d$
Three-way valve	$\dot{m}_5h_5 = \dot{m}_6h_6 + \dot{m}_7h_7$	$\dot{m}_5ex_5 = \dot{m}_6ex_6 + \dot{m}_7ex_7 + \dot{E}x_d$
Throttle valve 1	$\dot{m}_6h_6 = \dot{m}_{11}h_{11}$	$\dot{m}_6ex_6 = \dot{m}_{11}ex_{11} + \dot{E}x_d$
Chemical treatment	$\dot{m}_7h_7 = \dot{m}_8h_8$	$\dot{m}_7ex_7 = \dot{m}_8ex_8 + \dot{E}x_d$
Pump 2	$\dot{m}_7h_7 + \dot{W}_{in} = \dot{m}_8h_8$	$\dot{m}_7ex_7 + \dot{W}_{in} = \dot{m}_8ex_8 + \dot{E}x_d$
RO module	$\dot{m}_9h_9 = \dot{m}_{10}h_{10}$	$\dot{m}_9ex_9 = \dot{m}_{10}ex_{10} + \dot{E}x_d$
Mixing chamber	$\dot{m}_{10}h_{10} + \dot{m}_{11}h_{11} = \dot{m}_{12}h_{12}$	$\dot{m}_{10}ex_{10} + \dot{m}_{11}ex_{11} = \dot{m}_{12}ex_{12} + \dot{E}x_d$
Throttle valve 2	$\dot{m}_{12}h_{12} = \dot{m}_{13}h_{13}$	$\dot{m}_{12}ex_{12} = \dot{m}_{13}ex_{13} + \dot{E}x_d$

The overall energetic and exergetic efficiency equations of the designed case study can be written as follows:

$$\eta_{\text{ov}} = \frac{\dot{m}_{H_2}\text{LHV}_{H_2} + \dot{m}_{\text{SP12}}(h_{\text{SP13}} - h_{\text{SP12}}) + \dot{m}_{\text{fw}}h_{\text{fw}} + \dot{W}_{\text{C3}}}{\dot{q}_{\text{in,sol}}A_{\text{PV}} + \dot{m}_{\text{biomass}}\text{LHV}_{\text{biomass}} + \dot{m}_{\text{sw}}ex_{\text{sw}}} \tag{9.66}$$

$$\psi_{\text{ov}} = \frac{\dot{m}_{H_2}ex_{H_2} + \dot{m}_{\text{SP12}}(ex_{\text{SP13}} - ex_{\text{SP12}}) + \dot{W}_{\text{C3}} + \dot{m}_{\text{fw}}h_{\text{fw}}}{\dot{q}_{\text{in,sol}}\left(1 - \frac{T_0}{T_S}\right)A_{\text{PV}} + \dot{m}_{\text{biomass}}ex_{\text{biomass}} + \dot{m}_{\text{sw}}ex_{\text{sw}}} \tag{9.67}$$

9.5.3 Results and Discussion

The comprehensive analysis of several significant subsystems, such as solar PV panels, water electrolysis system, syngas composition, and WGSR is already conducted and presented in Chapters 3 and 8. This section only displays the significant results of the remaining subsystems of the designed case study 10.

Fig. 9.23 displays the effect of recovery ratio on energetic and exergetic efficiencies. The recovery ratio ranged from 0.6 to 0.7. It can be depicted that with the rise in recovery ratio from 0.6 to 0.7, the energetic efficiency increases from 0.598 to 0.698 and exergetic efficiency increases from 0.283 to 0.329.

It is significant to investigate the effect of input biomass, steam, and oxygen flowrates on the biomass gasification unit on the turbine work rate, hydrogen flowrate in streams SP15 and SP17 and on the output flowrates of steam, carbon dioxide, carbon monoxide, hydrogen, and oxygen. Fig. 9.24 displays the biomass input

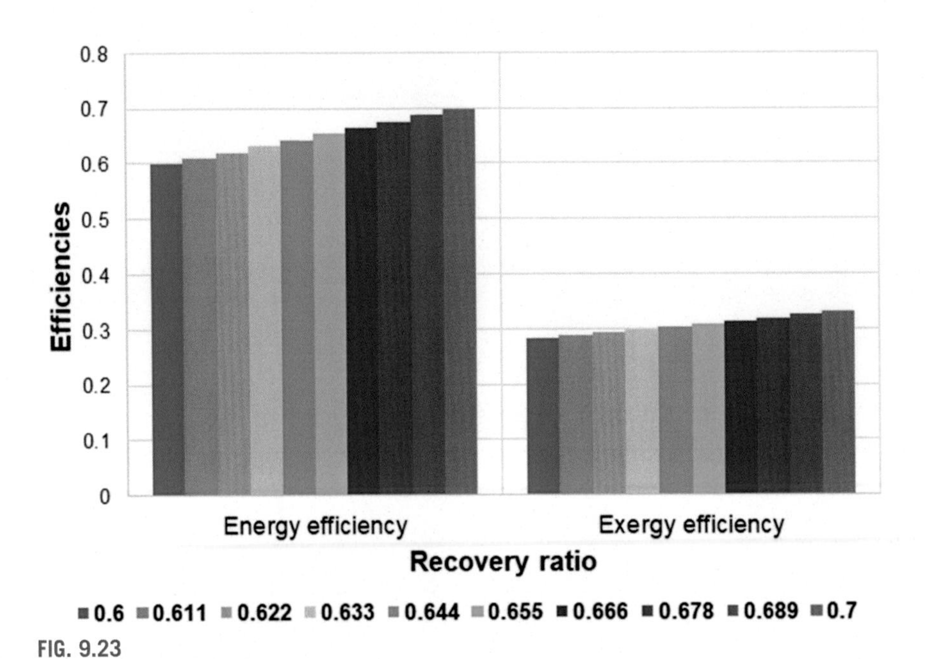

FIG. 9.23

Effect of recovery ratio on energetic and exergetic efficiencies.

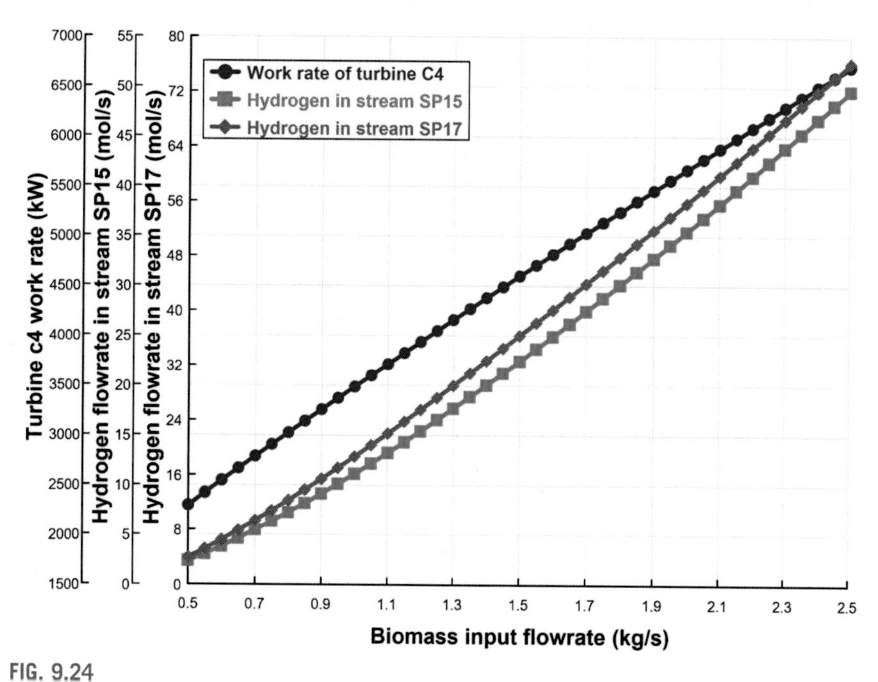

FIG. 9.24

Biomass input flowrate effect on the turbine work rate and hydrogen flowrate in streams SP15 and SP17.

flowrate effect on the turbine work rate and hydrogen flowrate in streams SP15 and SP17. Steam, oxygen, and biomass are the significant inputs of the entrained flow gasifier, and oxygen works as an oxidant that assists in gasifying rice husk and producing high-quality syngas. The biomass input flowrate ranged from 0.5 to 2.5 kg/s. The figure displays that the turbine work rate and hydrogen flowrate in streams SP15 and SP17 increase gradually with the rise in input biomass flowrate. It is evident that the hydrogen production rate in stream SP17 that is after WGSR is higher in comparison with the hydrogen production rate in stream SP15.

The effect of input steam, biomass, and oxygen flowrates is significant to be investigated against the output flowrates of steam, carbon dioxide, carbon monoxide, hydrogen, and oxygen. Fig. 9.25 exhibits the effect of steam input flowrate on the output flowrates of steam, carbon dioxide, carbon monoxide, hydrogen, and oxygen. Biomass, oxygen, and steam are the significant inputs of the biomass gasification unit. The steam input flowrate ranged from 0.1 to 1.5 kg/s. The figure displays that the flowrates of hydrogen, carbon monoxide, and steam increase with the rise in input steam flowrate while the flowrates of oxygen and carbon dioxide decrease with the rise in input steam flowrate. The steam flowrate is also expected to decrease as oxygen flowrate, but it increases because steam input flowrate increases continuously.

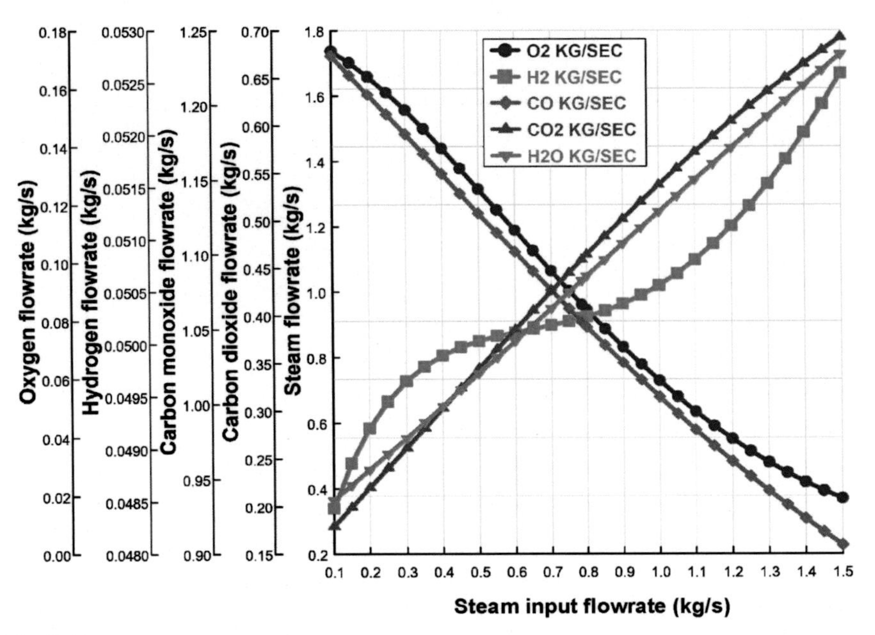

FIG. 9.25

Effect of steam input flowrate on the output flowrates of steam, carbon dioxide, carbon monoxide, hydrogen, and oxygen.

These sensitivity analyses show the significance of the optimum supply of all input parameters of steam, oxygen, and biomass. Fig. 9.27 displays the effect of biomass input flowrate on the output flowrates of steam, carbon dioxide, carbon monoxide, hydrogen, and oxygen. Biomass, steam, and oxygen are the significant inputs of the biomass gasification unit. The steam biomass flowrate ranged from 0.5 to 2.5 kg/s. The figure represents that the flowrates of steam, carbon dioxide, and oxygen increase in the beginning and start decreasing gradually with the rise in input biomass flowrate while the flowrates of hydrogen and carbon monoxide increase significantly with the rise in input biomass flowrate. The flowrates of hydrogen and carbon monoxide are desired to increase with the rise in biomass input flowrate which shows high-quality syngas.

Fig. 9.26 shows the effect of oxygen input flowrate on the output flowrates of steam, carbon dioxide, carbon monoxide, hydrogen, and oxygen. Biomass, steam, and oxygen are the significant inputs of the biomass gasification unit. Syngas composition is another significant parameter that is already investigated in Chapter 8. The steam input flowrate ranged from 0.5 to 1.5 kg/s. The figure displays that the flowrates of steam, carbon dioxide, and oxygen increase with the rise in input oxygen flowrate while the flowrates of hydrogen and carbon monoxide decrease with the rise in input steam flowrate. The flowrates of hydrogen and carbon monoxide are desired to increase as it is observed with the rise in input steam flowrate. This shows the significance of the optimum supply of all input parameters of steam, oxygen, and biomass.

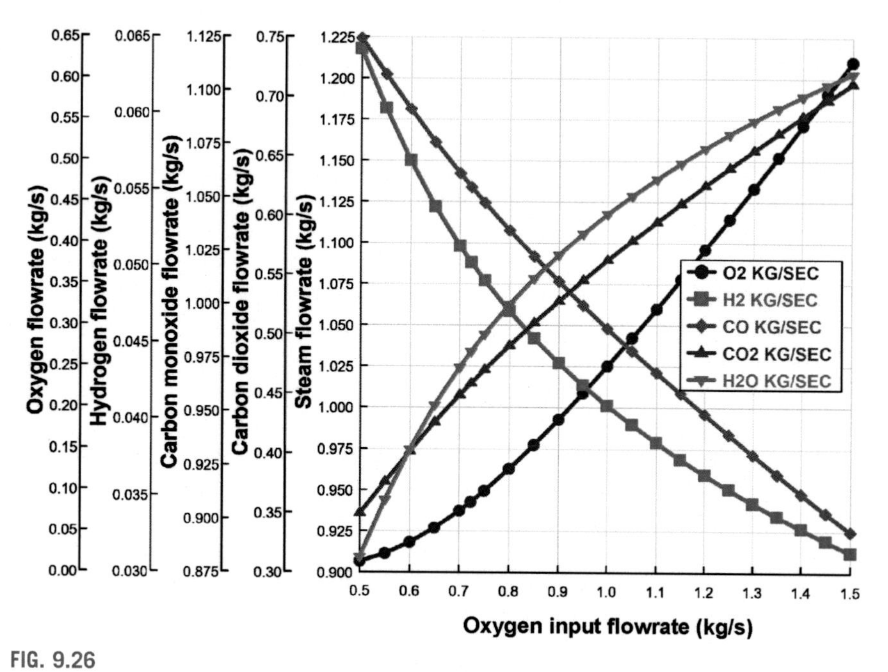

FIG. 9.26

Oxygen input flowrate effect on the output flowrates of steam, carbon dioxide, carbon monoxide, hydrogen, and oxygen.

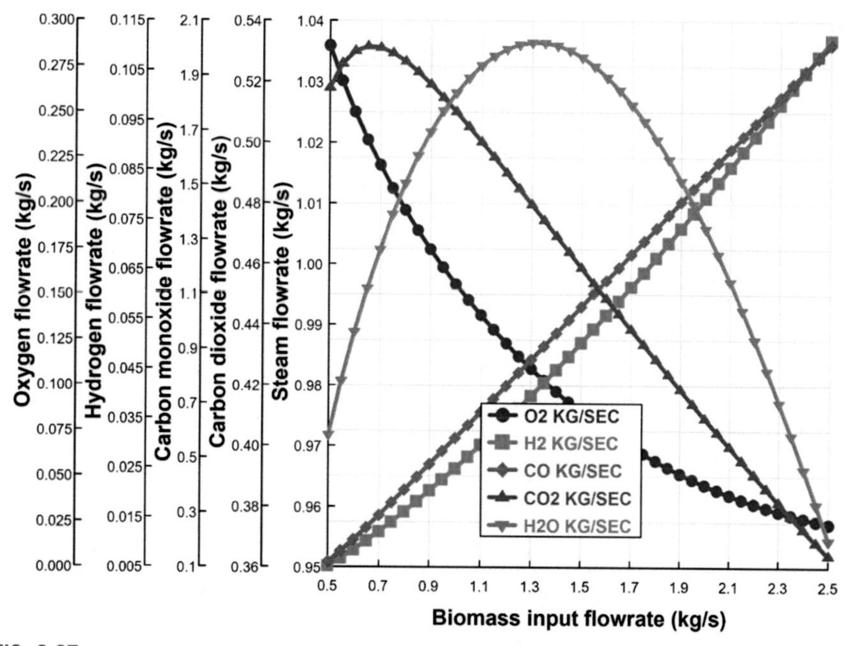

FIG. 9.27

Effect of biomass input flowrate on the output flowrates of steam, carbon dioxide, carbon monoxide, hydrogen, and oxygen.

9.6 Closing Remarks

The global energy systems are presently experiencing an extraordinary transformation: fundamentally a revolution in energy production, delivery, and consumption methods. The power industry is now integrating the analytical capabilities and collected energy data into standing energy infrastructures, renewable energy integration is rapidly varying the electricity generation blend, natural gas is substituting coal in larger capacities, transportation electrification is obvious by the increase in the number of plug-in road vehicles and sensors of energy management are presenting all types of appliances and devices. As this scheme advances, it causes the reconsideration in traditional standards of energy system operation and planning with a sturdy emphasis on how to attain the greatest efficient, reliable, and flexible energy system whereas offering energy services to the clients at a reasonable cost.

Renewable energy sources, such as wind, solar, hydro, geothermal, biomass, and ocean thermal energy conversion are probable to play a vital role in the global transition from conventional to renewable energy fuels to overcome numerous challenges, such as socio-economic challenges, environmental problems, GHG emissions, and carbon emission taxes. The integration of energy systems is projected to combine the energy carriers with infrastructures like hydrogen, communications, transportation, water, and combined heating and cooling to maximize the efficiency and effective utilization of renewable energy sources. The compatible integration of the energy components and subsystems is a significant prospect to track the optimal utilization of the newly established technologies and influence the global transition. Integrated energy systems can also be employed to recover the industrial waste heat using thermal management to produce useful outputs, such as hydrogen, electricity, ammonia, combined heating and cooling, domestic hot water, industrial drying, and many more.

This chapter presents three case studies undergoing the integration of different renewable energy sources to explore the performance of an integrated energy system and to utilize the energy effectively. Case study 8 is designed to explore the solar and ocean thermal energy conversion-assisted hydrogen production system. The solar thermal energy extracted from the solar heliostat field is employed for electricity generation and thermochemical Cu–Cl cycle-based hydrogen production. The electricity generated by both solar and OTEC systems is employed in the water electrolysis process for clean hydrogen production. Case study 9 investigates the integration of solar concentrating collector, wind, and geothermal energy-driven systems for hydrogen production. The solar thermal energy extracted from a solar concentrating collector is employed to generate electricity employing a double-stage Rankine cycle. The wind and geothermal energy system also generate power, and electricity produced by all three integrated sources is employed in the water electrolysis process for hydrogen production that is stored at high pressure using a multistage compression unit. Solar photovoltaic and biomass gasification systems are employed in case study 10 for producing hydrogen. The solar PV panels are employed to absorb the solar radiation intensity that produces photons using solar light to generate electricity and produced power is employed in the water electrolysis process for hydrogen production. The biomass gasification produces syngas that further passes through a water–gas shift reactor to produce additional hydrogen, and hydrogen is extracted using a separation unit and stored at high pressure. Renewable energy-assisted integrated energy systems can play a significant role in the global transition toward 100% renewable energy.

CHAPTER

10 Conclusions and Future Directions

Hydrogen can be produced using plentiful primary energy sources and employing numerous technical processes. Renewable energy-assisted water electrolysis is among the most promising options for hydrogen production. Renewable hydrogen is considered renewable energy, produced using different renewable energy sources such as wind, solar, hydro, geothermal, tidal, ocean thermal energy conversion, wave, and biomass and stored in the form of hydrogen gas. Renewable energy becomes permanently available by storing it in terms of zero-carbon hydrogen gas. At present, significant quantities of hydrogen are produced through natural gas that possesses numerous drawbacks of fossil fuel depletion, environmental problems, greenhouse gas (GHG) emissions, and carbon emissions taxes. A part of the electricity generated by solar, wind, biomass, and geothermal sources is also employed to produce hydrogen. Due to the challenges that occur with the usage of fossil fuels, a global transition from traditional to renewable energy is desirable and this book covers all the renewable energy sources and hydrogen production methods.

10.1 Conclusions

In this book, renewable hydrogen production technologies are explicated comprehensively considering all aspects of their development, analysis, operation, performance, assessment, and applications. This book encompasses the development and operation of renewable energy-driven hydrogen production technologies by summarizing the fundamentals and basic concepts to provide a complete understanding. The development of different renewable energy sources namely; wind, solar, hydro, geothermal, biomass, and ocean thermal energy conversion-assisted hydrogen production methods, is presented and discussed. The modeling of the significant hydrogen production methods, namely water electrolysis, thermochemical cycles, chlor-alkali electrochemical process, and direct renewable energy-assisted processes, is described in detail. The different electrolyzer types including proton exchange membrane electrolyzer, solid oxide (SO) electrolyzer, and alkaline (ALK) electrolyzer are modeled in detail using both simulation and experimental approaches to determine various electrochemical parameters.

Some of the significant uses of hydrogen are (1) it is used in oil refineries, (2) employed for ammonia synthesis, (3) used for the formation of synthetic fuels, (4) used for filling balloons because of being lighter, (5) used as an energy carrier,

Renewable Hydrogen Production. https://doi.org/10.1016/B978-0-323-85176-3.00006-8

(6) used as the energy storage system, (7) used for hydrogenation of oils and fats, (8) used as rocket fuel, (9) used for welding, (10) used for different chemical processes such as hydrodealkylation and hydrocracking, (11) used for the formation of production of hydrochloric acid, (12) used for metallic ores reduction, (13) used for cryogenics, (14) combined heat and power production systems, and (15) used as a fuel for hydrogen fuel cell vehicles. Thus, this book also covers the different fuel-cell types including proton exchange membrane fuel cells, SO fuel cell, ammonia fuel cells, ALK fuel cells, and phosphoric acid fuel cells and their detailed modeling to determine numerous electrochemical parameters that can assist to employ hydrogen in the transportation sector as hydrogen fuel-cell and hydrogen fuel-cell hybrid vehicles. Fig. 10.1 displays renewable hydrogen production methods and their applications in the transportation sector. It is displayed that the sources that are employed to generate electricity such as wind, solar, and biomass can directly be employed in a water electrolysis system that produces clean hydrogen that can be stored using storage systems. Renewable energy becomes permanently available by storing it in terms of zero-carbon hydrogen gas. The stored hydrogen gas can be employed in the transportation sector (aviation, heavy-duty trucks, fuel cell vehicles, and fuel cell trains) using hydrogen fuel cells, the industrial sector (synthesis of different chemical compounds such as ammonia and methanol), as fuel, as an energy storage medium, and as an energy carrier.

The currently existing global energy economy is not sustainable, and significant reasons are as follows. The global energy demand is growing significantly, which is majorly covered by fossil fuels that cause GHG emissions and facing depletion challenges as well. The supply and demand of fossil fuels are not balanced; thus, an alternative is required to be found. The emissions emitted by fossil fuels substantially

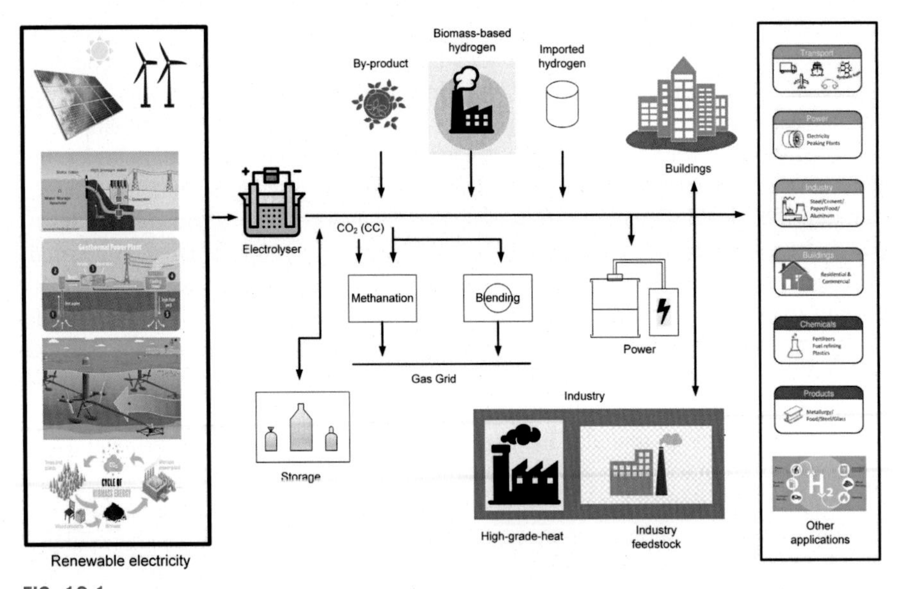

FIG. 10.1

Renewable hydrogen production methods and applications in transportation sector.

degrade the quality of air all around the globe. The resulting carbonaceous by-products are significantly causing climate and health impacts. The fossil fuel economy situates the nations and people under the excessive influence of energy suppliers and countries face a lack of economic independence. Hydrogen can eliminate these drawbacks in terms of the offered benefits. Hydrogen usage reduces environmental pollution greatly. The electricity can be generated by employing hydrogen and oxygen to the fuel cells, and only by-products of heat and water are released undergoing zero greenhouse gasses. This electricity can be employed to power vehicles, as energy carriers, as a heat source, as a fuel, and in many other applications. Hydrogen can be produced using numerous renewable energy sources either onsite where it is required or centrally and then distributed.

The water electrolysis process-based hydrogen production system separates water into its constituents of oxygen and hydrogen using electricity. Renewable energy can directly be employed to power the water electrolysis process, and the usage of renewable energy offers a sustainable system that is fossil fuel independent and undergoes zero-carbon emissions. The significant renewable sources that are currently employed to power electrolyzer for hydrogen production are wind, solar, hydro, and tidal energy. The hydrogen produced by the water electrolysis process is stored in the storage unit and can be employed in the fuel cells for electricity generation and undergoes the only by-products of heat and water.

10.2 Future Directions

Hydrogen energy center is constructing on the hydrogen benefits to comprehend sustainable energy economy. Considering the hydrogen production process, the electricity cost required for the water electrolysis process is a significant barrier to sustainable energy. Currently, the research scientific community is continuously investigating renewable energy sources to establish environmentally benign, clean, consistent, and sustainable hydrogen production. Wind power generation cost has been improved significantly in recent years and can generate electricity for $.04/kWh that is quite competitive with conventional electricity generation methods, and this cost is expected to be reduced further. If an adequate site for wind energy generation is selected, there is significant potential for hydrogen production capability. Tidal power is also getting momentum, and project demonstrations are found to be successful. Small-scale hydro systems have the potential for sites wherever limited production and usage are the determining factors. Improvements in solar photovoltaic cells are also lowering the electricity production cost continuously. If a suitable set of circumstances and the right climate are provided, solar energy can be a viable solution.

Renewable sources of energy are often restricted for commercial usage because of intermittent availability. The intermittent nature of some renewable energy sources such as solar and wind makes the storage systems mandatory to make these renewable energy sources consistent and sustainable. During the intervals of low wind speed and low solar radiation intensities, hydrogen can play a critical role as a storage medium to supply power throughout these intervals or to meet high energy demand. Renewable energy becomes permanently available by storing it in terms of zero-carbon hydrogen gas. Hydrogen can be employed as a mobile power source for the transportation sector by compressing and storing in small tanks similar to propane or gasoline.

To achieve net-zero CO_2 emissions by 2050, energy economy cannot only rely on renewable electricity. Decarbonizing certain industries such as transportation, industry, buildings and agriculture has been challenging and require novel solutions to complement clean and renewable electricity and hydrogen can directly be produced using renewable energy sources like solar, wind, geothermal, hydro, OTEC and biomass, that do not emit GHG emissions. The commercial deployment of the off-grid offshore hydrogen production industry demonstrates ground-breaking solutions with noteworthy potential around the globe. Hydrogen possesses the potential to decarbonize the marine industry by employing off-grid offshore wind energy and can also decrease the environmental impacts caused by current hydrocarbon-dependent transportation systems via carbon-free fuel including ammonia and hydrogen in the maritime and heavy-duty road vehicles. International hydrogen roadmaps have all highlighted the need for 'cluster/hub' models of development to drive deployment by matching fuel provision and end-users. Fig. 10.2 illustrates a roadmap to cover the period from the year 2020 up to 2050 for hydrogen energy systems and their potential deployment in various sectors, ranging from transportation to industrial.

Renewable energy sources such as hydro, geothermal, tidal, wave, biomass, and ocean thermal energy conversion offering reliable, consistent, and environmentally benign power production should be developed on the commercial scale to produce

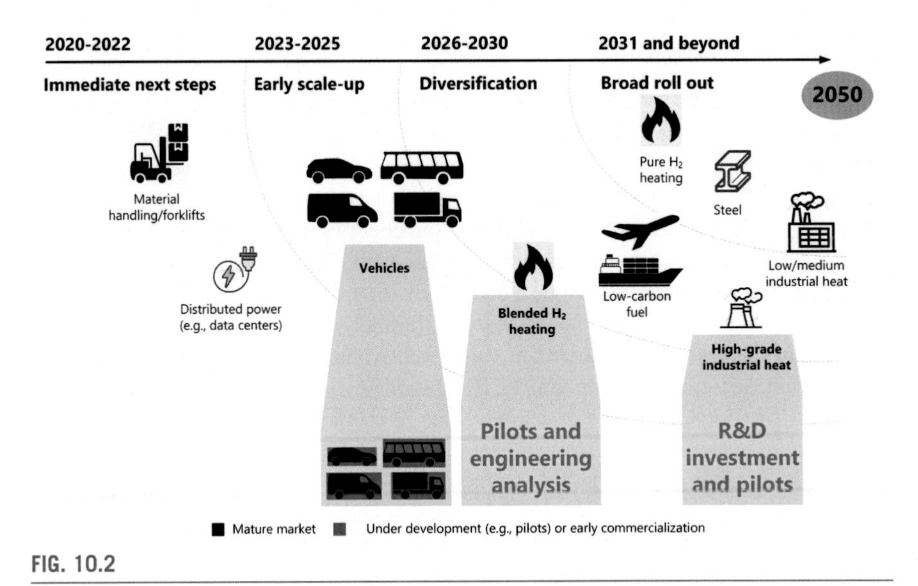

FIG. 10.2

A roadmap for hydrogen energy systems and their deployment.

clean hydrogen using the water electrolysis process. For the solar and wind energy sources, hydrogen storage systems that can store hydrogen during the intervals of high solar radiation intensities and wind speeds using excess electricity should be integrated, and the stored hydrogen can be employed in the fuel cells to meet the intervals of low or zero wind speed and low solar radiation intensities or intervals of high energy demand. With the continuous growth in hydrogen usage and technical advancements, the production, distribution, and product manufacturing costs will become very effective and affordable. The partnerships between businesses, universities, and governments can also play a significant role to build a sustainable energy economy and to commercially deploy the sustainable energy systems.

References

1. IEA: International Energy Agency. World energy statistics and balances. *IEA World Energy Statistics Balances*. 2018. https://doi.org/10.1787/enestats-data-en.
2. Zhang X, Dincer I. *Energy Solutions to Combat Global Warming*. Vol. 33. 2017.
3. International Energy Agency. *Data & Statistics - IEA*. [cited 24 April 2020]. Available from: https://www.iea.org/data-and-statistics?country=WORLD&fuel=CO2 emissions&indicator=CO2 emissions by sector.
4. Ishaq H, Dincer I. Comparative assessment of renewable energy-based hydrogen production methods. *Renew Sustain Energy Rev*. 2020;135:110192.
5. Al-Bassam AM, Conner JA, Manousiouthakis VI. Natural-Gas-derived hydrogen in the presence of carbon fuel taxes and concentrated solar power. *ACS Sustain Chem Eng*. 2018;6:3029–3038.
6. Kaiwen L, Bin Y, Tao Z. Economic analysis of hydrogen production from steam reforming process: a literature review. *Energy Sources B Energy Econ Plann*. 2018;13: 109–115.
7. Agrafiotis C, von Storch H, Roeb M, Sattler C. Solar thermal reforming of methane feedstocks for hydrogen and syngas production—a review. *Renew Sustain Energy Rev*. 2014;29:656–682.
8. *Global Demand for Pure Hydrogen, 1975–2018 – Charts – Data & Statistics - IEA*. [cited 25 April 2020]. Available from: https://www.iea.org/data-and-statistics/charts/global-demand-for-pure-hydrogen-1975-2018.
9. REN21. *Renewables global status report*; 2020. Available from: https://www.ren21.net/gsr-2020/.
10. Scientific History Institute. *Historical Biography: Robert Boyle*; 2017. Available from: https://www.sciencehistory.org/historical-profile/robert-boyle.
11. West JB. Henry Cavendish (1731–1810): hydrogen, carbon dioxide, water, and weighing the world. *Am J Physiol Lung Cell Mol Physiol*. 2014;307:1–6.
12. Levei R. Short communication: the electrolysis of water. *J Electroanal Chem*. 2008;44: 120–125.
13. Wisniak J. James Dewar - more than a flask. *Indian J Chem Technol*. 2003;10:424–434.
14. IUPAC Recommendations 2002. Measurement of pH. Definition, standards, and procedures. *Int Union Pure Appl Chem Meas*. 2002;74:2169–2200.
15. Chemistry LibreTech. *Acids and Bases - The Brønsted-Lowry Definition*; 2020. Available from: https://chem.libretexts.org/Bookshelves/Organic_Chemistry/Map%3A_Organic_Chemistry_(McMurry)/02%3A_Polar_Covalent_Bonds_Acids_and_Bases/2.07%3A_Acids_and_Bases-The_Brnsted-Lowry_Definition.
16. Garrett B. Deuterium: Harold C. Urey. *J Chem Educ*. 1962;39:583.
17. Brown HC. From little acorns to tall oaks: from boranes through organoboranes. *Science*. 1979;210:485–492.
18. Hughes J. The strath report: britain confronts the H - Bomb , 1954 – 1955. *Hist Technol*. 2003;19:257–275.
19. Olah GA. *My Search for Carbocations and Their Role in Chemistry (Nobel Lecture)*. 2003:201–213. https://doi.org/10.1142/9789812791405_0038.
20. Kubas G. Chemical bonding of hydrogen molecules of transition metal complexes. *Los Alamos Natl Lab*. 1990;2223:522–525.

21. Nellis WJ. Making metallic hydrogen. *Sci Am*. 2000;282:84–90.
22. *Uses and Benefits of Hydrogen as a Future Fuel - Cleantech Rising*. [cited 6 May 2020]. Available from: https://cleantechrising.com/water-for-fuel/.
23. *Sustainable Development*. [cited 6 May 2020]. Available from: https://www.admtl.com/en/adm/communities/sustainable-development.
24. Dincer I, Zamfirescu C. *Sustainable Hydrogen Production*. Amsterdam, Netherlands: Elsevier; 2016.
25. Lototskyy M, Yartys VA. Comparative analysis of the efficiencies of hydrogen storage systems utilising solid state H storage materials. *J Alloys Compd*. 2015;645: S365–S373.
26. Zhang J, Dong Y, Wang Y, Zhu Y, Duan Y. A novel route to synthesize hydrogen storage material ammonia borane via copper (II)-ammonia complex liquid phase oxidization. *Int J Energy Res*. 2018;42:4395–4401.
27. Hosseini M, Dincer I, Naterer GF, Rosen MA. Thermodynamic analysis of filling compressed gaseous hydrogen storage tanks. *Int J Hydrogen Energy*. 2012;37:5063–5071.
28. IEA: International Energy Agency. *Capacity of New Projects for Hydrogen Production for Energy and Climate Purposes, Electrolytic Hydrogen, 2000–2018 – Charts – Data & Statistics - IEA*; 2020 [cited 16 May 2020]. Available from: https://www.iea.org/data-and-statistics/charts/capacity-of-new-projects-for-hydrogen-production-for-energy-and-climate-purposes-electrolytic-hydrogen-2000-2018.
29. Colucci JA. Hydrogen production using autothermal reforming of biodiesel and other hydrocarbons for fuel cell applications. *Sol Energy*. 2006:483–484.
30. Yilanci A, Dincer I, Ozturk HK. A review on solar-hydrogen/fuel cell hybrid energy systems for stationary applications. *Prog Energy Combust Sci*. 2009;35:231–244.
31. Williams M. Fuel Cell Handbook. *U.S. Dep. Energy Off. Foss. Energy Natl. Energy Technol. Lab*. 2004;26 (Seventh Edition).
32. US wind energy selling at record low price of 2.5 cents per kWh. *Renew Energy World*. [cited 23 May 2020]. Available from: https://www.renewableenergyworld.com/2015/08/19/us-wind-energy-selling-at-record-low-price-of-2-5-cents-per-kwh/#gref.
33. Acar C, Dincer I. Comparative assessment of hydrogen production methods from renewable and non-renewable sources. *Int J Hydrogen Energy*. 2014;39:1–12.
34. Cetinkaya E, Dincer I, Naterer GF. Life cycle assessment of various hydrogen production methods. *Int J Hydrogen Energy*. 2012;37:2071–2080.
35. Ishaq H, Dincer I. The role of hydrogen in global transition to 100% renewable energy. In: *Accelerating the Transition to a 100% Renewable Energy Era*. Springer Nature Switzerland AG; 2020:275–307. https://doi.org/10.1007/978-3-030-40738-4_13.
36. Çelik D, Yıldız M. Investigation of hydrogen production methods in accordance with green chemistry principles. *Int J Hydrogen Energy*. 2017;42:23395–23401.
37. T-Raissi A, Block DL. Hydrogen: automotive fuel of the future. *IEEE Power Energy Mag*. 2004;2:40–45.
38. Dincer I, Acar C. A review on clean energy solutions for better sustainability. *Int J Energy Res*. 2015;39:585–606.
39. Shirasaki Y, Yasuda I. Membrane reactor for hydrogen production from natural gas at the Tokyo gas company : a case study. In: *Handbook of Membrane Reactors*. 2013.
40. Luk HT, Lei HM, Ng WY, Ju Y, Lam KF. Techno-economic analysis of distributed hydrogen production from natural gas. *Chin J Chem Eng*. 2012;20:489–496.

41. Ventura C, Azevedo JLT. Development of a numerical model for natural gas steam reforming and coupling with a furnace model. *Int J Hydrogen Energy.* 2010;35: 9776–9787.
42. Ishaq H, Dincer I. Analysis and optimization for energy, cost and carbon emission of a solar driven steam-autothermal hybrid methane reforming for hydrogen, ammonia and power production. *J Clean Prod.* 2019;234:242–257.
43. Al-Zareer M, Dincer I, Rosen MA. Analysis and assessment of a hydrogen production plant consisting of coal gasification, thermochemical water decomposition and hydrogen compression systems. *Energy Convers Manag.* 2018;157:600–618.
44. Muresan M, Cormos CC, Agachi PS. Techno-economical assessment of coal and biomass gasification-based hydrogen production supply chain system. *Chem Eng Res Des.* 2013;91:1527–1541.
45. Wang Z, Roberts RR, Naterer GF, Gabriel KS. Comparison of thermochemical, electrolytic, photoelectrolytic and photochemical solar-to-hydrogen production technologies. *Int J Hydrogen Energy.* 2012;37:16287–16301.
46. Zhang G, Wan X. A wind-hydrogen energy storage system model for massive wind energy curtailment. *Int J Hydrogen Energy.* 2014;39:1243–1252.
47. Ghazvini M, Sadeghzadeh M, Ahmadi MH, Moosavi S, Pourfayaz F. Geothermal energy use in hydrogen production: a review. *Int J Energy Res.* 2019;43:7823–7851.
48. Our World in Data. *Renewable Energy*; 2019. Available from: https://ourworldindata.org/renewable-energy.
49. Nikolaidis P, Poullikkas A. A comparative overview of hydrogen production processes. *Renew Sustain Energy Rev.* 2017;67:597–611.
50. Dincer I, Joshi AS. *Solar Based Hydrogen Production Systems*. New York Heidelberg Dordrecht London: Springer; 2013.
51. Naterer G, Suppiah S, Lewis M, et al. Recent Canadian advances in nuclear-based hydrogen production and the thermochemical Cu-Cl cycle. *Int J Hydrogen Energy.* 2009;34:2901–2917.
52. Naterer GF, Dincer I, Zamfirescu C. *Hydrogen Production from Nuclear Energy.* London New York: Springer Verlag; 2013. https://doi.org/10.1007/978-1-4471-4938-5.
53. *Demonstration Advances to Produce Hydrogen Using Molten Salt Reactor Nuclear Technology.* [cited 30 May 2020]. Available from: https://www.powermag.com/demonstration-advances-to-produce-hydrogen-using-molten-salt-reactor-nuclear-technology/.
54. Bolt A, Dincer I, Agelin-chaab M. Experimental study of hydrogen production process with aluminum and water. *Int J Hydrogen Energy.* 2020;45:14232–14244.
55. Sheikhbahaei V, Baniasadi E, Naterer GF. Experimental investigation of solar assisted hydrogen production from water and aluminum. *Int J Hydrogen Energy.* 2018;43: 9181–9191.
56. Chehade G, Lytle S, Ishaq H, Dincer I. Hydrogen production by microwave based plasma dissociation of water. *Fuel.* 2020;264:116831.
57. Mizeraczyk J, Jasinski M. Plasma processing methods for hydrogen production. *Phys J Appl Phys.* 2020;75:24702.
58. Rusanov VD, Fridman AA, Sholin GV. The physics of a chemically active plasma with nonequilibrium vibrational excitation of molecules. *Usp Fiz Nauk.* 1981;134.
59. Rashwan SS, Dincer I, Mohany A, Pollet BG. The Sono-Hydro-Gen process (ultrasound induced hydrogen production): challenges and opportunities. *Int J Hydrogen Energy.* 2019;44:14500–14526.

60. Rashwan SS, Dincer I, Mohany A. An investigation of ultrasonic based hydrogen production. *Energy*. 2020;205:118006.
61. Rabbani M. *Design, Analysis and Optimization of Novel Photo Electrochemical Hydrogen Production Systems* [Ph.D. thesis]. 2013.
62. Chandran RR, Chin DT. Reactor analysis of a chlor-alkali membrane cell. *Proc Electrochem Soc*. 1984;84−11:294−324.
63. Nakumura T. Hydrogen production from water utilizing. *Sol Energy*. 1977;19:467−475.
64. Roeb M, Neises M, Monnerie N, et al. Materials-related aspects of thermochemical water and carbon dioxide splitting: a review. *Materials*. 2012;5:2015−2054.
65. Ishaq H, Dincer I. A comparative evaluation of three Cu-Cl cycles for hydrogen production. *Int J Hydrogen Energy*. 2019;44:7958−7968.
66. Ni M, Leung MKH, Leung DYC. Energy and exergy analysis of hydrogen production by a proton exchange membrane (PEM) electrolyzer plant. *Energy Convers Manag*. 2008;49:2748−2756.
67. Ni M, Leung MKH, Leung DYC. Energy and exergy analysis of hydrogen production by solid oxide steam electrolyzer plant. *Int J Hydrogen Energy*. 2007;32:4648−4660.
68. Manabe A, Kashiwase M, Hashimoto T, et al. Basic study of alkaline water electrolysis. *Electrochim Acta*. 2013;100:249−256.
69. IRENA. *Hydrogen from Renewable Power: Technology Outlook for the Energy Transition*; 2018. Available from: www.irena.org.
70. Bagotski VS. *Fundamentals of Electrochemistry*. Honolen, New Jersey: Wiley Interscience; 2006.
71. Ozden E, Tari I. Energy-exergy and economic analyses of a hybrid solar-hydrogen renewable energy system in Ankara, Turkey. *Appl Therm Eng*. 2016. https://doi.org/10.1016/j.applthermaleng.2016.01.042.
72. Turner J, Sverdrup G, Mann MK, et al. Renewable hydrogen production. *Int J Energy Res*. 2008;32:379−407.
73. Ogden J. Introduction to a future hydrogen infrastructure. In: *Transition to Renewable Energy Systems*. 2013:795−811. https://doi.org/10.1002/9783527673872.ch38.
74. Duffie JA, Beckman WA. *Solar Engineering of Thermal Processes*. Hoboken, New Jersey: John Wiley & Sons, Inc.; 2013.
75. Janusz N, Tadeusz B, Wenxian L. Solar photoelectrochemical production of hydrogen. In: *Handb. Hydrog. Energy.*. 2014:455.
76. Chen HM, Chen CK, Liu RS, Zhang L, Zhang J, Wilkinson DP. Nano-architecture and material designs for water splitting photoelectrodes. *Chem Soc Rev*. 2012;41:5654−5671.
77. Acar C, Dincer I. Review and evaluation of hydrogen production options for better environment. *J Clean Prod*. 2019;218:835−849.
78. Acar C, Dinçer İ, Naterer GF. Review of photocatalytic water-splitting methods for sustainable hydrogen production. *Int J Energy Res*. 2016;40.
79. Jeon HS, Min BK. Solar-hydrogen production by a monolithic photovoltaic-electrolytic cell. *J Electrochem Sci Technol*. 2013;3:149−153.
80. Siddiqui O, Ishaq H, Chehade G, Dincer I. Experimental investigation of an integrated solar powered clean hydrogen to ammonia synthesis system. *Appl Therm Eng*. 2020;176:115443.
81. Georgiou MD. *Computer Simulations to Calculate Energy Flux and Image Size on a Receiver for Different Reflector Geometries Using Caustics* [MASc. thesis]. University of Illinois; 2012.

82. Ismail AA, Bahnemann DW. Photochemical splitting of water for hydrogen production by photocatalysis: a review. *Sol Energy Mater Sol Cells*. 2014;128:85–101.
83. Ishaq H, Dincer I, Naterer GF. Development and assessment of a solar, wind and hydrogen hybrid trigeneration system. *Int J Hydrogen Energy*. 2018;43:23148–23160.
84. Olateju B, Kumar A. Hydrogen production from wind energy in Western Canada for upgrading bitumen from oil sands. *Energy*. 2011;36:6326–6339.
85. Honnery D, Moriarty P. Estimating global hydrogen production from wind. *Int J Hydrogen Energy*. 2009;34:727–736.
86. Briguglio N, Andaloro L, Ferraro M, et al. Renewable energy for hydrogen production and sustainable urban mobility. *Int J Hydrogen Energy*. 2010;35:9996–10003.
87. Bose T, Agbossou K, Kolhe M, Hamelin J. *Stand-Alone Renewable Energy System Based on Hydrogen Production*. Institut de recherche sur l'hydrogène; 2004:1–9.
88. Our World in Data. *Wind Energy Generation, 1965 to 2018*. Available from: https://ourworldindata.org/search?q=wind.
89. *Wind Turbine Design*. [cited 1 July 2020]. Available from: https://en.wikipedia.org/wiki/Wind_turbine_design.
90. Sahin AD, Dincer I, Rosen MA. Thermodynamic analysis of wind energy. *Int J Energy Res*. 2006;30:553–566.
91. Osczevski RJ. Windward cooling: an overlooked factor in the calculation of wind chill. *Bull Am Meteorol Soc*. 2000;81:2975–2978.
92. Froude R. On the part played in propulsion by differences of fluid pressure. *Trans Inst Nav Archit*. 1889;30:390.
93. Ishaq H, Dincer I. Evaluation of a wind energy based system for co-generation of hydrogen and methanol production. *Int J Hydrogen Energy*. 2020. https://doi.org/10.1016/j.ijhydene.2020.01.037.
94. Ishaq H, Dincer I, Naterer GF. Performance investigation of an integrated wind energy system for co-generation of power and hydrogen. *Int J Hydrogen Energy*. 2018;43:9153–9164.
95. Siddiqui O, Dincer I. Design and analysis of a novel solar-wind based integrated energy system utilizing ammonia for energy storage. *Energy Convers Manag*. 2019;195:866–884.
96. Lund JW, Bertani R, Boyd TL. Worldwide geothermal energy utilization 2015. *GRC Trans*. 2015;39:79–91.
97. International Renewable Energy Agency (IRENA). *Geothermal Energy*; 2020 [cited 7 July 2020]. Available from: https://www.irena.org/geothermal.
98. Kalinci Y, Hepbasli A, Tavman I. Determination of optimum pipe diameter along with energetic and exergetic evaluation of geothermal district heating systems: modeling and application. *Energy Build*. 2008;40:742–755.
99. Elminshawy NAS, Siddiqui FR, Addas MF. Development of an active solar humidification-dehumidification (HDH) desalination system integrated with geothermal energy. *Energy Convers Manag*. 2016;126:608–621.
100. Karakilcik H, Erden M, Karakilcik M. Investigation of hydrogen production performance of chlor-alkali cell integrated into a power generation system based on geothermal resources. *Int J Hydrogen Energy*. 2019:14145–14150. https://doi.org/10.1016/j.ijhydene.2018.09.095.
101. International Energy Agency. *Direct Use of Geothermal Energy, World, 2012–2024*. Available from: https://www.iea.org/data-and-statistics/charts/direct-use-of-geothermal-energy-world-2012-2024.

102. Our World in Data. *Installed Geothermal Energy Capacity, 1990 to 2018*. Available from: https://ourworldindata.org/search?q=geothermal.
103. Yilmaz C, Kanoglu M. Thermodynamic evaluation of geothermal energy powered hydrogen production by PEM water electrolysis. *Energy*. 2014;69:592–602.
104. Kanoglu M, Yilmaz C, Abusoglu A. Geothermal energy use in hydrogen production. *J Therm Eng*. 2016;2:699–708.
105. Siddiqui O, Dincer I. Exergetic performance investigation of varying flashing from single to quadruple for geothermal power plants. *J Energy Resour Technol Trans ASME*. 2019;141:1–11.
106. Khaliq A. Exergy analysis of gas turbine trigeneration system for combined production of power heat and refrigeration. *Int J Refrig*. 2009;32:534–545.
107. Siddiqui O, Ishaq H, Dincer I. A novel solar and geothermal-based trigeneration system for electricity generation, hydrogen production and cooling. *Energy Convers Manag*. 2019;198:111812.
108. International Energy Agency. *Data and Statistics, Explore Energy Data by Category, Indicator, Country or Region*; 2020 [cited 16 July 2020]. Available from: https://www.iea.org/data-and-statistics?country=CANADA&fuel=Energy supply&indicator=Electricity generation by source.
109. Christopher K, Dimitrios R. A review on exergy comparison of hydrogen production methods from renewable energy sources. *Energy Environ Sci*. 2012;5:6640–6651.
110. Scherer L, Pfister S. Global water footprint assessment of hydropower. *Renew Energy*. 2016;99:711–720.
111. International Renewable Energy Agency. *Hydro Power*; 2020 [cited 26 July 2020]. Available from: https://www.irena.org/hydropower.
112. Ware A. Reliability-constrained hydropower valuation. *Energy Pol*. 2018;118:633–641.
113. Munoz-Hernandez G, Mansoor S, Jones D. *Modelling and Controlling Hydropower Plants*. Vol. 369. London Dordrecht Heidelberg New York: Springer; 2013.
114. Zarfl C, Lumsdon AE, Berlekamp J, Tydecks L, Tockner K. A global boom in hydropower dam construction. *Aquat Sci*. 2014;77:161–170.
115. Liu Y, Packey DJ. Combined-cycle hydropower systems - the potential of applying hydrokinetic turbines in the tailwaters of existing conventional hydropower stations. *Renew Energy*. 2014;66:228–231.
116. Singh VK, Singal SK. Operation of hydro power plants-a review. *Renew Sustain Energy Rev*. 2017;69:610–619.
117. Tarnay DS. Hydrogen production at hydro-power plants. *Int J Hydrogen Energy*. 1985;10:577–584.
118. USGS Science for a Changing World. *Hydroelectric Power: How it Works*; 2020 [cited 28 July 2020]. Available from: https://www.usgs.gov/special-topic/water-science-school/science/hydroelectric-power-how-it-works?qt-science_center_objects=0#qt-science_center_objects.
119. Bergant A, Simpson AR, Tijsseling AS. Water hammer with column separation: a historical review. *J Fluid Struct*. 2006;22:135–171.
120. Fard RN, Tedeschi E. Integration of distributed energy resources into offshore and subsea grids. *CPSS Trans Power Electron Appl*. 2018;3:36–45.
121. Mofor L, Goldsmith J, Jones F. Ocean energy: technology readiness, patents, deployment status and outlook. *Int Renew Energy Agency*. 2014. https://doi.org/10.1007/978-3-540-77932-2.

122. International Renewable Energy Agency. *Ocean Energy Data*; 2020 [cited 11 August 2020]. Available from: https://www.irena.org/ocean.
123. Goward Brown AJ, Neill SP, Lewis MJ. Tidal energy extraction in three-dimensional ocean models. *Renew Energy*. 2017;114:244–257.
124. Valizadeh R, Abbaspour M, Rahni MT. A low cost Hydrokinetic Wells turbine system for oceanic surface waves energy harvesting. *Renew Energy*. 2020;156:610–623.
125. Kim DY, Kim YT. Preliminary design and performance analysis of a radial inflow turbine for ocean thermal energy conversion. *Renew Energy*. 2017;106:255–263.
126. Parkinson SC, Dragoon K, Reikard G, García-Medina G, Özkan-Haller HT, Brekken TKA. Integrating ocean wave energy at large-scales: a study of the US Pacific Northwest. *Renew Energy*. 2015;76:551–559.
127. Ishaq H, Dincer I. A comparative evaluation of OTEC, solar and wind energy based systems for clean hydrogen production. *J Clean Prod*. 2019:118736. https://doi.org/10.1016/j.jclepro.2019.118736.
128. Our World in Data. *CO_2 Emissions per Capita, 2017*; 2017 [cited 24 April 2020]. Available from: https://ourworldindata.org/search?q=CO2+emissions.
129. International Energy Agency. *Data & Statistics CO_2 Emissions by Sector - IEA*. [cited 13 April 2020]. Available from: https://www.iea.org/data-and-statistics?country=WORLD&fuel=CO2 emissions&indicator=CO2 emissions by sector.
130. Schipfer F, Kranzl L. *IEA Bioenergy Task 40 Sustainable International Bioenergy Trade Securing Supply and Demand: Country Report Austria 2014*. Vol. 40. 2015:29.
131. International Renewable Energy Agency. *Bioenergy*; 2020 [cited 22 August 2020]. Available from: https://www.irena.org/bioenergy.
132. Chang ACC, Chang HF, Lin FJ, Lin KH, Chen CH. Biomass gasification for hydrogen production. *Int J Hydrogen Energy*. 2011;36:14252–14260.
133. Acharya B, Dutta A, Basu P. An investigation into steam gasification of biomass for hydrogen enriched gas production in presence of CaO. *Int J Hydrogen Energy*. 2010; 35:1582–1589.
134. Kumar A, Jones DD, Hanna MA. Thermochemical biomass gasification: a review of the current status of the technology. *Energies*. 2009;2:556–581.
135. Andersson J, Lundgren J. Techno-economic analysis of ammonia production via integrated biomass gasification. *Appl Energy*. 2014;130:484–490.
136. Chuayboon S, Abanades S, Rodat S. Comprehensive performance assessment of a continuous solar-driven biomass gasifier. *Fuel Process Technol*. 2018;182:1–14.
137. Our World in Data. *Biofuel Energy Production 2019*; 2020 [cited 23 August 2020]. Available from: https://ourworldindata.org/search?q=biofuel.
138. Yin C. Prediction of higher heating values of biomass from proximate and ultimate analyses. *Fuel*. 2011;90:1128–1132.
139. Yan L, Yue G, He B. Exergy analysis of a coal/biomass co-hydrogasification based chemical looping power generation system. *Energy*. 2015;93:1778–1787.
140. Ud Din Z, Zainal ZA. Biomass integrated gasification–SOFC systems: technology overview. *Renew Sustain Energy Rev*. 2016;53:1356–1376.
141. Dincer I, Acar C. Review and evaluation of hydrogen production methods for better sustainability. *Int J Hydrogen Energy*. 2014;40:11094–11111.
142. REN21 community. Renewables 2019 global status report. *Resources*. 2019;8.
143. Bocci E, Zuccari F, Dell'Era A. Renewable and hydrogen energy integrated house. *Int J Hydrogen Energy*. 2011;36:7963–7968.

144. National Renewable Energy Laboratory. Renewable electricity generation and storage technologies. *Renew. Electr. Futur. Study.* 2012;2:2.
145. RAEL. *Energy Efficiency & Financing Districts for Local Governments. Renew. Appropr. Energy Lab. Energy Resour. Gr. Univ. California, Berkeley.* 2009.
146. Al-Zareer M, Dincer I, Rosen MA. Influence of selected gasification parameters on syngas composition from biomass gasification. *J Energy Resour Technol.* 2018. https://doi.org/10.1115/1.4039601.
147. Augustine AS, Ma YH, Kazantzis NK. High pressure palladium membrane reactor for the high temperature water-gas shift reaction. *Int J Hydrogen Energy.* 2011;36: 5350–5360.
148. Our World in Data. *Hydropower Generation, 2018*; 2020 [cited 16 July 2020]. Available from: https://ourworldindata.org/search?q=hydro.

Appendix

Table A-1 CO_2 Emissions per Capita.

CO_2 emissions (tonnes per capita)	Australia	Canada	China	Germany	Japan	Netherlands	Saudi Arabia	South Africa	United Arab Emirates	United Kingdom	United States
1990	16.34	16.75	2.06	12.80	9.28	10.85	11.36	8.32	27.79	10.51	20.28
1991	16.20	16.21	2.14	12.76	9.34	11.34	15.85	8.47	28.77	10.62	19.89
1992	16.29	16.52	2.21	12.08	9.39	11.25	16.40	7.65	27.68	10.32	20.10
1993	16.35	16.30	2.33	11.89	9.31	11.19	17.54	7.95	29.70	10.05	20.32
1994	16.44	16.64	2.45	11.62	9.72	11.15	16.79	8.21	31.18	9.96	20.47
1995	16.90	16.90	2.63	11.57	9.80	11.19	12.53	8.59	28.59	9.78	20.47
1996	17.10	17.24	2.73	11.79	9.87	11.72	13.50	8.47	15.72	10.12	20.93
1997	17.37	17.57	2.71	11.44	9.80	11.20	11.06	8.82	15.21	9.64	20.96
1998	17.93	17.70	2.58	11.34	9.47	11.20	10.42	8.49	28.43	9.69	20.88
1999	18.24	18.06	2.55	11.01	9.74	10.80	11.13	8.31	25.95	9.54	20.91
2000	18.37	18.63	2.61	11.06	9.90	10.80	14.27	8.26	35.43	9.62	21.28
2001	18.56	18.20	2.65	11.25	9.77	11.05	13.92	8.00	30.25	9.74	20.72
2002	18.60	18.23	2.91	11.04	9.99	10.95	14.87	7.56	23.88	9.44	20.67
2003	18.77	18.58	3.41	11.04	10.04	11.10	14.48	8.46	28.24	9.59	20.66
2004	19.21	18.34	3.90	10.87	9.98	11.14	16.99	9.30	27.37	9.57	20.87
2005	19.10	17.86	4.37	10.62	10.03	10.85	16.56	8.50	25.01	9.45	20.78
2006	19.09	17.48	4.80	10.78	9.83	10.51	17.55	9.03	23.25	9.34	20.32
2007	19.12	17.99	5.13	10.48	10.12	10.46	15.31	9.30	22.06	9.11	20.40
2008	18.98	17.21	5.49	10.53	9.56	10.59	16.67	9.80	22.36	8.78	19.56
2009	18.80	16.00	5.74	9.76	9.02	10.24	17.47	9.84	21.55	7.89	17.95
2010	18.39	16.26	6.25	10.31	9.42	10.92	18.88	9.16	19.12	8.09	18.47
2011	17.98	16.24	6.87	10.02	9.82	10.10	17.62	8.97	18.76	7.36	17.91
2012	17.84	16.15	7.01	10.06	10.14	9.82	19.36	8.81	19.54	7.59	17.13
2013	17.20	16.14	7.08	10.25	10.24	9.78	18.06	8.64	18.68	7.39	17.49
2014	16.76	15.95	7.06	9.74	9.86	9.32	19.56	8.95	23.02	6.75	17.53
2015	16.93	15.73	6.96	9.76	9.56	9.74	19.67	8.34	24.85	6.46	16.94
2016	17.13	15.38	6.91	9.79	9.43	9.74	19.57	8.36	25.18	6.06	16.48
2017	16.90	15.64	6.98	9.73	9.45	9.63	19.28	8.05	24.66	5.81	16.24

Data from [130].

Table A-2 Global Solar Energy Generation by Different Regions (Terawatt-Hours).

Year	Africa	Asia Pacific	CIS	Canada	China	Europe	India	Japan	Middle East	United Arab Emirates	United Kingdom	United States
2000	0.02	0.43	0.00	0.02	0.02	0.13	0.01	0.34	0.00	0.00	0.00	0.52
2001	0.02	0.61	0.00	0.02	0.03	0.17	0.01	0.50	0.00	0.00	0.00	0.57
2002	0.03	0.83	0.00	0.02	0.05	0.28	0.01	0.69	0.00	0.00	0.00	0.60
2003	0.03	1.12	0.00	0.02	0.06	0.47	0.02	0.95	0.00	0.00	0.00	0.61
2004	0.04	1.46	0.00	0.01	0.08	0.75	0.02	1.27	0.00	0.00	0.00	0.70
2005	0.04	1.85	0.00	0.02	0.08	1.49	0.02	1.63	0.00	0.00	0.01	0.75
2006	0.05	2.29	0.00	0.02	0.10	2.52	0.01	2.00	0.00	0.00	0.01	0.82
2007	0.06	2.75	0.00	0.03	0.11	3.82	0.06	2.31	0.00	0.00	0.01	1.10
2008	0.08	3.33	0.00	0.04	0.15	7.50	0.06	2.59	0.00	0.00	0.02	1.63
2009	0.11	4.38	0.00	0.11	0.28	14.19	0.08	3.05	0.04	0.02	0.02	2.08
2010	0.22	6.74	0.00	0.25	0.70	23.26	0.11	3.98	0.10	0.02	0.04	3.01
2011	0.46	12.15	0.00	0.57	2.61	46.71	0.83	5.44	0.23	0.02	0.24	4.74
2012	0.55	17.58	0.01	0.88	3.59	71.37	2.10	7.37	0.43	0.02	1.35	9.04
2013	0.82	32.34	0.02	1.50	8.37	86.94	3.43	12.91	0.77	0.10	2.01	16.04
2014	1.83	62.87	0.18	2.12	23.51	98.33	4.91	23.55	1.50	0.30	4.05	29.22
2015	3.57	99.19	0.41	2.90	43.56	109.88	6.57	34.54	2.39	0.30	7.53	39.43
2016	4.91	141.76	0.64	3.03	61.69	113.95	11.56	48.54	3.46	0.33	10.41	55.42
2017	6.61	227.20	0.77	3.29	117.80	124.54	21.52	61.83	4.41	0.53	11.52	78.06
2018	9.03	314.21	0.88	3.55	177.50	139.05	30.73	71.69	6.12	0.95	12.92	97.12

Data from [130].

Table A-3 Global Solar PV Energy Consumption by Different Regions (Gigawatt-Hours).

Year	Total Africa	Total Asia Pacific	Total CIS	Total Europe	Total Middle East	Total North America	Total South and Central America
1990	0.00	4.00	0.00	12.50	0.00	371.79	0.00
1991	0.00	9.45	0.00	15.50	0.00	480.25	0.00
1992	0.00	27.60	0.00	27.09	0.00	413.90	0.00
1993	0.00	43.10	0.00	32.38	0.00	481.21	0.00
1994	0.00	53.40	0.00	37.69	0.00	508.93	0.00
1995	0.00	70.66	0.00	46.84	0.00	523.31	0.00
1996	0.00	98.16	0.00	53.81	0.00	553.24	0.00
1997	0.00	147.02	0.00	63.93	0.00	545.64	0.00
1998	0.00	202.92	0.00	87.09	0.00	541.76	0.02
1999	0.00	286.27	0.00	91.20	0.00	538.61	0.13
2000	18.40	428.43	0.00	134.47	0.00	541.93	2.17
2001	24.70	608.47	0.00	170.17	0.00	601.93	3.81
2002	30.80	829.25	0.00	276.78	0.00	632.33	6.20
2003	33.90	1116.54	0.00	468.73	0.00	641.44	8.57
2004	39.10	1464.06	0.00	751.13	0.00	721.40	11.12
2005	44.20	1852.81	0.00	1486.41	0.00	776.37	17.91
2006	51.90	2291.28	0.00	2524.12	0.10	857.12	8.21
2007	63.63	2748.58	0.00	3815.73	0.10	1133.36	10.52
2008	84.92	3330.60	0.00	7501.10	3.50	1687.48	14.54
2009	107.87	4383.50	0.01	14192.83	44.80	2212.26	23.76
2010	223.86	6738.75	0.24	23263.39	97.94	3298.24	60.74
2011	460.53	12154.40	1.82	46713.48	227.48	5351.50	126.64
2012	553.43	17583.03	6.34	71869.61	429.93	9987.01	335.08
2013	824.28	32338.91	18.84	86935.09	766.22	17644.50	530.74
2014	1826.26	62874.75	177.38	98832.51	1499.20	31556.96	1143.46
2015	3568.28	99193.73	411.04	109880.18	2388.00	42567.26	2730.73
2016	4911.32	141755.23	635.75	113950.48	3455.74	58703.79	4965.80
2017	6610.57	227196.64	767.30	124542.38	4409.89	82535.27	7455.61
2018	9029.09	314208.55	881.30	139052.06	6121.15	102907.23	12431.53

Data from [130].

Table A-4 Global Wind Energy Consumption and Installed Wind Capacity.

Year	Wind Generation (gigawatt-hours)	Wind Capacity (gigawatts)
1980	10.5	–
1981	10.5	–
1982	18.5	–
1983	32.79495	–
1984	44.75556	–
1985	64.2202	–
1986	138.8313	–
1987	195.3768	–
1988	331.5798	–
1989	2649.777	–
1990	3632.471	–
1991	4086.707	–
1992	4733.212	–
1993	5697.569	–
1994	7122.93	–
1995	8261.923	4.778
1996	9204.601	6.07
1997	12017.82	7.623075
1998	15921.26	9.936175
1999	21216.17	13.42656
2000	31420.94	17.3037
2001	38390.95	23.9764
2002	52331.76	30.9795
2003	62916.93	38.3917
2004	85117.16	46.9174
2005	104085.9	58.4518
2006	132859.1	73.1655
2007	170682.6	91.5111
2008	220572.1	115.3629
2009	275949.3	150.1813
2010	341614.5	180.9412
2011	436786.4	220.1294
2012	523809.4	267.1129
2013	645302.2	300.3026
2014	712031.7	349.699
2015	831384.5	416.7388
2016	956873.5	467.5776
2017	1,127,990	515.1749
2018	1,269,953	564.347

Table A-5 Geothermal Capacity in Different World Regions (Megawatts) [104].

Year	Total Africa	Total Asia Pacific	Total CIS	Total Europe	Total North America	Total South and Central America
1990	45	1537.7	11	617.6	3656.5	130
1995	45	2168.8	11	710.9	3913.96	230
2000	65.3	3344.7	23	794	3797	360.8
2001	65.3	3603.7	21	807	3841	365.8
2002	65.3	3559.7	70	902	3872	417
2003	65.3	3592.6	70	940	3996	435
2004	135.3	3607.6	56	875.2	4054	428
2005	135.3	3722.7	79	933.1	4089	449
2006	135.3	3719.6	87	1142.1	4131	474.2
2007	135.3	3887.6	90	1207.9	4193	527.4
2008	135.3	4104.7	80	1304.9	4206	527.4
2009	170.3	4284.7	81	1380.9	4386	526.8
2010	205.3	4384	81	1430.9	4463	526.8
2011	205.3	4423.1	81	1557.2	4386.8	525.4
2012	212.8	4508.1	81	1608.2	4548	655.6
2013	212.8	4597.4	79	1771.2	4588	640.6
2014	373.4	4888.8	78	1904.3	4570	637.1
2015	626.2	4938.6	78	2125.3	4717.8	640
2016	670.2	5138.6	78	2321.5	4730.6	629.5
2017	680.2	5308.6	78	2619.9	4658.1	689
2018	670.3	5487.6	78	2883.7	4768.845	713

Table A-6 Installed Hydro Capacity Around the Globe From 1990 to 2018.

Years	Africa	Asia Pacific	CIS	Canada	Central America	China	Europe	Japan	Middle East	North America	South and Central America	United Kingdom	United States
1990	57.25	401.67	210.95	295.68	11.64	126.74	502.46	89.31	15.28	612.45	360.99	5.12	292.28
1991	60.65	413.57	212.15	307.28	11.33	124.69	513.09	97.49	11.02	617.23	385.43	4.54	287.33
1992	57.86	403.96	216.00	315.17	10.79	130.69	529.66	82.55	17.83	593.41	392.79	5.35	251.43
1993	56.68	443.31	222.58	322.21	12.02	151.85	550.47	95.59	20.05	628.12	423.06	4.30	279.25
1994	57.60	446.11	226.33	327.86	11.62	167.43	557.20	67.27	16.55	607.47	448.44	5.12	259.34
1995	59.77	486.73	220.31	334.06	11.46	190.58	564.89	82.12	18.72	672.88	465.69	4.65	311.22
1996	62.32	476.74	198.81	354.59	13.23	187.97	548.09	80.52	18.70	733.48	485.69	3.40	347.55
1997	63.85	479.23	197.85	348.68	12.77	195.98	567.68	89.80	16.09	731.01	513.91	4.40	355.97
1998	66.65	510.30	198.33	330.87	12.85	198.89	593.90	92.51	17.21	677.55	526.61	5.12	322.09
1999	70.68	502.87	203.16	345.00	15.00	196.58	598.18	86.42	11.80	694.32	527.29	5.34	316.61
2000	75.25	524.43	208.43	356.76	15.37	222.41	517.62	87.25	10.73	662.60	555.65	5.09	272.76
2001	80.86	573.09	217.31	331.52	13.71	277.43	512.01	84.17	11.62	570.20	521.40	4.05	210.24
2002	85.18	580.68	207.32	349.27	14.46	287.97	563.91	82.38	16.82	632.30	547.36	4.79	258.17
2003	82.87	595.31	205.72	336.14	15.09	283.68	529.52	93.43	18.58	625.96	570.65	3.23	269.97
2004	87.41	683.72	225.80	338.40	16.04	353.54	570.10	93.03	23.13	626.13	589.03	4.84	262.55
2005	89.07	727.42	223.52	361.96	17.12	397.02	571.67	77.11	23.92	656.09	624.15	4.92	266.43
2006	92.24	800.65	222.01	352.89	18.00	435.79	561.85	87.76	28.25	668.88	652.68	4.59	285.54
2007	95.34	852.30	227.41	367.62	17.94	485.26	563.94	74.70	26.71	638.01	675.74	5.08	243.04
2008	97.16	996.68	207.67	377.49	20.10	636.96	592.99	74.44	13.69	667.77	680.30	5.14	251.05
2009	99.76	965.84	218.32	368.69	18.83	615.64	589.69	68.80	12.12	666.94	697.53	5.23	271.53
2010	107.43	1094.88	216.84	351.38	21.24	711.38	649.72	86.91	17.39	645.82	700.88	3.59	257.27
2011	110.45	1115.87	212.75	375.72	20.92	688.05	569.61	80.77	18.33	728.09	744.10	5.69	316.10
2012	110.95	1279.09	213.03	380.27	22.31	862.79	625.76	76.06	21.41	686.21	730.32	5.31	274.03
2013	117.67	1365.04	229.43	391.79	21.83	909.61	661.56	78.00	23.36	686.36	709.81	4.70	266.55
2014	123.76	1510.90	221.14	382.50	21.49	1051.15	644.01	79.97	19.94	677.14	686.25	5.89	255.75
2015	120.27	1565.09	215.46	382.19	22.43	1114.52	635.84	83.84	16.77	659.54	671.46	6.30	246.45
2016	116.69	1626.95	234.65	387.17	22.81	1153.27	651.36	79.96	20.17	681.63	686.30	5.62	263.76
2017	124.58	1649.45	240.15	396.50	27.17	1165.07	584.89	79.24	20.79	725.16	720.44	5.93	296.81
2018	132.84	1718.51	244.84	387.25	26.91	1202.43	642.07	81.00	15.19	708.35	731.31	5.46	288.71

Data from [150].

Index

'*Note*: Page numbers followed by "f" indicate figures and "t" indicate tables.'

I

K

L

M

N

O

P

R

S

T

U

V

W